MW00609770

PSYCHOLOGY *according to* SHAKESPEARE

What You Can Learn about Human Nature from Shakespeare's Great Plays

PHILIP G. ZIMBARDO, PhD
ROBERT L. JOHNSON, PhD

Guilford, Connecticut

Prometheus Books
An imprint of Globe Pequot, the trade division of The Rowman & Littlefield Publishing Group, Inc.
4501 Forbes Blvd., Ste. 200
Lanham, MD 20706
www.rowman.com

Distributed by NATIONAL BOOK NETWORK

British Library Cataloguing in Publication Information Available

Library of Congress Cataloging-in-Publication Data

Names: Zimbardo, Philip G., author. | Johnson, Robert L. (Robert Lee), author.
Title: Psychology according to Shakespeare / Philip G. Zimbardo, PhD, Robert L. Johnson, PhD.
Description: Lanham, MD : Prometheus Books, [2024] | Includes bibliographical references and index. | Summary: "William Shakespeare has undergone psychological analyses ever since Freud diagnosed Hamlet with an Oedipus complex. But now, two psychologists propose to turn the tables by telling how Shakespeare himself understood human behavior and the innermost workings of the human mind"—Provided by publisher.
Identifiers: LCCN 2023044422 (print) | LCCN 2023044423 (ebook) | ISBN 9781633889606 (cloth) | ISBN 9781633889613 (epub)
Subjects: LCSH: Psychology. | Human behavior—Psychological aspects. | Shakespeare, William, 1564–1616.
Classification: LCC BF121 .Z535 2024 (print) | LCC BF121 (ebook) | DDC 150—dc23/eng/20231218
LC record available at https://lccn.loc.gov/2023044422
LC ebook record available at https://lccn.loc.gov/2023044423

™ The paper used in this publication meets the minimum requirements of American National Standard for Information Sciences—Permanence of Paper for Printed Library Materials, ANSI/NISO Z39.48-1992

CONTENTS

PROLOGUE

Why Shakespeare and Psychology?

For more than a century—ever since Freud diagnosed Hamlet with an Oedipus complex—Will Shakespeare has been the object of psychological analyses. Now we propose to turn the tables: to tell how the Bard himself understood the brain, behavior, and the inner workings of the human mind. And that's why we two psychologists have ventured into the world of the Bard, hoping that we can illuminate the psychological aspects of Shakespeare's work that have never before been gathered in one volume. Here is a brief overview of what we found:

Even today, there exists no better depiction of a psychopath than *Richard III*, no more poignant portrayal of dementia than *King Lear*, nor a more unforgettable illustration of obsessive-compulsive disorder than Lady Macbeth's attempts to wash away the damn'd blood spot. What has not been revealed so completely before, however, are the many forms of mental illness Shakespeare described in terms that we can associate with our twenty-first-century conception of mental disorders as described in the psychiatric manual known as the *DSM-5*.

Nor was Shakespeare's psychology limited to mental disorders. His fascination with human nature ranged across the psychological spectrum, from brain anatomy to personality, cognition, emotion, perception, lifespan development, and states of consciousness. To illustrate, we have stories to tell involving astrology, potions, poisons, the four fluids called "humors," anatomical dissections of freshly hanged criminals, and a mental hospital called Bedlam—all showing how his perspective was grounded in the medicine and culture of his time.

And yet Will Shakespeare's intellect, curiosity, and temperament allowed him to glimpse ideas and issues that would become important in psychological science centuries later. So, here we retell of his fashioning the felicitous phrase *nature-nurture* for Prospero to utter in frustration with Caliban; we also tell how the nature-nurture dichotomy became central in psychology's quest to understand the tension between heredity and environment. Then, in *Measure for Measure*, he made audiences puzzle over the related issue: Which exerts the greater influence on human behavior: internal traits or the external situation? And in *Hamlet*, he explored the push-pull between reason and emotion in the mind of a seemingly dithering prince.

Ever the keen observer of human behavior, ever the proto-psychologist, Mr. Shakespeare (see figure P.1) serves up tales highlighting flaws of human cognition. In *Othello*, we watch the general's losing battle with *confirmation bias* and the "green-eyed monster." At Elsinore, we hear of Hamlet's struggle with the *rationalist delusion.* And at Cawdor Castle, we witness Macbeth's undoing, caught in the trap of the *sunk cost heuristic*—all some four hundred years before scientists Amos Tversky and Daniel Kahneman rediscovered these same concepts. Arguably, our man from Stratford even understood the concept that we now call *cognitive dissonance.*

Figure P.1 William Shakespeare portrait by the American artist George Henry Hall, 1896. *Folger Digital Image Collection.*

Aside from bringing together the Bard's own psychology, this book will show how his interest in mind and behavior ranged across the full spectrum of modern psychology, including topics that we now call *biopsychology*, *neuroscience*, *social psychology*, *thinking and intelligence*, *motivation and emotion*, and *reason vs intuition*. Granted, the psychological concepts he knew have evolved over the intervening centuries—for example, the Elizabethan notion of *sensus communis* eventually became "consciousness," and the old idea of the humors morphed into our current understanding of hormones and neurotransmitters. Nevertheless, some of Mr. Shakespeare's concerns seem especially timely today, as in the subplot of queer vs straight issues complicating the story of *Troilus and Cressida* and in Shylock's telling of prejudices inflicted on ethnic minorities.

To put our interest in Shakespeare more broadly but more simply, we see his works and modern psychology as providing different perspectives on the same ideas—two ways of looking at mind and behavior. It's analogous to the two perspectives one sees in the Necker cube—a hoary illusion borrowed from the psychology of visual perception[1] (figure P.2). The "cube" itself is, of course, merely a two-dimensional drawing on a flat surface that the brain interprets as a three-dimensional object. If you examine the cube for a while, you will suddenly see it "flip" so that you see it from a new angle. Try it before reading on!

We would extend that notion even further, adding all the sciences, arts, and humanities. In doing so, we are suggesting that human knowledge is like a multidimensional web that interconnects ideas across disciplines with links that we have only begun to discover. It is a profound idea, yet one rarely taught—or perhaps, rarely learned. It is a concept that has been nearly lost as knowledge has exploded and disciplines have become more and more specialized and fragmented. Accordingly, we have written *Psychology according to Shakespeare* to nurture the notion of connections among diverse fields of knowledge.

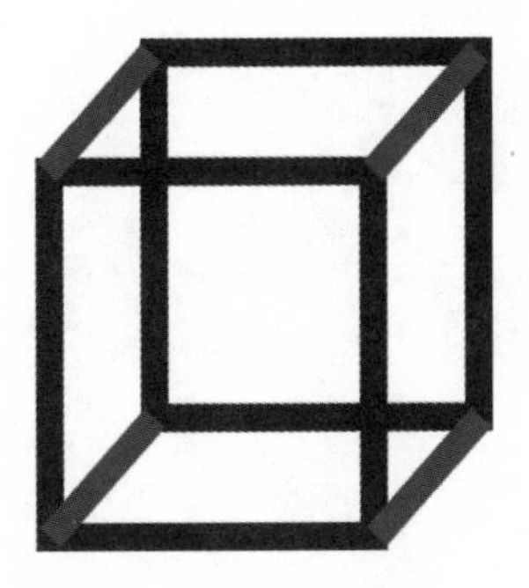

Figure P.2 The Necker cube.

THE FOUR BIG ISSUES CONNECTING SHAKESPEARE AND PSYCHOLOGY

In writing this book, we have found that Shakespeare and psychology have a shared interest in four big issues:

- *Heredity vs environment*—Which influences us more: nature or nurture, biology or experience?
- *The person vs the situation*—Do our responses to new situations depend more on our internal influences, such as our personality traits and biology, or on external influences, such as the demands of the social situations in which we find ourselves?
- *The nature of the human mind*—Are we driven more by our conscious or unconscious thought processes?
- *Reason vs emotion*—Have we *Homo sapiens* evolved to be rational beings who sometimes get swept away by emotional events? Or are we really emotional creatures who fancy ourselves as reasonable, but who only occasionally resort to reason?

Accordingly, after the introductory chapter, the remainder of the book is organized in four parts, with each part focusing on one of these issues.

Philip G. Zimbardo and Robert L. Johnson

INTRODUCTION

Shakespeare's Psychology and the Roots of Genius

We sit in The Anchor with tall glasses of ale, musing about our stormy relationship with William Shakespeare. We are here because he finally won us over—and so has The Anchor. Still full of life after four hundred years, this venerable watering hole has served tipplers like us since Shakespeare trod the stage of the nearby Globe Theater on the south bank of the Thames (figure I.1). Its location makes it likely that he, too, drained a few pints here, perhaps scanning the crowd for interesting characters. Hangouts such as this were haunts for Elizabethan locals whom he may have turned into such memorable characters as Falstaff, Mistress Quickly, Doll Tearsheet, Bardolph, Pistol, and Bottom. We spot modern candidates for several of them.

At a distance of four centuries, Shakespeare was not an easy fellow to befriend, and we suspect that we are not alone. Our teachers urged us to adore him for the beauty of his language—for words that, as Hamlet said, cascade "trippingly from the tongue." Instead, we tripped over the "forsooths," "anons," "withals," and "sirrhas." Oh, he had a clever way with words—yet that was not how we finally connected with him.

For Elizabethan audiences, of course, language was not the obstacle it presents for us. They were familiar with occupations such as *tinker*, *fletcher*, *draper*, and *herald*. They knew sports such as bowls and fencing, and everyday Elizabethan-English words, such as *nonce*, *vouchsafe*, *dandle*, and *cuckold*, that are no longer common. Nor did they struggle with the wordplay that slips over our heads because the pronunciation has changed.[1] To Londoners of Shakespeare's day, "hour" and "whore" sounded alike, so those attending *As You Like It* must have howled when they heard this:

JAQUES: 'Tis but an hour ago since it was nine,
And after one hour more 'twill be eleven;
And so, from hour to hour, we ripe and ripe,
And then, from hour to hour, we rot and rot;
And thereby hangs a tale.

—*As You Like It*, 2.7.25–30[2]

Alas, the pun is usually lost on moderns, as are the meanings of other words that have changed: "brave" once meant *handsome*, "honest" meant *pure*, "quaint" was *beautiful* or *ornate*, "still" was a synonym for *always* or *forever*, while "want" meant *need* or *lack*. Likewise, Shakespeare's audiences had no trouble understanding the references when Hamlet tells Horatio:

> I am but mad north-north-west. When the wind is southerly,
> I know a hawk from a handsaw.
>
> —*Hamlet*, 2.2.402–3

Figure I.1 The Anchor, a venerable London pub, perched on the bank of the Thames River in Southwark. It was rebuilt in 1676 after the Great Fire. Shakespeare almost certainly had a pint or two here because it was only a stone's throw from The Globe, the home of the Bard's theater company.

We do get the gist; we know Hamlet is saying that he has become distraught, but not insane. Yet we are left wondering what it is about a *hawk* that could be difficult to distinguish from a *handsaw.* Elizabethan audiences probably just smiled, knowing that this was a pun borrowed from the sport of hunting with hawks and falcons: in those days, another meaning of hawk was "pickaxe," while handsaw was a pun on "hernshaw," another word for heron.[3]

Happily for the two of us—both psychologists—we have found much more in Shakespeare than his language. How else to explain why so many of the plays have been completely rewritten for the modern stage and screen? To name a few: *West Side Story* is *Romeo and Juliet* set in New York City; *The Lion King* was inspired by *Hamlet*; *Scotland PA* is a spoof on *Macbeth*; the acclaimed Japanese film *Ran* derives from *King Lear*; and *Kiss Me Kate* was Cole Porter's musical update of *The Taming of the Shrew.* In the same vein, the Oregon Shakespeare Festival has embarked on a project to translate Shakespeare's plays into modern English and present them alongside those in their original form. All these adaptations suggest that what appeals most about Shakespeare's plays are the stories and the larger-than-life characters—really, *ourselves* writ large—who must deal with the same problems and emotions that we in the audience face, but on a grander scale. So, we see Romeo and Juliet falling in love across the boundaries of family enmity, the wives of Windsor deciding how to deal with a merry old lecher, and Macbeth being dragged down by ambition and pride. This is what we read into a line from Ben Jonson's eulogy for his friend Will: "He was not of an age, but for all time."[4]

THE PSYCHOLOGY IN SHAKESPEARE

It was at a performance of *Cymbeline* that one of us realized that the story of two young princes, abducted from their cradles and raised in the wilderness of Wales, was also Will Shakespeare's exploration of the nature-nurture issue. It was an epiphany! Shakespeare was a psychologist—of sorts! That insight led us to a close reading of the canon—all thirty-eight plays—where we found a slew of psychological insights.[5] Shakespeare seemed to have anticipated everything from Freudian ego defense mechanisms to cognitive dissonance theory.

In our view, his best plays are also case studies disguised as tales that open wide a window into the recesses of the human mind. So, Othello is the model of a jealous general whose life unravels in a lethal rage. Lady Macbeth's portrayal is a mental autopsy of obsessive-compulsive disorder, as she famously tries to wash the "damn'd spot" of blood from her hands. And Richard III is the psychopath who murders his way to the English throne (at least, in Shakespeare's spin on

English history). With such characterizations, the Bard's work represents a great turning point in literature, as well as in our understanding of the human mind.[6] Like no writer before him, he grasped the idea that people have private mental lives that he could reveal on stage, showing us their hidden thoughts, feelings, and motives. To be sure, others had dealt cursorily with matters of the mind, but none before had developed the art of getting the audience so fully inside the heads of their characters.

Perhaps the most famous example comes from *Hamlet*, in which the prince returns home from college to discover that Uncle Claudius may have murdered his father and married his mother—who was likely a willing coconspirator. Clearly, "there is something rotten in the state of Denmark"[7] because the uncle now wears the crown; Hamlet's inheritance and status in the nobility of Denmark are uncertain. Nor are those the least of his woes; Hamlet also realizes that, as the logical heir, his life is in jeopardy. What's a prince to do? Should he take revenge on his uncle . . . and his mother? Or should he decide "not to be," that is, to escape the problem by suicide? Hamlet's perfectly ordered world has turned upside down, and he sees no clear way to set it upright again. To refresh your memory, here is the prince's private debate with himself:

HAMLET: To be or not to be—that is the question:
Whether 'tis nobler in the mind to suffer
The slings and arrows of outrageous fortune,
Or to take arms against a sea of troubles
And by opposing, end them. To die, to sleep—
No more—and by a sleep to say we end
The heartache and the thousand natural shocks
That flesh is heir to—'tis a consummation
Devoutly to be wished. To die, to sleep—
To sleep, perchance to dream. Ay, there's the rub
For in that sleep of death what dreams may come
When we have shuffled off this mortal coil,
Must give us pause: there's the respect
That makes calamity of so long life;
For who would bear the whips and scorns of time,
The oppressor's wrong, the proud man's contumely,
The pangs of despis'd love, the law's delay,
The insolence of office and the spurns
That patient merit of the unworthy takes,
When he himself might his quietus make
With a bare bodkin? who would fardels bear,

> To grunt and sweat under a weary life,
> But that the dread of something after death,
> The undiscover'd country from whose bourn
> No traveller returns, puzzles the will
> And makes us rather bear those ills we have
> Than fly to others that we know not of?
> Thus conscience does make cowards of us all.
>
> —*Hamlet*, 3.1.64–91

[Spoiler alert: Hamlet decides "to be"—although he unexpectedly finds himself *not* to be for long, as nearly everybody dies, and the Norwegians take over Denmark.]

Hamlet's words resonate only in his mind, of course, because to express them openly would be treason. Nonetheless, Shakespeare found a way for us to hear the prince's ruminations. His solution: a theatrical device called a *soliloquy*, a speech that only the audience hears—much like an extended "aside." While Shakespeare was not the first to employ the soliloquy, he was the first to use it to draw us so completely into the minds of his protagonists.

Nor was William Shakespeare the first writer to comment extensively on mind and behavior. That honor may go to Plato, two millennia earlier. Likewise, if you want to know about the strong emotions that led to the Trojan War, read Homer.[8] If you want to know about the mindset of young aristocrats in plague-scourged Italy, read Boccaccio. And if you want to know about love and lust in medieval England, read Chaucer.

Shakespeare was simply the first playwright to bring so much psychology to the stage. Curiously, his contemporary, the Spanish writer Miguel de Cervantes, used much the same approach in his novel, *Don Quixote*, which takes readers deep into the mind of his Woeful Knight.[9] Although (we think) Shakespeare and Cervantes never met, the fact that they simultaneously pioneered such similar psycho-literary devices seems no coincidence. Something seems to have been in the air during these times, and as Hamlet taught us, it is the artist's job to hold up a mirror to nature.[10]

The point is that Will Shakespeare—and Cervantes, too—was as much a psychologist as he was a wordsmith, even though psychology as a discipline, or even a word, did not yet exist.[11] The redoubtable Shakespeare scholar Harold Bloom concurred. The central conceit of his tome *Shakespeare: The Invention of the Human* is that the Bard was a psychologist—a student of mind and behavior. And indeed, Shakespeare was an astute observer of human nature, human problems, personalities, and foibles; he was the quiet fellow at the bar observing everyone else.[12] Never before had anyone dealt so broadly with the whole range of human emotions, desires, conflicts, and social interactions. And by doing so, we have discovered, he anticipated much of modern psychology, from observations on the brain to jealousy, rage, power, lust, flaws of memory, love, and its sometimes mirror image, mental illness.

Bloom was not the first to see the psychology in Shakespeare. For years, textbooks have found the Bard's works to be a rich source of quotes that illustrate mental processes, as when Hamlet says, "There is nothing either good or bad, but thinking makes it so" (*Hamlet*, 2.2.268–70), or when Iago speaks of jealousy as "the green-eyed monster which doth mock the meat it feeds on" (*Othello*, 3.3.196). Among the first to comment on the Shakespeare-psychology connection, psychoanalyst Ernest Jones proposed over a century ago that Hamlet loved his mother, perhaps a little too much.[13] Most recently, psychologist Steven Pinker has argued that Shakespeare pioneered the psychology of morality in *Measure for Measure*.

Will Shakespeare's psychology is all the more remarkable because there is so much of it. As we will show, his plays are filled with his explorations of motivation, emotion, hallucinations, delusions, personalities, social interactions . . . and even what we now call psychotherapy. But from the perspective of his audiences, what Shakespeare seemed to do so well—what no one had done so well before—was to show us ourselves. That is, he identified the common themes of our private mental experiences and reflected them back to us on stage.

WHY DOES SHAKESPEARE ENDURE?

In Shakespeare's London, theaters had to compete with exhibitions of bear-baiting, cock fights, and public hangings. Jousting tournaments and armed combat contests still occurred, even though they were remnants of the Middle Ages.[14] Besides, there were public performances by musicians, acrobats, dancers, and puppeteers. Other pastimes included bowling and team sports resembling soccer, tennis, and badminton.[15] Nevertheless, plays were popular entertainments among Londoners in Shakespeare's time, and his company did well, filling the house with audiences eager to see characters like Falstaff, Hamlet, Rosalind, Juliet, and quasi-historical figures including Kings Henry V, Richard III, and Elizabeth's father, Henry VIII.

Using his ability to connect with diverse segments of his audiences, our playwright was a pioneer in the psychology of marketing. For the "groundlings"—ordinary folk who paid a penny (about the price of a pint of ale at The Anchor) to stand on the ground in front of the stage—he gave them characters like themselves: the yeomen, assorted "rustics," whores, and cutpurses, along with Falstaff, Doll Tearsheet, and the Dromio twins, all on stage and larger than life. He also gave them familiar old tales, but always with a new and often bawdy twist. Here, from *The Taming of the Shrew*, is an example that modern readers can easily understand, as Petruchio and his reluctantly betrothed Kate are matching wits:

> PETRUCHIO: Come, come, you wasp! I'faith you are too angry.
> KATHERINE: If I be waspish, best beware my sting.
> PETRUCHIO: My remedy is then to pluck it out.
> KATHERINE: Ay, if the fool could find where it lies.
> PETRUCHIO: Who knows not where a wasp does wear his sting?
> In his tail.
> KATHERINE: In his tongue.
> PETRUCHIO: Whose tongue?
> KATHERINE: Yours, if you talk of tales, and so farewell.
> PETRUCHIO: What, with my tongue in your tail?
>
> —*Taming of the Shrew*, 2.1.222–31

To be sure, Shakespeare was a cunning linguist.

To entice spectators with some education (and who could afford the shilling required for a bench seat), he included more sophisticated references to current events, the heroes and villains from English history, and the Greek and Roman classics, particularly Homer and Ovid. As for nobility in the audience, their ancestors strutted and fretted in the history plays, portrayed in the best possible light, of course.[16] Especially mindful of his Queen, Shakespeare took pains to depict Elizabeth's forebears approvingly, particularly her father, the much-married Henry VIII. The Bard was not so kind to the Tudor family's rivals, however, even when it meant trifling with the truth. So, the family nemesis, Richard III, became Shakespeare's arch-villain, even though he was actually a good king, as English kings come and go.[17]

To fill the seats, Shakespeare had still other tricks up his ruffled sleeve. Appealing to a populace under the constant threat of war, his historical plays reassured the citizenry that all would be well for England and correspondingly bad for the dreaded Spaniards and the despised French. For those seeking the romance of exotic places, he set plays in Denmark, Italy, Sicily, Spain, France, Greece, Egypt, Turkey, the Balkans, and even the New World. For those in love with love, he offered up *Much Ado about Nothing*, *Love's Labour's Lost*, *Midsummer Night's Dream*, and a dozen more. For those who adored gardens and plants, his plays mention nearly two hundred flowers, shrubs, herbs, and trees. And for fans of a little mayhem or a good murder, the plays snuff seventy-four characters in sundry ways, including stabbing, drowning, beheading, poisoning, hanging, burying to the neck, dismemberment, snakebite, indigestion, being torn apart by a bear, and baked in a pie.

Of course, our Shakespeare never knew that his works would remain popular with audiences down to the present day. We infer this because he published none of his own plays, yet they did endure. And their most pervasive influence may have been on our language. Yes, we're admitting that we have come to appreciate his use of language, particularly his *shaping* of the English language. All told, the fellow is said to have coined several thousand words (or at least penned them for the first time), among them *auspicious*, *baseless*, *castigate*, *dwindle*, *discontent*, *circumstantial*, *monumental*, *sanctimonious*, *madcap*, *moonbeam*, *bloodstained*, *cold-blooded*, *lackluster*, *premeditated*, *watchdog*, and *obscene*.[18] Consider, too, the many Shakespearean expressions we use so often that they have become clichés, among them:

- "It's Greek to me" (*Julius Caesar*, 1.2.295)
- "All that glisters is not gold" (*Merchant of Venice*, 2.7.73)
- "In a pickle" (*The Tempest*, 5.1.337–38)
- "Clothes make the man" (*Hamlet*, 1.3.78)[19]
- "A laughing stock" (*The Merry Wives of Windsor*, 3.1.85)
- "Wear my heart upon my sleeve" (*Othello*, 1.1.44)
- "Break the ice" (*The Taming of the Shrew*, 1.2.271)
- "Bated breath" (*Merchant of Venice*, 1.3.116)
- "Brave new world" (*The Tempest*, 5.1.217)
- "Dead as a doornail" (*Henry VI, Part 2*, 4.10.38)
- "Eaten me out of house and home" (*Henry IV, Part 2*, 2.1.76)
- "Foregone conclusion" (*Othello*, 3.3.486)
- "Good riddance" (*Troilus and Cressida*, 2.1.124)
- "In my heart of hearts" (*Hamlet*, 3.2.78)[20]
- "In my mind's eye" (*Hamlet*, 1.1.124 and 1.2.193)
- "Kill with kindness" (*The Taming of the Shrew*, 4.1.188)
- "One fell swoop" (*Macbeth*, 4.3.258)

Even the seemingly modern "knock-knock" jokes may have started with a comic-relief scene in *Macbeth*, in which the Porter exclaims, "Knock, knock! Who's there?" (2.3.3).

Yet for us, as psychologists, the first connection was the psychology itself. As we pushed through the canon, we found not just the soliloquies and case studies but plays shot through with observations on dozens of topics still prominent in modern psychology. One example comes again from *Hamlet*, where our ambivalent hero plots to smoke out his father's murderers, whom he suspects are his Uncle Claudius and his mother, Queen Gertrude. Hamlet craftily hires a theatrical troupe to stage a play called *The Murder of Gonzago*,[21] in which a monarch is poisoned in much the same way as Hamlet suspects that his father was killed. When the player-queen later proclaims everlasting love for the husband she has just dispatched, Hamlet turns to his mother and says, "Madam, how do you like this play?" Whereupon Queen Gertrude replies, "The lady doth protest too much, methinks." For Hamlet, it was an example of what Freud would later call *reaction formation*, an exaggerated denial of one's own motives.[22]

Another example comes from *Julius Caesar*, where Cassius tries to persuade Brutus to join a plot to assassinate Caesar and save the Roman Republic. Cassius says:

> Men at some time are masters of their fates:
> The fault, dear Brutus, is not in our stars,
> But in ourselves, that we are underlings.
>
> —*Julius Caesar*, 1.2.145–47

With these words, Shakespeare identifies the all-too-human tendency to blame our fears and failings on forces beyond our control, when the fault really lies in ourselves. Three centuries later, Freud dubbed this *displacement*; now psychologists would call it the *self-serving bias*.[23]

We will see, too, that Shakespeare focused entire plays on psychological themes. Here, briefly, are some examples to be fleshed out in coming chapters:

- *Henry V* is ostensibly about the conquest of France by England's greatest hero. But it is also about the psychology of motivation and leadership: how a military leader prepares his troops for battle against seemingly impossible odds.
- *The Merchant of Venice* is a tale of cultural conflict that also explores the psychology of antisemitic prejudice.
- *King Lear* deals with the problems of inheritance. In addition, it reveals the treacherous journey through the decline of old age. As early as the mid-nineteenth century, doctors recognized Lear's condition as a form of senile dementia.[24]
- *Othello*, we have seen, explores how the emotions of jealousy and jealous rage can become the "green-eyed monster" that disastrously overrules reason. Likewise, in *The Winter's Tale*, jealous rage also ruins relationships.

- *The Two Noble Kinsmen* and *The Two Gentlemen of Verona* are studies of gender roles and male bonding, from an Elizabethan perspective.
- *The Tempest*, like *Cymbeline*, explores the nature-nurture (heredity vs. environment) issue. *The Tempest* is also the play for which Shakespeare coined the term *nature-nurture.*
- And of course, many plays in the Shakespearean canon deal with the emotion of love, in one way or another. In *A Midsummer Night's Dream*, we learn that "the course of true love never did run smooth." Then in *Romeo and Juliet*, *Antony and Cleopatra*, *Love's Labour's Lost*, *Much Ado About Nothing*, *As You Like It*, and *Taming of the Shrew*, we discover just how rough that course can be.

We cannot know whether Mr. Shakespeare deliberately and consciously articulated these issues to himself. Nevertheless, we can see them as themes weaving through his plays. Because they are so fundamental to our understanding of mind and behavior, they must have resonated, at some level, with Elizabethan audiences, as well as with us today, and so must have factored into his success as a playwright.

WHERE DID THE BARD GET HIS PSYCHOLOGICAL PERSPECTIVE?

A central argument of this book is that Shakespeare was a psychological observer of a world in transition between medieval and early modern times, identifying habits of mind and patterns in human behavior and culture that would later enter the knowledge base of modern psychology. In that sense, Antonio of *The Tempest* was anticipating psychology when he declared, "What's past is prologue."

Even so, we must be wary of committing Bardolatry by giving Shakespeare more credit than he is due.[25] As a product of his time, he was a Renaissance man with roots still deep in the medieval world. Accordingly, we should not be the least surprised that Shakespeare did not always get his psychology right by modern standards. For example, he often attributes temperaments to the positions of the planets or to the "humors" (body fluids), a view dating back to Galen and Hippocrates. Moreover, his plays frequently rely on devices based on superstition: sorcerers, witches, ghosts, prophetic dreams, astrology, and omens. The point is, while he was not always right, he was obviously fascinated by human behavior and was the first to reveal inner mental processes so extensively on stage.

Does the Bard have anything to teach us moderns about human behavior and mental processes? Let us count the ways. For one, his perspective on behavior and mind was wider and more holistic than we find among most twenty-first-century psychologists, whose focus typically falls on narrow issues such as color vision, prejudice, IQ, or mental illness. For another, Shakespeare explores territory that remains largely ignored by today's psychological science. To give just a few examples, *Much Ado About Nothing* and *Othello* both deal with jealousy and revenge, the *Henry IV* plays explore the desire for power, and *The Merchant of Venice* considers several of the Seven Deadly Sins.[26] Most of these topics scarcely appear in psychological literature,[27] yet we know that they have been factors in real-world tragedy from the time of the ancients—and an untapped reservoir of ideas for new psychological research.

A third way that Mr. Shakespeare can teach us something about psychology is more surprising. He seems to have intuited some notions about the workings of the brain that, centuries later, neuroscientists would confirm using laboratory experiments and brain-scanning devices. We will take a closer look at some of these insights in upcoming chapters: one we now know as the *functional shift* and another we dub the *Lady Macbeth effect.* Most surprising of all (to your authors) was the discovery that Shakespeare's *Measure for Measure* anticipated both the design and the results of Philip Zimbardo's famous Stanford Prison Experiment. More about all of these *anon.*[28]

In the following pages, then, we will highlight many psychological principles enunciated by Shakespeare. But we will also take the opportunity to point out some of the most important ideas in today's scientific psychology and to discuss how those notions differ from the views held in Shakespeare's time. This book, then, is about Shakespeare's psychology and modern psychology—and about Shakespeare himself.

We have organized this volume according to issues, such as *nature vs. nurture* and *reason vs. emotion*, that were of obvious importance to the playwright and are still debated in modern psychology. In the following pages, then, we will discuss both his perspective and the views that twenty-first-century psychologists hold on these topics. With this approach, we hope to suggest a new way of looking at the old fellow and his works.

For those readers who are not specialists in psychology, we hope to introduce you to a few new psychological concepts, as well as to suggest how you may have appeared through his psychological lens, had you shared a pint with him in The Anchor. Even more broadly, we hope to open a dialogue between psychology and literature, where the twain do not often meet.

One further note of caution: Shakespeare did not necessarily hold all the opinions that he put in the mouths of his characters. So, we cannot say that we are always inferring a true picture of the Bard's personal beliefs either from his writings or from the *zeitgeist* (worldview) of the times. We can safely assume, however, that what he wrote represented some mixture of his own views and those common in Elizabethan and Jacobean England. In this sense, Shakespeare can be read as presenting the psychology of an age.

THE CONTEXT OF CREATIVITY AND GENIUS

What was it about the man and his times that might have combined to produce the best-loved writer in the English language, having an unmatched facility with words and a penetrating curiosity about the human mind?[29] Answering that question is a tall order since we know rather little about Will Shakespeare, the man.[30]

He came from Stratford-upon-Avon, a small town roughly ninety miles northwest of London, bringing scarcely more than his talent. As the son of a leather and wool merchant, he had no family pedigree or fortune. Unlike rival playwright, Christopher Marlowe, Will Shakespeare never attended a university, although his grammar-school education did provide a solid grounding in Latin, classical literature, and Greek mythology.[31] In sum, nothing in young Will's lineage or upbringing signaled his emergence as a great literary figure.[32] Still, we might suppose that teachers or some other mentor must have noticed and encouraged his talents. No records, however, survive to document this or any other details of his education—which, again, is not remarkable because he was merely a commoner from the rural English Midlands. Nor do we have evidence that he ever traveled abroad, except through books, to visit the faraway places in which he set many of his works.[33]

It wasn't until his mid-twenties that Shakespeare left Stratford for London. The date was circa 1590, and Elizabeth I was the queen. So, what was London like when Will Shakespeare burst upon its theater scene and into history? The English Renaissance was in full bloom; it was a time of excitement and new ideas, but also of turmoil and fear. The very theaters that would make Will Shakespeare famous, along with fellow playwrights Marlowe, Jonson, and Kyd, were shuttered periodically due to the plague—the Black Death that swept again and again over the populace. All the while, England was still reeling from the religious strife created a generation earlier when Elizabeth's father, King Henry VIII, had dared to defy Rome, outlaw Catholicism, and form his own Church of England. The reason? Henry wanted authority to divorce his wife Catherine of Aragon and marry her maid of honor,[34] Ann Boleyn, who eventually became the second of his six wives—and Elizabeth's mother.

Then in 1599, Spain declared war, and England braced itself for another onslaught of the powerful Spanish Armada, which Londoners feared could sail up the Thames River any day. Our playwright seems to have responded well to the energy of those turbulent times, for 1599 was, perhaps, his most productive year.[35] And because his plays were a commercial success, we can conclude that the public wanted what he had to offer, particularly the rousing call to patriotism that was his play *Henry V.* In addition, they must have wanted pure entertainment, perhaps to divert their thoughts from the dangers that seemed to be lurking everywhere. Shakespeare gave it to them in the three other plays that flowed from his pen that year: *Hamlet*, *Julius Caesar*, and *As You Like It*—plays that also just happened to emphasize the legitimacy of the Crown and the established order.

Despite these successes, we should not imagine Shakespeare as a standalone genius. It could not have been mere coincidence that so many other great figures, the likes of novelist Miguel de Cervantes,[36] philosopher Francis Bacon,[37] Galileo Galilei,[38] and mathematician/philosopher René Descartes, flourished at the same time.[39] (See figure I.2.) Genius does not exist in a vacuum.[40] Rather, it requires the right person appearing in the right place and context, usually connecting with others of like minds. For these thinkers, artists, pioneers, and explorers, the context was the cultural stew made from new ideas resulting from the rise of science, the loss of confidence in religion, the end of feudalism, and a flowering of art and literature, accompanied by increased literacy and constant threats of war and disease. In the wake of all this lay the new age, the Renaissance, with its new ways of understanding the world.

William Shakespeare was one of the shapers of the new age to come. But he was also a product of his heritage and the unsettled times in which he lived. Clearly, he "stood on the shoulders of giants" (to borrow a phrase from Isaac Newton, who borrowed it from someone else).[41] As writers often do, Shakespeare found inspiration in classical sources: He lifted the story of *Troilus and Cressida* from Homer (who also dabbled a bit in the psychology of *hubris*). He reworked stories from Chaucer and Boccaccio to write *The Two Noble Kinsmen*. And, for his English history plays, he borrowed extensively from Holinshed's *Chronicles*. Some of his psychology probably came from Plato and Aristotle, both of whom young Will would have studied in grammar school and both of whom offered complex psychological theories, such as Plato's famous Allegory of the Cave.[42] But where Shakespeare's special interest in mental processes came from remains a mystery.

Neither can we know whether the playwright was aware of how transitional his times were. Yet there he was, in the crosscurrents of two worlds: the Middle Ages and the Renaissance. And so, his plays reflect the conflicting

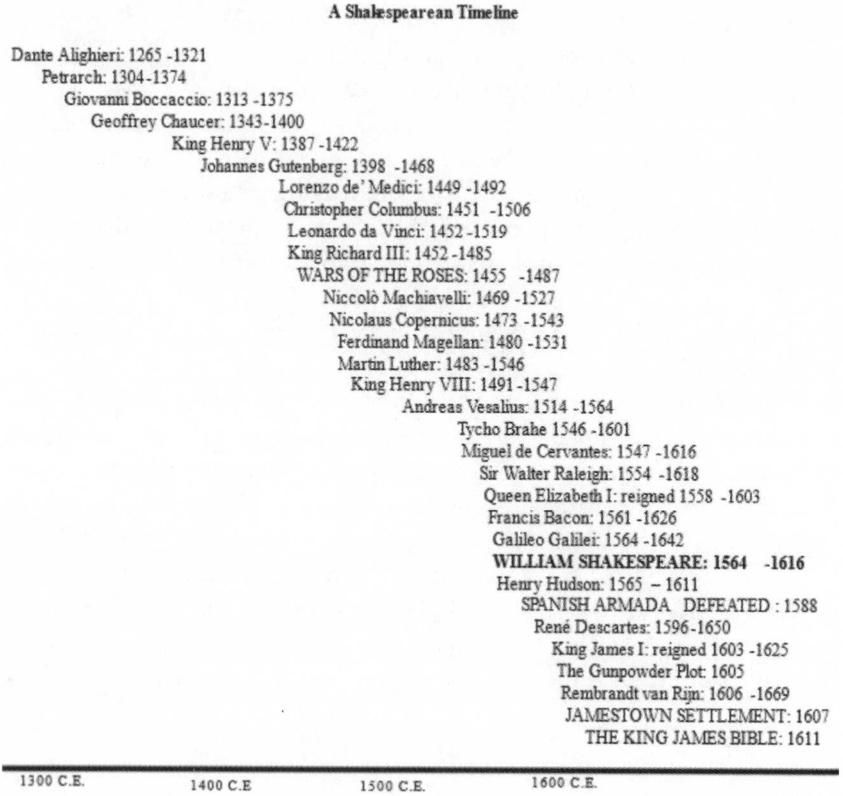

A Shakespearean Timeline

Dante Alighieri: 1265 -1321
Petrarch: 1304-1374
Giovanni Boccaccio: 1313 -1375
Geoffrey Chaucer: 1343-1400
King Henry V: 1387 -1422
Johannes Gutenberg: 1398 -1468
Lorenzo de' Medici: 1449 -1492
Christopher Columbus: 1451 -1506
Leonardo da Vinci: 1452 -1519
King Richard III: 1452 -1485
WARS OF THE ROSES: 1455 -1487
Niccolò Machiavelli: 1469 -1527
Nicolaus Copernicus: 1473 -1543
Ferdinand Magellan: 1480 -1531
Martin Luther: 1483 -1546
King Henry VIII: 1491 -1547
Andreas Vesalius: 1514 -1564
Tycho Brahe 1546 -1601
Miguel de Cervantes: 1547 -1616
Sir Walter Raleigh: 1554 -1618
Queen Elizabeth I: reigned 1558 -1603
Francis Bacon: 1561 -1626
Galileo Galilei: 1564 -1642
WILLIAM SHAKESPEARE: 1564 -1616
Henry Hudson: 1565 – 1611
SPANISH ARMADA DEFEATED : 1588
René Descartes: 1596-1650
King James I: reigned 1603 -1625
The Gunpowder Plot: 1605
Rembrandt van Rijn: 1606 -1669
JAMESTOWN SETTLEMENT: 1607
THE KING JAMES BIBLE: 1611

1300 C.E. 1400 C.E 1500 C.E. 1600 C.E.

Figure I.2 A Shakespearean timeline.

cultural forces pressing on him in the late 1500s and early 1600s, a world as chaotic and uncertain as Hamlet's private world—or our own. For centuries, nearly everyone, from the British Isles to Asia Minor, had accepted the classical view of the universe: an Earth governed by divinely ordained monarchs and the heavens governed by God, assisted by an assortment of angels and opposed by Satan and his legions of demons. The Earth itself stood still at the center of this universe. Around it moved concentric layer upon layer of crystalline spheres to which the stars and planets were fixed. This now-quaint notion comes down to us as "the music of the spheres."[43] At last, in Shakespeare's time, scholars began questioning these ancient beliefs. Among them, Copernicus, Galileo, and Bacon were pioneering a whole new way of thinking that we now call *science*.

Until quite recently, critics judged that Shakespeare ignored much of the scientific revolution going on around him.[44] What really interested him were tales of faraway lands, told by travelers to Greece, Italy, and Asia Minor and, especially, the reports from explorers like Christopher Columbus, who discovered a vast New World inhabited by people whose appearance and customs were unlike anything previously known. We see this in *The Tempest*, based on a true and harrowing tale of a shipwreck that occurred in 1609 when a vessel headed for the English colony in Virginia was thought to have been lost in a storm. Miraculously, it landed on a beach in the Bermoothes (now known as the Bermuda Islands), spent the winter, and eventually reached Jamestown—a year late. In *The Tempest*, Shakespeare "improved" this story by weaving a tale about a magician, a "brave new world," and its strange inhabitants, set against a backdrop described as follows by the sprite, Ariel:

> Safely in harbor
> Is the King's ship; in the deep nook where once
> Thou call'dst me up at midnight to fetch dew
> From the still-vexed [stormy] Bermoothes[45]
>
> —*The Tempest*, 1.2.269–72

But the New World of the Americas was not the only new world. Shakespeare also seems to have been alluding to changes that the Age of Exploration and the English Renaissance were bringing to the familiar Old World, as reflected in both *The Tempest* and *Hamlet*—and possibly in other plays featuring social upheavals, including *Richard II*, *Richard III*, *Julius Caesar*, and *Coriolanus*. It wasn't just the maps and the heavens that were changing; the earthly order, too, was in flux. As trade became more important than landed wealth, earldoms and dukedoms were losing their importance in the social hierarchy of England. No longer did the population cluster for protection mainly around castles and their lords. In their place, cities were becoming the centers of civilization, creating all the promises and woes of urbanization, along with a new and powerful merchant class. Those cities, in turn, offered the ideal breeding grounds for the plague, against which the Church, with its message of pestilence as divine retribution against a sinful populace, proved feckless in staunching wave after wave of death.

Then came the Reformation, raising still more questions about Church authority and doctrine, even as the Inquisition struggled to stamp out these dangerous new doubts. But the force of new ideas and new technology was relentless. Galileo's telescope was important, just as was Gutenberg's invention of moveable metal type, which resulted in books, including the Bible, becoming widely available to an increasingly literate population educated in the public "grammar" schools.[46] Readers now had more access to information from which to form their own opinions about religion and politics. And they were eager to read for pleasure, which was also a boon to writers of the day.

As the old world-order gave way under the flood of daring new ideas and technologies, the old elites—the Church and the nobility—let some of their knowledge and power slip into the hands of the individual. Renaissance-style *humanism* and *individualism* were germinating, and with them, new ideas about human nature and nurture. As you can imagine, these new ideas were controversial—even potentially lethal, especially when they involved the Church or the Crown. Wisely, for his safety, Shakespeare deftly sidestepped such hot-button issues of his day.[47] Nevertheless, he echoed new notions of the Renaissance when he wrote:

> HAMLET: What a piece of work is a man! how noble in reason! how infinite in faculties, in form and moving how express and admirable, in action how like an angel, in apprehension how like a god: the beauty of the world! the paragon of animals—and yet, to me, what is this a quintessence of dust?
>
> —*Hamlet*, 2.2.327

WHO WAS WILLIAM SHAKESPEARE?

We will not engage with the "authorship controversy." To our minds, James Shapiro, Jonathan Bate, and Stephen Greenblatt have settled the issue.[48] We do recognize him instantly in portraits featuring his balding pate (figure I.3). But for us, the Big Questions about Mr. Shakespeare are these: What sort of person was he? How did he get along with his friends? Did he have enemies? What was he like in public and in private? What were his interests, faults, motives, goals, politics, and ideals? If we could share a pitcher of ale with him in The Anchor, what would we talk about? These are more personal questions than whether Shakespeare authored Shakespeare.

Unfortunately, psychology and literary criticism have too often approached Shakespeare in a much different way: by putting him, along with his characters, on the couch. Some three hundred years after Shakespeare, Sigmund Freud's system of *psychoanalysis* captured the popular imagination with its doctrine of unconscious sexual and aggressive urges that drive our thoughts and behaviors. Shakespeare became one of its first targets when Freud's friend Ernest Jones proposed that Oedipal lust drove Hamlet mad. By extension, Jones opined, Shakespeare must have struggled with erotic desires for his mother.[49] Freud himself attributed Macbeth's murderous proclivities to being childless,

Figure I.3 Portrait of William Shakespeare—thought to have been painted, perhaps from life, by John Taylor, a contemporary of the playwright. *Wikimedia Commons.*

suggesting, moreover, that Shakespeare had been prompted by unconscious motives to write the play when his own son died.[50] Not to be outdone at their own game, some literary critics soon appropriated psychoanalysis as the lens through which they viewed their world.[51] One, for example, interpreted Benedick's flirtatious battles with Beatrice, in *Much Ado About Nothing*, as much ado about castration anxiety.[52] Another proposed that the entire Shakespeare canon reflects a tension between the Bard's dominant and aggressive masculine side and his gentle, but feared, feminine side.[53] We are not in sympathy with these views; we find nothing in Shakespeare's works that suggests he was anticipating the Freudian model of the unconscious mind. Rather, we will argue that Shakespeare's view of unconscious processes was much more in tune with psychology's more recent conception of the conscious and unconscious minds espoused by two psychological luminaries: one is Daniel Kahneman, in his remarkable book on these two partitions of the mind, *Thinking Fast and Slow*;

the other is Jonathan Haidt, in his book *The Righteous Mind*, explaining new research on the moral dimensions of our minds.[54]

Searching for the psychology in Shakespeare is also a game that can be played according to many sets of rules. In recent years, a critical method known as *deconstruction* has become fashionable among those who shape opinions about literature. Like Freudian analysis, deconstruction is a filter through which literature is viewed. It aims to uncover psychological, political, and cultural assumptions implicit in literary works, assumptions of which the author may not even be aware. For example, a deconstructionist might ask: What does the portrayal of Othello the Moor tell us about Shakespeare's attitude toward minorities or Islam? What was Shakespeare's view of women? Was he an early feminist, as suggested by his portrayal of the clever Portia in *Merchant of Venice*—or was he a misogynist, as suggested in *Two Gentlemen of Verona*, where one "gentleman" takes it upon himself to forgive his best friend for the rape of the woman he loves? Does his *Winter's Tale* treatment of a king's son (disguised as a shepherd) suggest that his assumptions about the nobility are elitist? Deconstructionists have even guessed at the organization of Shakespeare's brain, using methods borrowed from cognitive psychology.[55] This approach is closer to what we propose. However (again, in our opinion), attempts at using psychology in the deconstruction of Shakespeare can lapse too quickly into psychoanalysis, perhaps since literary critics may lack a strong background in the broader science of psychology—which we are modestly prepared to supply! On the other hand, we would happily learn more about literature from them. Increased dialogue between psychology and the humanities is one of the principal goals of this book.

Any speculation about Shakespeare's thoughts and motives inevitably contains a measure of conjecture. Yet we have discovered a small trove of comments on his personal characteristics by his contemporaries. These do suggest that he was liked and respected by most peers in the theatrical world. It is another irony, then, that we must make inferences about Shakespeare's thoughts, motives, and desires, given that he was the playwright who first revealed the hidden world of the mind in so much detail.

Alas, psychological science has not developed tools ideally suited to the task of deducing an individual's frame of mind from literary texts. Rather, it specializes in settling questions that can be answered by gathering empirical evidence, usually on large numbers of people. So, while we can make our "educated guesses," these should not be mistaken for cold, objective facts. What we *can* do with some confidence, however, is to address the more limited questions posed by this book. We can ask: Which concepts in modern psychology did Shakespeare anticipate? Of which was he apparently oblivious? What did he get wrong? And, to turn those questions on their heads: What blind spots in modern psychology can Shakespeare help us see?

As for the Bard himself, we can safely infer from his writings that he was a man of great intellect and boundless curiosity. We can also learn from his work something of his interests and passions. And then there is a small trove of comments about him from his friends and acquaintances. We will follow those paths further in an upcoming chapter, dealing with what Shakespeare had to say about human personality—and what the psychology of personality might be able to say about him.

CREATIVITY, GENIUS, AND MULTIPLE INTELLIGENCES

Now, having disparaged previous attempts to apply psychology to Shakespeare, we beg your indulgence for a short demonstration of how we plan to go about understanding the man a little better. To do so, we first draw on some ideas proposed by psychologist Howard Gardner, hoping to show how his insights into the psychology of creativity may help us understand how young Will Shakespeare became the beloved Bard.

We know that our playwright-poet had a prodigious intellect. But was it predisposed toward writing for the theater? Or could such a talent just as well have been applied to sculpting like Michelangelo? Composing music like Bach? Or beating Galileo to the discovery of Jupiter's moons? Such questions may sound far-fetched, yet they get at an important issue in the psychology of creativity: Is creativity a single trait that can be applied to any opportunity the world presents? Or is creativity an umbrella term for many different abilities, each specific to a particular field, such as art, literature, or science?

In a provocative book, *Creating Minds*,[56] Howard Gardner explored the origin of creative genius through the lives of seven individuals—Sigmund Freud, Albert Einstein, Pablo Picasso, Igor Stravinsky, T. S. Eliot, Martha Graham, and Mohandas Gandhi—whose work shaped the twentieth century and exemplifies Gardner's theory of *multiple intelligences*.[57] While Dr. Gardner finds some common patterns running through the lives of these greats, he argues that each one represents a distinct type of intelligence: He picked Einstein for his logical-mathematical ability and Picasso for his visual-spatial intellect. Stravinsky's gift was musical, while Eliot's genius arose from his verbal-linguistic ability; for Graham, the dancer, talent lay in bodily kinesthetic control. Then for *intra*personal intelligence (understanding the mind), Gardner chose Freud, and for *inter*personal intelligence (understanding human interaction), he selected Gandhi. After comparing the talents and histories of these seven geniuses, Gardner concluded that creative intelligence does come in different forms.[58] That is, Einstein would probably have failed as a painter (although he was a pretty good violinist), Picasso could not have invented relativity, and Gandhi's dancing doubtless never would have eclipsed Graham's, no matter how much he practiced.

So, what could this suggest about the Bard from Stratford? He might have been a flop as a scientist, but he clearly stands alongside Eliot, a writer with high verbal-linguistic intelligence. We suspect Gardner would also agree that Shakespeare exhibited genius in two other categories. Like Freud, he showed great insight into private mental processes: the hallmark of intrapersonal intelligence, which showed itself so clearly in Hamlet's soliloquy. And like Gandhi, Shakespeare understood interpersonal relationships, as we see in his stories of love (such as *Romeo and Juliet*), of war (such as *Henry V*), and of fantasy (such as *The Tempest*). In Gardner's framework, then, Will Shakespeare was a genius three times over, demonstrated by his skill with words, his understanding of the mind, and his understanding of human interaction.

There is more: Gardner found certain similarities, as well as differences, among the highly creative individuals representing his seven intelligences. All of his exemplars followed a similar script of leaving their hometowns, their families, and their friends to go where the action was. In Shakespeare's case, we see the same pattern in his move from small-town Stratford to the theater scene in London. Moreover, Gardner found that his subjects all had mentors—or a series of mentors—moving on when their skills outstripped those of their teachers. We have no evidence that young Shakespeare ever had such a mentor, although it would be surprising if some teacher in Stratford had not encouraged his talents.

Gardner also found that each of his exemplars experienced major bursts of creativity at approximately ten-year intervals. This is hard to judge in Shakespeare's case because his life as a writer was relatively short—about twenty-five years—although longer than average in those days. As we have seen, the Bard's most productive year, 1599 (the year he wrote *Hamlet*, *Julius Caesar*, *Henry V*, and *As You Like It*), came ten years after his first play (possibly *Two Gentlemen of Verona*).[59] Then, in 1608, the year his mother died and nearly ten years after his prolific year of 1599, Shakespeare's life changed again. He returned to Stratford, where he wrote six or seven new plays. During his career as a poet and playwright, he was continually productive at a prodigious rate, but his work reflects periodic changes in style, seemingly at about ten-year intervals. During the final period of his writing career, when he returned to Stratford, critics suggest, his writing style changed, becoming darker and more complex as he more freely combined comedy with tragedy and romance in such works as *Pericles*, *Cymbeline*, *Winter's Tale*, and *The Tempest*.[60]

Gardner's seven subjects all benefitted from what he called a "matrix of support" consisting of highly creative individuals. While the time of greatest creativity could be a time of isolation or loneliness, each of his genius exemplars had a close friend or circle of friends upon whom they could rely for social support—much as their mentors had served them in their early years. For Freud it was Fleiss; for Picasso it was Braque; for Einstein it was a small study group nicknamed The Olympiad. And, we suggest, for Shakespeare it was probably a small group of fellow playwrights and collaborators, including Marlowe, Kyd, Jonson, and Fletcher.

One final and noteworthy finding in Gardner's research suggests a darker side of genius. Every one of his creative subjects made a "Faustian bargain"—meaning that they metaphorically sold their souls for their greatness, leaving a string of broken relationships behind, as they emerged on the world stage. Was it the same with Shakespeare? Although we know little about his personal life, scholars have suggested that his relationship with his wife, Ann Hathaway, was strained. Church documents show that he was eighteen and she twenty-six and pregnant when they were married in 1582. There is no real documentation about their relationship thereafter, although he moved to London about ten years after the marriage, leaving Anne and their children some five leagues behind in Stratford. This suggests, at least, a distant relationship, punctuated by long absences from his wife and family. A curious and controversial line in his will reinforces this view: "I give unto my wife my second-best bed with the furniture." This strikes some as an odd bequest for a beloved wife.[61] Perhaps Will was not always a faithful, caring husband, and perhaps his Faustian characters—Macbeth, Hamlet, Henry V, Pericles, and Richard III—were not modeled after those he met at The Anchor, The George, or The Boar's Head. Could they have been modeled after Will himself? The answer to that, of course, would be psychological speculation, not psychological science.

Part I

NATURE VS. NURTURE

Your politics, your food preferences, your personality, your career, your educational background—everything you have ever thought or done—all were influenced by some combination of your heredity and environment. A pair of older terms, *nature* and *nurture*, carries essentially the same meaning. "Nature" refers to our physical selves, our biology, and our inherited traits, while "nurture" encompasses our upbringing, learning, and all other life experiences.

As you will see, Sir Francis Galton borrowed the terms *nature* and *nurture* from William Shakespeare and introduced them into the emerging field of psychology during the late 1800s. But which one, nature or nurture, exerts the more powerful influence on us? It's a question that Shakespeare explored in *The Tempest* and *As You Like It*—plays that we will feature in the two chapters comprising part I of this book. Even today, nature vs. nurture remains an unsettled issue—one that lies at the heart of modern psychology.

· 1 ·

NATURE-NURTURE, NEUROSCIENCE, AND THE BRAIN OF THE BARD

The Tempest

Milan, 1611[1]—Prospero was the Duke of Milan until his dastardly brother Antonio usurped his title and position, setting Prospero and his young daughter Miranda adrift in the sea. Prospero's faithful servant Gonzalo, however, had managed to provision their boat with food, water, and books on wizardry. Eventually, the two castaways made landfall on an island inhabited by strange creatures, whom Prospero enslaved, using magical powers he had learned from the books. One such creature, the mercurial sprite Ariel, could fly anywhere on an errand in the blink of an eye. Another, the ill-tempered monster Caliban, was the son of the late witch Sycorax, who had previously ruled the island. And so Miranda came of age among these strange beings, having no interaction with another civilized human, aside from her father—until he divined a boatload of their former countrymen passing nearby. Among them was his brother Antonio.

As the play begins, Prospero conjures a storm (a *tempest!*), causing the passing ship to founder and wash its cargo of nobles and their entourage onto the island. Among them are Antonio's friend, King Alonzo of Naples, and his brother Sebastian, as well as the young, handsome, and eligible Prince Ferdinand.

The beautiful young Miranda fears that all on the ship have perished (figure 1.1), but she later encounters handsome young Prince Ferdinand on the shore, and the two predictably fall in love at first glance. While Miranda and Ferdinand moon over each other, the remaining characters engage in antics and schemes, some comical and some sinister—but all dealing with both the fears and fascinations arising from cultural and class differences among them.

Figure 1.1 *Miranda—The Tempest,* a painting by John William Waterhouse (1849–1916). *Wikimedia Commons.*

This being a Shakespearean romantic comedy, the play climaxes with (nearly) all trespasses being forgiven and strained relationships reconciled. Prospero frees Ariel and Caliban; Miranda and Ferdinand are betrothed; Alonzo's ship magically reappears, not having been wrecked, after all; and the entire party prepares for a return to Milan, where Prospero will resume the duties of his dukedom.

The Tempest flowed from Shakespeare's pen more than a hundred years after Columbus's legendary voyages across the Atlantic. It was a time when reports from the Americas continued to fill in a still-fragmentary picture of the New World. The English, eager for news of these mysterious lands, had a special interest in their Virginia colony, named for the Virgin Queen, Elizabeth. But it was tales of the "savages," mistaken by Columbus for East Indians, that particularly piqued their curiosity. One of those stories involved a ship wrecked in the "vexed Bermoothes" on its way to Jamestown.[2] We can assume that our Bard found the account fascinating, too, for he wove it into *The Tempest.*

When the play debuted, England was a heavyweight competitor in the Age of Exploration, staking claims on new lands in Asia and America, thanks to the likes of Henry Hudson, Walter Raleigh, and Francis Drake. *The Tempest,* then, connected with a public interest in the exotic cultures English explorers had "discovered." It also played on speculation among intellectuals about *human nature*—that is, about what people might have been like in a "state of nature," unaffected by the constraints of civilization.[3] In the Romantic view, championed by Montaigne and Rousseau, our uncivilized forebears were "noble savages," benign and innocent Ariel-like creatures.[4] In opposition, the dour Thomas Hobbes and his crowd saw the "state of nature" in more Caliban-ish terms, proclaiming that life for uncivilized peoples must be "solitary, poor, nasty, brutish, and short."[5]

Either way, both Hobbes and Rousseau were using the term *nature* in much the same way we might speak of "heredity" or the biological basis of behavior. Likewise, they might use *nurture* to mean what we think of as "culture" or "experience" or "learning." So, one of the much-debated issues of their era (shortly after Shakespeare's death) was the relative contributions of nature and nurture to "civilized" beings (such as themselves). The nature-nurture issue is still debated today, although we are most likely to frame it in terms of heredity vs. environment.

NATURE AND NURTURE IN THE BARD'S WORLD AND BEFORE

You might well wonder what Shakespeare could have known about the nature-nurture issue at a decades-earlier time when his world had just stepped out of the Middle Ages. Yet he was far from the first to ponder the question. That honor may go back nearly two thousand years to Psamtik I, one of the last pharaohs of Egypt, who reportedly conducted a nature-nurture experiment to determine the origin of language.[6] The historian Herodotus tells us that Psamtik gave two newborn children to a shepherd, with instructions never to speak to them but just to listen for the first intelligible word they uttered. Old Psamtik believed this utterance would reveal the original language of humankind. The word turned out to be *bekos,* the Phrygian term for "bread," leading the pharaoh to conclude that the Phrygian tongue was the original human language.[7] Whatever the truth may be of the Psamtik tale, it tells us that the nature-nurture issue carries a long history.

To give Mr. Shakespeare his due, we note that he likely coined the very term *nature-nurture* expressly for a scene in *The Tempest*, where Prospero vents his frustration with Caliban:[8]

> PROSPERO: A devil, a born devil, on whose nature
> Nurture can never stick; on whom my pains,
> Humanely taken, all, all lost, quite lost.
>
> — *The Tempest*, 4.1.211

Some credit also goes to the nineteenth-century English scientist, Sir Francis Galton, for borrowing the alliterative word pair *nature-nurture* from Shakespeare and bringing it into the fledgling field of psychology.[9] While Galton never acknowledged Shakespeare as his source, we do know that he was the first scientist to use the term "nature-nurture," and we also know that he was a Shakespeare fan. Said Galton:

> The phrase "nature and nurture" is a convenient jingle of words, for it separates under two distinct heads the innumerable elements of which personality is composed. Nature is all that a man brings with himself into the world; nurture is every influence without that affects him after his birth.[10]

Galton, incidentally, had a darker side related to the nature-nurture issue. As a founding father of the eugenics movement, he sought to discourage breeding among those with "inferior" heredity, such as the poor, the insane, the "mentally defective," and those of "inferior" races.[11] You can judge for yourself whether Galton, as a member of the British aristocracy and the academic elite, had a self-interested bias in the matter of nature and nurture.

While Shakespeare seems to be ultimately responsible for the felicitous pairing of *nature* and *nurture*, he sometimes gets extra credit for things he did *not* say about human nature. In an oft-quoted line from *Troilus and Cressida*, the Bard puts these words into Ulysses's mouth: "One touch of nature makes the whole world kin." The playwright's fans frequently mistake this statement for a Shakespearean embrace of "universal brotherhood."[12] But it is important to read it in context, where Ulysses expresses a very different sentiment, and a rather cynical view of human nature, at that.[13]

> ULYSSES: One touch of nature makes the whole world kin,
> That all with one consent praise new-born gawds,
> Though they are made and moulded of things past,
> And give to dust that is a little gilt
> More laud than gilt o'erdusted.
>
> —*Troilus and Cressida*, 3.3.155

The language Shakespeare used here probably causes our confusion. "Nature" here refers to what *we* would call our *biological* nature. And so, he is describing one unfortunate way in which we are all alike: We are easily distracted by anything new and shiny ("gawds") and without substance. And so he is lamenting that a person's most recent exploits grab our attention and push his past deeds, virtues, and foibles out of memory.[14] It is not a call for a Kumbaya moment but rather a reminder that human memory is short and that fame is fleeting—surely a specter always in the back of a playwright's mind.

HOW ELIZABETHANS UNDERSTOOD HEREDITY

People in Renaissance England had an agriculturist's knowledge of inherited traits.[15] They knew that their children, like domesticated animals, could inherit the parents' physical and mental characteristics, along with their temperament and other personality traits, even though they knew little of the underlying biology.[16] Importantly, this minimal understanding of heredity extended to "noble traits" and was used to justify passing the Crown from one generation of the nobility to another, a practice they also believed was sanctified by God.[17]

Shakespeare slipped the notion of God-ordained inheritance into *The Winter's Tale*, where a gentleman tells us how he recognized the long-lost Perdita as royalty, even though she had been abandoned in infancy and raised by a shepherd:

> GENTLEMAN: The majesty of the creature in resemblance of the mother, the affection of nobleness which nature shows above her breeding and many other evidences proclaim her, with all certainty, to be the King's daughter.
>
> —*The Winter's Tale*, 5.2.38

Nor was this a one-off reference to inherited aristocratic temperament. In *Cymbeline*, the courtier Belarius kidnaps the king's two infant sons and rears them as his own in the wilds of ancient Britain. Years later, Belarius expresses wonder at their noble bearing, despite having been raised in the wilderness. This time Shakespeare crafts the argument in terms of plant breeding:

> BELARIUS: Thou divine Nature, how thyself thou blazon'st
> In these two princely boys! . . . Tis wonder
> That an invisible instinct should frame them
> To loyalty unlearn'd, honour untaught,

> Civility not seen from other, Valour
> That wildly grows in them, but yields a crop
> As if it had been sow'd!
>
> —*Cymbeline*, 4.2.218

The theme of a hereditary noble temperament also threads through the history plays, where Shakespeare spins accounts of English history to justify the line of succession leading to the Tudors.[18] In the *Henry VI* trilogy, for example, he explains heredity to the audience—this time in the language of an orchardist grafting an inferior limb on a superior stalk.

> SUFFOLK: Blunt-witted lord, ignoble in demeanor!
> If ever lady wrong'd her lord so much,
> Thy mother took into her blameful bed
> Some stern untutor'd churl; and noble stock
> Was graft with crabtree slip, whose fruit thou art
> And never of the Nevil's noble race.
>
> —*Henry VI, Part 2*, 3.2.218

So, how did educated people in the early 1600s think that the parents' traits were transmitted to their children? They knew nothing, of course, about sperm cells or ova, let alone genes and chromosomes.[19] Medical authorities taught that hereditary traits flowed in the blood, from which they were passed somehow to the offspring, although they had no explanation as to how the blood of both parents comingled to do the job. Two and a half centuries later, scientists, including the great Charles Darwin, continued to believe that the blood carried our hereditary traits.[20]

Blood, for several reasons, held a special place in people's thinking. It was, of course, a fluid known to be vital for life, but it was also a central element in the Eucharist, where people believed it carried the essence of Christ. In ordinary folk, blood was considered the most important of the four *humors* (fluids) that determined temperament and personality, for it carried the other three humors—black bile, yellow bile, and phlegm—throughout the body.[21] This "blood theory" lay at the heart of Renaissance medicine, which held that humors out of balance could cause disease, and so the practice of "bleeding" was intended to vent the unwanted humors from the blood and cure the patient.

A clearer understanding of inheritance would emerge in the late 1800s when a monk named Mendel began tinkering with his pea plants to determine the laws of cross-breeding and hybridization. Then, only a few years after

Mendel, other scientists armed with microscopes discovered chromosomes and showed them to be the vectors of inheritance. Finally, in the mid-twentieth century, it would be Watson and Crick's discovery of DNA's "double helix" and their subsequent unraveling of the genetic code that finally revealed the mechanism of heredity.[22]

Still, the blood theory dies hard. It was not long ago that many states had a "one-drop rule," meaning that a person with "one drop" of Negro "blood" was considered "colored" (and therefore assigned to second-class citizenship). Even today, many people remain nearly as vague as the Elizabethans were about the mechanisms of heredity. As if to prove the point, we still use the old metaphor when talking about "blood relatives," "bloodlines," "blue bloods," and "hot bloods."

WHY THE NATURE–NURTURE ISSUE WAS IMPORTANT FOR THE BARD

When Shakespeare had to choose between nature and nurture, he chose nature—as in his description of Caliban—and he had practical reasons for doing so. As we have seen, his audiences generally believed that the rigid class structure of English society was prescribed by God. The monarchy did too, of course, so writing plays that supported the Tudors' version of English history helped the Bard keep his head. It also made good financial sense, since the Crown, under Elizabeth, was a patron and financial underwriter of the arts, and Shakespeare's company was a favorite at her court. She was never formally their patron, but after her death, King James *did* become a patron of the company, which changed its name to The King's Men.[23]

With all his emphasis on nature, it is noteworthy that Shakespeare often speaks sarcastically or comically about *learning* (which is merely another way of referring to nurture or experience)—for example, in *The Taming of the Shrew*, *The Merry Wives of Windsor*, *Henry VI, Pt. 2*, and *Love's Labour's Lost*. This is particularly surprising coming from someone who so obviously prized learning in his own life as much as did our playwright. Perhaps he was pandering to many in his audience who had no schooling or detested the schooling they had. Or perhaps he had become cynical about the dominance of nature over nurture, for he knew that noble status was not always inherited but could be acquired—for a price. Like his father, Will sought to rise above his mean station in life, and indeed he did. Legal records show that, in his later years, he renewed his father's unsuccessful application for a coat of arms, which brought with it official standing among the gentry.[24]

A related problem that Shakespeare must have pondered was the mixture of nature and nurture to which he owed his success as a playwright, poet, actor, and businessman. What combination of heredity and environment was at work in his own brain? No other member of the Shakespeare family became a distinguished literary figure—indeed, many were not even literate! We, too, must wonder: Was his obvious talent produced by a freak mutation—a "literary gene" since lost? Or could it have been the encouragement of a mentor, unknown to us? Shakespeare himself is silent on the issue. But if he ever wondered about the contributions of nature and nurture in his own life, he was not alone. For years, scholars have argued whether a fellow with his unremarkable ancestry and a mere grammar-school education could have written the works attributed to his name.[25] Was Shakespeare's genius more nature or more nurture? That is the question.

THE NATURE–NURTURE ISSUE TODAY

There are two reasons why no scientist these days would deliberately isolate children from human language, as Pharaoh Psamtik allegedly did. One is that it would be unethical. The other: psychologists now have far better tools to parse nature and nurture.

Here again, Will Shakespeare seems to have anticipated the methods of modern biological psychology in one of his earliest plays, *The Comedy of Errors*. It's a slapstick farce centering on a pair of twins, separated years earlier (yes, in a shipwreck!), each having one of another twin duo as his servant.[26] The humor derives from a series of mistaken-identity episodes that eventually result in their reunion. Except for the shipwreck, *The Comedy of Errors* is a veritable quick-start guide for the modern psychological study of heredity and environment which examines the similarities in personality, appearance, and aptitudes of twins separated early in life.

Perhaps it was his experience as the father of a twin pair, Hamnet and Judith, that planted the idea, for Shakespeare wrote *two* plays featuring confusion between "identical" twins. The other was *Twelfth Night*, and the twins were Viola and Sebastian (also separated in a shipwreck).[27] But *identical*? How could a girl and boy be identical twins? As we have seen, medicine in those days knew little about the biology of reproduction and inheritance, and physicians were particularly ignorant about twinning. Elizabethans made no distinction between identical versus fraternal twins and had no idea that male-female twins must come from different pairs of eggs and sperm cells. (Audiences also got the irony that both Viola and Sebastian were played by male actors.) The

Elizabethans regarded all twins, regardless of their genders, as coming from essentially the same obscure source: a mother's overreaction to sperm.[28] Moreover, literature of the time often portrayed twins as sinister, perhaps because they were uncommon in the Elizabethan world (mainly because they presented difficulties associated with low birth weight, prematurity, and complications of delivery in an era of high infant mortality). Their comparative rarity also made twins objects of fascination and mystery, a fact that Shakespeare must have recognized as he used twins as a plot device to generate comedy and confusion.

Today, twins reunited after being separated at birth and raised in different families, or even in different cultures, allow psychologists to study the interaction of heredity and environment. Such was the case with the identical twins Oskar Stohr and Jack Yufe.[29] Oskar and Jack were born on the Caribbean island of Trinidad but were sent on different paths in infancy when their parents divorced. Jack remained in Trinidad with his father, where he stood out as the red-headed Jewish boy in a Caribbean culture of mixed African and Native American roots. Eventually, he became an Israeli army officer. Meanwhile, his twin, Oskar, went with his mother to Germany, just as the Nazis were coming to power. Fearful for their safety, she cautioned Oskar never to tell that his father was Jewish. He heeded her advice and took it a step further by joining the Hitler Youth movement, apparently to disguise his heritage. Says Thomas Bouchard, the psychologist who studied them, "Jack and his brother clearly have the greatest differences in background I've ever seen among identical twins reared apart."[30]

When he first saw them, Dr. Bouchard was not surprised that the pair displayed many similarities in their looks, down to their receding hairlines and even their neatly trimmed mustaches. After all, they were identical twins. When reunited in Minneapolis, home of the Minnesota Twin Study Project, Jack and Oskar both arrived wearing wire-rimmed glasses, identical shirts (with two pockets and epaulets), and white jackets. But it was their "likes" rather than their looks that Bouchard found most remarkable. Both had a taste for sweet liqueurs and spicy foods; they each had a habit of reading the ending of a book first; both liked to sneeze loudly, as a joke; and they both exhibited a similar "hot" temperament. (They never quite managed to bond with each other, largely because of their political views.) Perhaps most striking were certain shared mannerisms, such as habits of dipping buttered toast in coffee, flushing the toilet both before and after using it, scratching their heads with a ring finger, and storing rubber bands on their wrists.

Which of these were mere coincidences and which were due to their identical heredity?[31] Dr. Bouchard went after the problem with brain scans and a battery of psychological tests. Their brain waves looked nearly identical, as did their scores on intelligence tests. And when he gave them a personality test, the scores were as similar as the scores of the same individual taking the test twice. Yet none of this information can answer the nature-nurture question for Jack and Oskar—or any particular set of twins. What Bouchard was looking for were patterns that appear broadly across many twin pairs. Those can be said with more confidence to have a strong hereditary influence, especially if found in pairs raised in quite different environments, as were Stohr and Yufe. In the final analysis, twin research demonstrates that "identical" twins, who have the same genetic makeup, are usually much more alike than ordinary siblings on measures of intelligence, interests, abilities, personality, and temperament. But they are never absolutely identical. Twin studies show that both heredity and environment always play a role in determining our characteristics. Even small differences found between twins are magnified by enriched environments and minimized by impoverished ones. This echoes in adopted children from impoverished or neglecting environments who often flourish, with demonstrable gains in IQ points, when transferred to more nurturing homes. The Big Idea coming out of Bouchard's work is that heredity and environment always *interact*—always work together—to produce complex traits.

THE BRAIN AS THE NEXUS OF NATURE AND NURTURE

If nature and nurture interact, one might ask: How do they work their psychological effects? The short answer is that the influence of both plays out in the brain. The long answer is that our twenty-first-century understanding of these effects on the brain comes from knowledge that gradually emerged on two other fronts over the four hundred years since Shakespeare's time: (a) a much deeper and science-based understanding of the brain and (b) the extension of Charles Darwin's ideas into the field now called *evolutionary psychology*. To see how these developments came about, let's start with the understanding of the brain in the Bard's day.

At the turn of the seventeenth century, there was no *science* of medicine. Rather, medical knowledge consisted largely of lore dating back to antiquity. While physicians of the day did correctly understand the brain to be the organ of thought and reason, they knew nothing of its component neurons and pathways, with their capacity to carry messages and process information, nor did they know that different brain areas controlled different senses, emotions, and behaviors. But they were learning about the brain from dissections being performed by physicians and artists, such as Leonardo. And thereby hangs another tale—with an intended pun, as you will see.

Some three centuries before Shakespeare, the Church relaxed its prohibition on dissection of human bodies, allowing doctors to perform anatomical studies for the first time since the classical Roman era. Most of the subjects were freshly hanged criminals because other potential donors were reluctant to arrive at Heaven's gates with their worldly bodies mutilated. Thus, it fell to the courts to supply the cadavers by sentencing prisoners to be "hung by the neck until dead" and then to be "anatomized." A barber-surgeon often did the dissecting, under the supervising eyes of a professor of anatomy, who could then lecture about the procedure without having to soil his hands with substances exuded from the corpse.[32]

Many such dissections occurred in special "theaters" constructed at medical schools (figure 1.2), but others were staged in makeshift venues set up as close as possible to the gallows. Anatomists preferred to have their subjects as fresh and warm as possible. This was in part because, after about three days, especially in the summer, the stench could become overwhelming. Another reason for proximity to the gallows was to prevent "body-snatching," for a fresh corpse could fetch as much as £10—roughly a year's wages for a common laborer.[33]

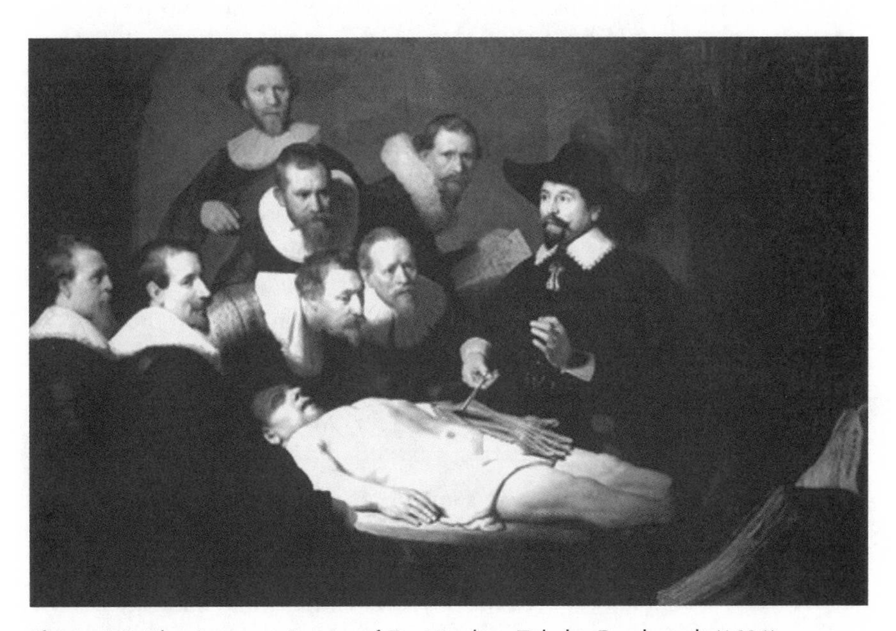

Figure 1.2 *The Anatomy Lesson of Dr. Nicolaes Tulp* by Rembrandt (1631). *Wikimedia Commons.*

All over Europe, dissections and autopsies became performance art, open to the public (customarily, for a small fee). England was slightly behind the times because of the ill will between the pope and Henry VIII; public dissections did not become common there until late in Shakespeare's career. One source of English corpses was the notorious site of public hangings in London, known as the Tyburn gallows, located just a few yards from what is now Marble Arch in Marylebone—not far from The Globe, the theatrical home to Shakespeare's troupe.

We have no hard evidence that Will ever witnessed a public dissection, but he was a curious man, and his works are filled with medical references to the liver, spleen, heart, and brain, along with the four "humors" believed to be responsible for differences in human temperaments. He was obviously fascinated by the brain, for his plays contain some 153 references to the organ.[34]

But how did the brain work? One couldn't tell its functions simply by studying its form. It would not be until the late 1800s that the roles of specific brain structures would begin to be understood. But that didn't stop scholars and physicians in the pre-science era from guessing the brain's functions from its form. In Shakespeare's day, medicine was still in the sway of the second-century Greek physician Galen, who taught that our sensations mingled in the *ventricles* (hollow spaces) of the brain, where they somehow melded with memory and intellect (also called *sensus communis* or "common sense") to form our perceptions of the world. That is the image Shakespeare invokes when, in *Love's Labour's Lost*, we hear the same Holofernes speak of the brain's ventricles, and of the membrane enveloping the brain, known as *pia mater* (see figure 1.3).[35] And then we hear him explaining that the hollow ventricles are the reservoir in which the brain stores memories, which mingle together to produce our thoughts:

> HOLOFERNES: This is a gift that I have, simple, simple—a foolish extravagant spirit, full of forms, figures, shapes, objects, ideas, apprehensions, motions, revolutions. These are begot in the ventricle of memory, nourished in the womb of pia mater, and delivered upon the mellowing of occasion.
>
> — *Love's Labour's Lost*, 4.2.82

It is tempting to think that at least some of our playwright's medical knowledge may have come by way of his son-in-law, Dr. John Hall, a physician who married Will's daughter Susanna.[36] However, that relationship most likely does not account for Holofernes's references to the ventricles and pia mater. Why? Because Susanna was a maiden of about twelve years old when her father wrote the play.[37]

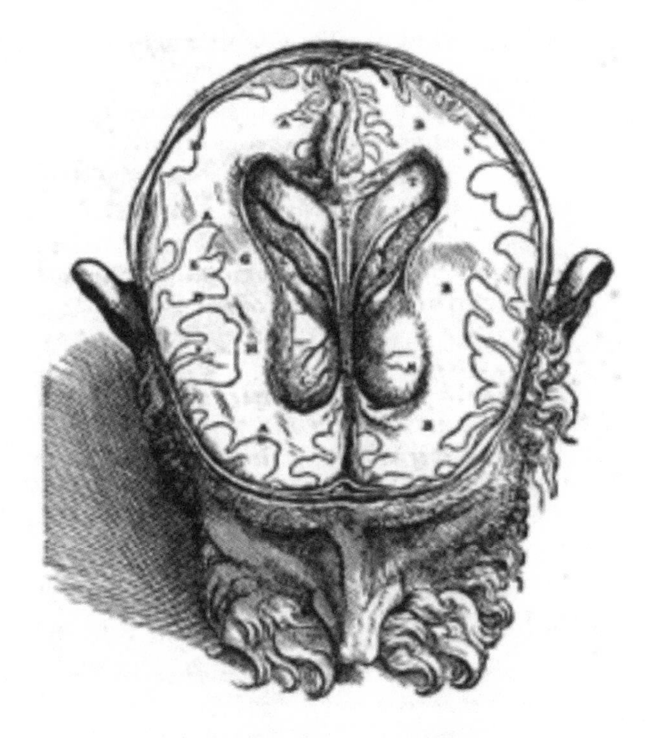

Figure 1.3 Vesalius's drawing of the brain, with the ventricles shown in cross-sectional (transverse) view. *Wikimedia Commons.*

Doctors of medicine in Elizabethan times did correctly infer that diseases of the brain could cause mental illness (although they were likely to attribute it to an imbalance of the humors). So, we hear Lady Macbeth declare that human problems may stem from thinking "so brainsickly of things."[38] The same idea echoes in *The Tempest*, where Prospero (to Ferdinand) ascribes his confusion to a "troubled" brain:

PROSPERO: Sir, I am vexed.
Bear with my weakness. My old brain is troubled.
Be not disturbed with my infirmity.
If you be pleased, retire into my cell
And there repose. A turn or two I'll walk
To still my beating mind.

—*The Tempest*, 4.1.163

COMMON SENSE AND THE BINDING PROBLEM

How does the brain manage to combine the elements into a meaningful thought or image? Shakespeare's sources believed that the *sensus communis* did the job by allowing the brain to accept only sensible constructs, but they did not understand the underlying mechanism. Nor do we, even today. Neuroscientists call it *the binding problem.*

To illustrate, vision requires combining several streams of incoming visual information from, say, a rose. Those streams separately tell the brain about light and dark, lines, edges, colors, and perhaps movement in the wind. Each feeds into a different region of the visual cortex, which detects and identifies the individual properties of the rose. Then, somehow, the brain recombines these components of the visual image and links them with memories, emotions, and language. ("It's a yellow rose!") Meanwhile, other senses may be processing the rose's scent or the prick of its thorn and associating these sensations with the visual image. It is all done quickly and with apparent ease, yet we have no good explanation of how the brain manages this feat. Now consider these lines that Shakespeare wrote for Juliet:

> What's Montague? It is nor hand, nor foot,
> Nor arm, nor face, nor any other part
> Belonging to a man. O, be some other name!
> What's in a name? that which we call a rose
> By any other name would smell as sweet;
> So Romeo would, were he not Romeo call'd,
> Retain that dear perfection which he owes
> Without that title.
>
> —*Romeo and Juliet*, 2.2.41

While Shakespeare never tried to explain just how the brain combined the separate attributes of a sensory image—how it constructs a coherent image of a Montague or a rose—he clearly knew, four centuries ago, that it must do so.

What happens after the brain "binds" all the elements of the rose into one percept? It is only then that the image enters into consciousness. Therein lies an even deeper mystery: How does the brain create consciousness? It is the old problem of the *sensus communis* dressed in the language of biopsychology and neuroscience.[39]

THE "THEATER" OF CONSCIOUSNESS

Consider what we experience in the conscious mind. Neuroscience now tells us that conscious thoughts, sensations, desires, and memories consist entirely of electrochemical signals shuttling among neurons because those signals are the brain's only language. When we "see" light, no light actually enters the brain because light penetrates only as far as the retinas of our eyes. Beyond that point, light must be converted into the electrochemical signals the brain understands. So what we "see" is nothing more than the activity of nerve cells. The same goes for sound waves, which nerve cells must also translate into the neural language of brain cells. And so it goes for the chemical essences of odors and tastes. Consciousness seems like nothing so much as a mental version of The Globe, where an image of the world is playing on the brain's stage. In reality, however, all we really experience are the electrochemical messages shuttling from one neuron to another through our neural network.[40]

Yet the theater of the mind is not the most profound mystery of consciousness. An even deeper question asks: Who or what is observing these thoughts and images inside our heads? In what sense are *we* the audience? If consciousness is like a screen where the brain projects the continuous movie of our lives and dreams, does that imply a little entity who watches the movie . . . in an even smaller venue inside her head? Quickly such theatrics turn into an absurdity—an infinite regress. While much has been written on this problem, so far no one has found a convincing solution.

Shakespeare never used the word "consciousness" and never seems to have puzzled over its deepest mysteries. The word itself did not appear in print until a half-century after his death. Yet his frequent use of sleep, dreaming, and drug-induced states as plot devices leaves no doubt that he was interested in what we call *states of consciousness*, for these mental states turn up in so many of his plays, including *A Midsummer Night's Dream*, *Romeo and Juliet*, *King Lear*, *The Tempest*, *Cymbeline*, and *Hamlet*. Moreover, he sometimes used "conscience" with a double meaning that included "awareness," as in Hamlet's sardonic wordplay:

> HAMLET: Thus conscience doth make cowards of us all;
> And thus the native hue of resolution
> Is sicklied o'er with the pale cast of thought.
> And enterprises of great pith and moment
> With this regard their currents turn awry,
> And lose the name of action.
>
> —*Hamlet*, 3.1.90

What can we conclude about Shakespeare's conception of consciousness—or "conscience"? It seems that the Bard was telling us here that he conceived of it more as an actor than an anatomical structure—as a sort of player in the theater of the brain.

And so, we have come nearly full circle to Elizabethan times, but with still no understanding of the deepest mystery of all in psychology. Yet, we have made great advances in our understanding of the brain. No part of the great neural knot remains *terra incognita.*[41] Indeed, neuroscience has found a function for every part of the brain, even those lying deep underneath the gray matter of the cortex. Yet we still know neither the locus of the conscious mind nor how it works.[42]

PLUCKING OUT A ROOTED SORROW: NOW AND THEN

In our twenty-first-century view, the brain consists largely of cells that process information, much like a computer does, so we talk of its "wiring" and of the brain's neural "network" of connections that link its components.[43] The metaphor is, perhaps, less apt for describing communication at the level of individual neurons, which depends on rapid chemical processes occurring in tiny sacs, known as *synapses* that connect one nerve cell to another—a notion arguably a descendent of the old humor theory.[44] Medicine has learned to exploit this chemistry with drugs that enhance or suppress the messages shuttling across the synapses. Such processes are fundamental, as well, to our current understanding of thought, memory, motivation, and emotion, along with the malfunctions that can cause mental illness.[45]

The current metaphor for the brain has always been the most complex device known at the time. A century ago, it was a telephone network. Even earlier, people conceived of the brain in terms of steampunk-like automatons, driven by pulleys, springs, chains, and brass gears, while Freud saw the brain in terms of the hydraulics and steam pressures that drove the Industrial Revolution. In Shakespeare's time, people likened the brain to a clockwork, causing them to wonder what makes us "tick." We hear this metaphor in *The Tempest* when Sebastian sarcastically says of old Gonzalo, "Look he's winding up the watch of his wit. By and by it will strike" (2.1.14).

Behind these metaphors lay the hope of discovering the mechanisms that generate thoughts, feelings, actions, and perhaps the locus of consciousness, the self, or even the soul. As we have seen, Leonardo, Vesalius, and others attempted to locate such mechanisms by dissecting the brain. Some anatomists countered that such ideas were nonsense because the whole brain was involved with all our thoughts and actions, an argument based on the observation that extent of a brain injury was more important than the site of the injury.

Then, in the early 1800s, a group calling themselves *phrenologists* proposed that specific patches of cerebral cortex control distinct aspects of a person's personality and intellect. Further, their eccentric leader, Franz Joseph Gall, claimed that by measuring the skull's irregularities he could "read" a person's mental strengths and weaknesses, which he supposed reflected physical development in the underlying brain. A gullible public quickly embraced these ideas, and so did many in the medical community.

Eventually, however, phrenology was debunked by scientific evidence. Even so, it turns out that Gall wasn't altogether wrong. Different areas of the brain *do* have different functions—just not the ones that Gall and his troops thought were associated with the skull's bumps and depressions. The evidence showed the brain to be a collection of modules specialized for specific tasks, such as vision, hearing, and movement.[46]

Some of the first empirical evidence for localization of function in the brain came in 1848 from a dramatic accident in which Phineas Gage, a railroad worker, was tamping a charge of blasting powder with a thirteen-pound steel bar when the charge exploded, driving the bar up through the left cheek and out through the top of his skull (figure 1.4). Miraculously, Gage did not die.

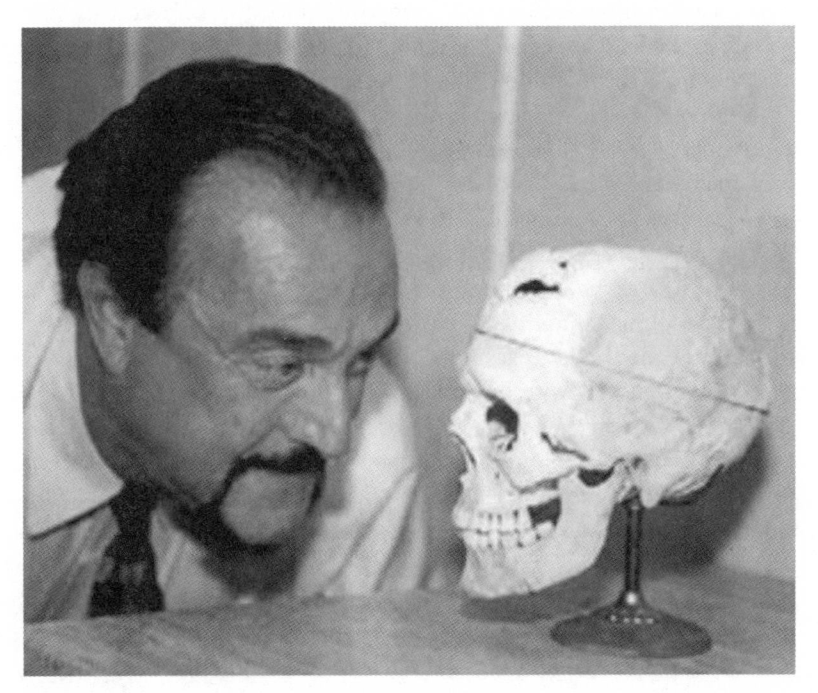

Figure 1.4 Dr. Philip Zimbardo examines Phineas Gage's skull at the University of Akron's National Museum of Psychology.

But upon recovery, he exhibited a profound personality change as a result of the self-inflicted "lobotomy" that destroyed the left frontal lobe of his brain. The previously even-tempered Gage became profane and obnoxious. The physician who treated him later observed, "He is fitful, irreverent, indulging at times in the grossest profanity, which was not previously his custom."[47] Obviously, the left frontal lobe of the brain has a role in curbing some of our worst tendencies.

Then, in the 1860s, more evidence came from the autopsy table. There Dr. Paul Broca found similar patterns of lesions farther back in the left frontal lobes of patients who had experienced severe speech difficulties before death.[48] Later work confirmed Broca's suggestion that a certain patch of cortex in that region was involved in the *production* (but not the *understanding*) of speech. Further studies demonstrated that the two brain hemispheres play different roles not only in language but in emotions and mathematical ability.[49]

In the 1870s, studies of combat casualties in the Franco-Prussian War produced more evidence for localization. In a rather grisly experiment, army surgeons Fritsch and Hitzig used an electric probe on the exposed brains of mortally wounded soldiers to show that a narrow strip of cortex on each side of the brain is responsible for movements on the opposite side of the body.[50] Subsequent observations demonstrated that each hemisphere processes auditory and tactile sensations, also on the opposite side.[51]

Back in the laboratory, animal studies (mostly on rats but also frogs, monkeys, and chimpanzees) confirmed and extended these observations. Such investigations would have been unthinkable in Shakespeare's day because humans were considered a special creation of God, distinct from other creatures. The concept that people share a common ancestry with all living things arose much later when Darwin published his theory of evolution in his groundbreaking book, *On the Origin of Species.*[52]

Finally, in the mid-twentieth century, it was Canadian surgeon Wilder Penfield who continued to fill in the cortical map. Penfield pioneered a method that used local anesthesia to keep his patients fully conscious, but free of pain, during brain surgery. This technique allowed patients to tell Dr. Penfield what they were experiencing when he stimulated their brains with a mild electric probe—and so he was able to identify and remove diseased tissue and avoid removing healthy brain tissue. For brain science, the result was a detailed map of the brain's gray-matter cloak, including previously mysterious "silent areas."

In the most striking of his discoveries, Penfield found that he could evoke memories by stimulating the temporal lobes—some memories so vivid that patients thought they were actual events in the operating room. One heard a popular song played by an orchestra, another heard singing. Still others had *déjà vu* experiences: "Something brings back a memory. I can see Seven-Up

Bottling Company . . . Harrison Bakery," and "Yes doctor, yes. Now I hear people laughing—my friends in South Africa!"[53]

Thanks to Penfield and his probe, medicine was able at last to distinguish diseased from healthy brain tissue and to perform surgically what Macbeth had requested metaphorically when he implored the doctor to cure his wife's mental distress:

> MACBETH: Cure her of that.
> Canst thou not minister to a mind diseased,
> Pluck from the memory a rooted sorrow,
> Raze out the written troubles of the brain
> And with some sweet oblivious antidote
> Cleanse the stuffed bosom of that perilous stuff
> Which weighs upon the heart?
>
> —*Macbeth*, 5.3.49

THE FUNCTIONAL SHIFT: SHAKESPEARE AS AN INTUITIVE NEUROSCIENTIST

We have seen that Shakespeare's knowledge of the brain was remarkable for a layman of his time, although it was sometimes quite wrong by modern standards. Nevertheless, there was one thing about the brain that Shakespeare knew intuitively but which seems altogether astonishing today. It's a neurological trick the brain plays when hearing a word used in a grammatically unusual way. We can illustrate with some examples (below in **bold**). The first comes from *The Tempest*, as Caliban plots to have his master, Prospero, murdered.

> CALIBAN: Why, as I told thee, 'tis a custom with him,
> I' th' afternoon to sleep: there thou mayst **brain** him,
> Having first seized his books, or with a log
> Batter his skull, or paunch him with a stake,
> Or cut his wezand with thy knife.
>
> —*The Tempest*, 2.2.96

Here Caliban uses "brain" not as a noun, as we usually do, but as a verb, meaning to strike a lethal blow to the head. Linguists call such a grammatically uncommon use of words a *functional shift.*

In another example from *The Tempest*, the servant Trinculo uses "brained" in an unexpected way to mean the quality of one's intellect.

TRINCULO: Servant-monster! the folly of this island! They say there's but five upon this isle: we are three of them; if th' other two be **brained** like us, the state totters.

—*The Tempest*, 3.2.5

What happens when the brain senses these grammatically unusual words? To find out, curious neuroscientists who had noticed Shakespeare's functional shifts concocted sentences containing the very words that Shakespeare had grammatically changed. Here are a few examples used in the experiment, again with the functionally shifted words in **bold**:[54]

- He was no longer alone in the world; he was **wived** to a kind and beautiful woman.
- At last, the dancer is ready; he now **foots** his course with grace and vigor.
- You do not have to shout; please **season** your anger before talking to me.
- You're secretly my enemy: you **monster** my deeds, making them look most foul.

Next, while volunteers read the sentences aloud, the scientists eavesdropped on their brains with an fMRI scanning machine.[55] The scans showed spikes in neural activity during processing of the functionally shifted words—but remarkably *not* in brain areas normally used to process speech! Rather, the spikes emanated from loci in the brain known to focus attention. The scientists concluded: "These findings show how Shakespeare's grammatical exploration forces the listener to take a more active role in integrating the meaning of what is said."[56] In other words, Shakespeare knew how to make an audience stay awake during a performance.[57]

ANTICIPATION OF EVOLUTIONARY PSYCHOLOGY

Our current understanding of nature and nurture grows out of developments in biopsychology, neuroscience, and a newcomer field called *evolutionary psychology*. It is to the latter that we now turn, to persuade you that Shakespeare had an intuitive grasp of Darwin's fundamental idea that our behavior must serve the twin needs of survival and reproduction: the needs most crucial for sending offspring into the next generation.[58] Our thesis is that Shakespeare was a folk psychologist[59] who unwittingly anticipated evolutionary psychology with his tales wrapped around the very urges of "nature" that most often get people in trouble: aggression, jealousy, greed, pride, gluttony, and especially sex.[60]

Why, for example, do men and women typically have such different attitudes toward sexuality?[61] Evolutionary psychologists explain that males and females have faced, over the millennia, very different evolutionary pressures. Because men don't become pregnant, their best strategy (from the evolutionary perspective of multiplying their genes) is to be eager breeders: to have sex as often as they can and with as many fertile mates as possible. Meanwhile, a woman hoping to spread her genes faces a much larger biological investment in pregnancy and a far greater biological limitation on the number of offspring she can produce. Further, women typically invest much more time and energy in child-rearing than do men. A woman's optimum strategy, then, is to be a reluctant breeder, withholding sex from all but the best possible mates: those who are the healthiest, command the most resources, and seem most likely to protect their children. Many human customs and problems flow directly from these gender differences.[62]

While all that may be true, if you *really* want to understand how evolution has shaped the human mind and behavior, look to *literature* says evolutionary psychologist David Barash and his coauthor, daughter Natalie Barash.[63] Read the works of great writers like Homer, Boccaccio, Virginia Woolf, and, yes, William Shakespeare. There you will find tales wrought by gender differences, sexual duplicity, jealousy, treachery, and revenge—stories showing how humans cope with the problems of surviving and pushing their genes into the next generation.

Consider the great recurring literary theme of infidelity. *The Iliad*, for example, is an epic poem describing a ten-year war fought over Helen, a married woman who ran off with another man, Prince Paris of Troy. Likewise, the Greek play *Medea* tells of a woman who kills her children to get revenge on a husband who had cast her aside for another woman. And so it goes in Boccaccio's *Decameron*, in Chaucer's *Canterbury Tales*, and down to the present day in the Sondheim musical *Into the Woods*, where Prince Charming and the Baker's wife share a one-night stand. See the pattern?

Turning back to Shakespeare, we find in *Othello* a man who, in a fit of jealous rage, smothers his wife with a pillow. Why? Because he (falsely) believes she is having an affair with his lieutenant Cassio. Then in *Cymbeline*, Shakespeare gives us the randy and dandy Iachimo, who conspires to make Posthumus believe that his wife, Imogen, has cuckolded him. Similarly, in *The Winter's Tale*, a happy marriage and a lifelong friendship sour when the King of Sicily mistakenly believes that his queen has cheated with his friend, the King of Bohemia, and so the child she is carrying is not his. And in *Much Ado About*

Nothing, the plot turns on a deception, as Don John and his malefactors make Claudio believe that his betrothed, Hero, has been unfaithful the night before their wedding.

Notice that these stories all involve suspected infidelity and the resultant jealousy, or as Shakespeare called it, "the green-eyed monster." The Barash team tells us that these classic stories are read and reread not just because Shakespeare was a wordsmith or Boccaccio's tales were racy or Helen was attractive enough to launch a thousand ships toward Troy. While all that may be true, these works have become classics because there is also a bit of our Darwinian selves in all of them.

Not every literary classic involves infidelity and jealousy, of course. Other stories, such as *Huckleberry Finn*, *Snow White*, and *The Lion King* connect with different instincts. But they, too, touch on deeply felt fears, desires, and emotions that also bear on survival and reproduction. We don't mean to echo Freud here, for we are not speaking of an evil and Oedipal unconscious. Rather, as the Barashes say, the classics of literature all deal with problems of survival and reproduction that humans have faced ever since earning the surname *Sapiens.* Each is a lesson in evolutionary psychology, of how the mightiest can fall, leaving the survivors to reproduce. Shakespeare, Homer, Cervantes, and the rest all unknowingly anticipated evolutionary psychology with their tales that touch the deepest sources of our humanity.

So, we see that evolutionary psychology is not all about sex, but rather about motives and emotions that have been built into all of us because of their survival value for our ancestors. From this perspective, then, the history plays are largely about strivings for power. *The Merchant of Venice* can be read as a play dealing with a cultural wariness of the "other." And *King Lear* addresses the sources of family ties and parent-child bonding that we now understand as having, in part, a biological basis.[64]

THE BRAIN OF THE BARD

Neuroscientists would love to study Shakespeare's own brain, but it has long since turned to dust. Perhaps, if they could just examine his skull, they might make some reasonable inferences about the brain it once contained.[65] Indeed, the inside of the skull does record faint impressions of a brain's cortical convolutions (which were said in Shakespeare's time to resemble the twists and coils of the intestines).[66] We must caution you that, in matters of the brain, size does not much matter, for brain volume has essentially no relationship either to intelligence or creativity. So why would it be interesting to see Shakespeare's brain—or those of Einstein, Van Gogh, or da Vinci? We do have Einstein's brain, and anatomists report that, while it is not exceptionally large, it has features—unusual connections—that may account for his famed ability to visualize physical processes. Even Einstein's brain doesn't carry much weight, however. It was only about two pounds—and it's an inert specimen, quite incapable now of the creativity it had in life. So, could studying the brains of exceptional living individuals tell us anything about someone like William Shakespeare?

Neuroscientist Vilayanur Ramachandran reports that highly creative people are far more likely than the rest of us to have *synesthesia*, a syndrome that produces "cross-talk" among the brain's sensory areas.[67] A synesthete may tell you, for example, that Wednesdays are "green" or that a pear tastes "pointy." Such people are not hallucinating. Ramachandran has shown that the different parts of the synesthetes' brains responsible for these "mistaken" sensations are indeed communicating with each other. Moreover, synesthetes realize that these cross-sensations exist in their minds, not in the external world. That is, they know that the sensory color of Wednesdays and the pointiness of pears is a product of their imaginations and not hallucinations.

The condition runs in families, so there is apparently a hereditary component. And we should note, synesthesia occurs some seven or eight times more often in poets, painters, and novelists than in, say, accountants or dentists or plumbers. Some small degree of synesthesia is quite normal, says Ramachandran. It is the reason why we might call a cheese "sharp" or speak of a shirt with "loud" colors. In synesthetes, however, the cross-talk among the senses is enhanced and seems to enrich their experiences.

If you have *grapheme-color synesthesia*, for example, you might associate letters or numbers with colors. This makes sense when we realize that the part of the visual cortex that processes letters and numbers lies adjacent to the region that processes colors. Moreover, neuroscientists have found grapheme-color synesthetes to have more neural connections between the color- and number-processing brain areas than do the rest of us. Using brain-scanning technology, Ramachandran was able to show that both the color and the number areas "light up" when grapheme-color synesthetes see either a black-and-white number or a color. In "normal" people, this does not happen.

We need one more piece of the puzzle before relating all this back to Shakespeare. In *conceptual synesthesia*, the cross-talk associations involve both sensations and concepts. You can probably guess where this is heading, and it is here where the theory becomes speculative. Ramachandran believes that conceptual synesthetes are particularly good at creating metaphors, as when Shakespeare writes, "It is the east and Juliet is the sun" (*Romeo and Juliet*, 2.2.1). Did Shakespeare have conceptual synesthesia? At this point, we can only begin to answer that question by assessing more highly creative artists and writers alive today for conceptual synesthesia, and if so, perhaps to find ways of encouraging its development.

As for Shakespeare's skull, it is likely that we will never be able to examine it—for two reasons. First, there is the matter of the curse carved on his tombstone (figure 1.5):

GOOD FRIEND FOR JESUS SAKE FOREBEAR,
TO DIGG THE DVST ENCLOSED HERE.
BLESTE BE Y^{E} MAN Y^{T} SPARES THESE STONES,
AND CURSED BE HE Y^{T} MOVES MY BONES

The second reason involves a study of the gravesite using ground-penetrating radar.[68] The radar images show that Shakespeare's body is there—minus his head. One can only hope that his skull will someday be found and that the grave robbers who moved the skull will get their cursed comeuppance.

Figure 1.5 Shakespeare's tomb at Holy Trinity Church in Stratford upon Avon. *Wikimedia Commons.*

· *2* ·

THE AGES AND STAGES OF MAN (AND WOMAN)

As You Like It

The Forest of Arden (circa 1599)[1]—Duke Senior's mean brother Frederick has deposed him, after which the ex-duke has taken refuge in the forest, living quite comfortably with his entourage. As the play begins, Frederick banishes Senior's daughter Rosalind, whereupon her close friend and cousin, Celia (Frederick's daughter), decides to accompany her, as does Touchtone, the court fool. And so, they leave Fredrick's court together in search of Rosalind's heartthrob, Orlando, who just happens also to have left the court because his brother, Oliver, is trying to have him killed. These events, then, are the backdrop against which the antics of Shakespeare's melodramatic *As You Like It* unfold.

To make their travel safer, the two girls disguise themselves as boys (played on the Elizabethan stage by boys); Rosalind becomes "Ganymede" and Celia decides to be "Aliena." Coincidentally, all end up in the Forest of Arden, as does one other lesser character on whose wise words we will focus in this chapter. His name is Jaques, and we will return to him in a moment.

The principal plot centers on Rosalind and Orlando, who are hopelessly in love with each other, although neither realizes, at first, how reciprocal the other's feelings are—nor does either initially know that the other has come to the forest. Soon, however, Rosalind finds notes hung in the tree branches, proclaiming Orlando's love for her. This discovery gives canny Rosalind the upper hand in the budding relationship. So, in her disguise as the boy Ganymede, she finds the lovesick Orlando and proposes a sort of behavior therapy to cure him of his lovesick malady: Orlando will role-play wooing Ganymede, as though "he" were Rosalind (which *she* is).[2]

Amid this revelry, Rosalind, Celia, and Touchstone encounter a bevy of odd characters and become unintentionally involved in multiple love triangles. As for Orlando, he faithfully attends his therapy sessions with Ganymede.

Meanwhile, back at the court, the dastardly Duke Frederick realizes that Orlando is missing and orders the fratricidal brother Oliver to search for him. Happily, Orlando finds Oliver first—and saves his life by improbably rescuing him from a lion. In return, Oliver experiences a rapid change of heart and apologizes to his brother for having mistreated him all these years. Likewise, all the contentious love triangles sort themselves out, and there are just enough pairs that almost everyone can get married. A wave of happiness miraculously spreads to the court, where Duke Frederick repents for mistreating his brother and says he will join a religious order, Jaques vows to go with him, and Duke Senior is restored to his former status.

JAQUES'S "SEVEN AGES OF MAN"

For Elizabethan audiences, *As You Like It* was a comedic romp in the woods with a happy ending. But for our purposes, a famous speech by Jaques hints at Shakespeare's understanding of how people grow, change, and mature across the lifespan—the study of which we moderns now call *developmental psychology*. Jaques's version of development, however, calls for some explanation.

You see, the play is primarily a spoof on courtly love customs of the medieval past, which suggests that Shakespeare knew that times had changed since the Middle Ages—that cultural changes had influenced the way people interact. He shows us this, with dramatic license, in the exaggerated changes of heart in the usurping Duke Frederick and the murderous brother Oliver.

The wise and cynical observer of these intrigues is Jaques. By comparison with Orlando and Rosalind, Jaques is a relatively minor character, barely essential to the plot. Yet it is he who delivers the play's most oft-quoted lines, which we hear shortly after Orlando arrives at Duke Senior's forest camp, desperately hungry and disheveled. Jaques's famous lines are prefaced by the duke's observation that Orlando's woeful condition shows that things really could be worse than their own lonely banishment in the Forest of Arden:[3] (2.7.142)

DUKE SENIOR: Thou seest we are not all alone unhappy.
This wide and universal theater
Presents more woeful pageants than the scene
Wherein we play in.

Hearing this cue, Jaques responds with his declamation on the "Seven Ages of Man":[4]

JAQUES: All the world's a stage,
And all the men and women merely players.
They have their exits and their entrances,
And one man in his time plays many parts,
His acts being seven ages. At first the **infant**,
Mewling and puking in the nurse's arms.
Then the whining **schoolboy** with his satchel
And shining morning face, creeping like snail
Unwillingly to school. And then the **lover**,
Sighing like furnace, with a woeful ballad
Made to his mistress' eyebrow. Then a **soldier**,
Full of strange oaths and bearded like the pard,[5]
Jealous in honor, sudden and quick in quarrel,
Seeking the bubble reputation
Even in the cannon's mouth. And then the **justice**,
In fair round belly with good capon lined,[6]
With eyes severe and beard of formal cut,
Full of wise saws and modern instances;
And so he plays his part. The sixth age shifts
Into the **lean and slippered pantaloon**[7]
With spectacles on nose and pouch on side,
His youthful hose, well saved, a world too wide
For his shrunk shank, and his big manly voice,
Turning again toward childish treble, pipes
And whistles in his sound. Last scene of all,
That ends this strange eventful history,
Is **second childishness** and mere oblivion,
Sans teeth, sans eyes, sans taste, sans everything.

—*As You Like It*, 2.7.146

The idea of stages or "ages" of development was not original with Shakespeare. Aristotle proposed merely three: *youth*, *prime-of-life*, and *old age*. The Roman poets Ovid and Horace saw four: Horace called them "the child," "the beardless youth," "manhood," and "the gray-haired old man," who was also "a reprover and censurer of the younger generation."[8] Perhaps it was Hippocrates who first listed seven ages: *the infant* (birth to age seven), *the boy* (until fourteen), *the youth* (till "thrice seven"), *the young man* ("four sevens"), *the man* (until forty-nine), *the elderly man* (to fifty-six), and *the aged man* (fifty-seven until . . . you know). These groupings map well into Jaques's "Seven Ages" speech, so Shakespeare likely borrowed the model from Hippocrates.[9]

Let's consider, for a moment, how Shakespeare presented these "ages" in some of his other plays (with the names of the characters he created in **bold**).

The Infant

None of the Bard's works gives us so much as a single "mewling and puking" child, so it may appear that Will Shakespeare simply overlooked the First Age of Man. But why? The most likely explanation is that he deemed stories about children to be of little interest to adult theatergoers. He did, however, have three children of his own, so it is likely that he experienced at least some mewling and puking firsthand. He and Anne had a daughter, Susanna (born six months after Will and Anne were married), and a pair of twins, Judith and Hamnet. The latter died at age eleven, and we have hints that Will was devastated by the loss.[10]

Childhood was a threatening time in the Elizabethan world, especially for children in families of limited means—which accounted for nearly the whole population. Death swept away almost half the children of England, and that is reflected in the plays: as the Shakespeare scholar Ann Blake points out, most children in his works meet a violent end.[11]

While preschool-age children *did* appear in Shakespeare's London productions, their parts were always minor and usually anonymous—often as fairies, as in *The Merry Wives of Windsor* and *A Midsummer Night's Dream*. Among the few roles and references to children who are named, we find **Fleance**,[12] Banquo's son in *Macbeth*; the butterfly-maiming **Martius** in *Coriolanus*; and the sad and fragile **Mamillius** in *The Winter's Tale*. A closer look, says Blake, reveals some thirty-odd roles for children in the plays, albeit mostly bit parts.[13] Usually, the Bard does not give us clues enough to determine whether they are in the age of The Infant or the next.

The Schoolboy

Here, again, the roles are few and the parts minor. The most prominent ones at this stage are the ill-fated Princes in the Tower, who appear (and then disappear—never to be seen again) in *Richard III*.[14] The older of the two was **Edward**, briefly King Edward V, who was sent to the Tower of London when he was twelve, accompanied by his younger brother, **Richard of Shrewsbury**, age nine. Both were in this second of the Seven Ages. Unfortunately, for our purposes of discovering Shakespeare's views on child development, he gives us only a glimpse of the two princes' personalities.

We do know that, in Shakespeare's own boyhood, he almost certainly attended one of the new "grammar schools," located in Stratford. It offered a rigorous education in classical subjects to any boy in the community who could read and write. These schools were an enlightened English innovation, aimed at boosting the educational level of the general populace. Children of the nobles, however, were educated at home. And, as we might expect, the grammar schools accepted boys, not girls. The rigorous curriculum centered on learning Latin and Greek (which is why they were called "grammar schools") and on reading the classic works of Ovid, Horace, and Virgil. Learning was largely by rote—for ten to twelve hours a day, six days a week.[15]

The Lover

Shakespeare is all (or nearly all) about love, and so he describes the young fellow at the age of the Lover as, "Sighing like furnace, with a woeful ballad / Made to his mistress' eyebrow" (*As You Like It*, 2.7.146). But we must remember that the speaker, Jaques, is a jaundiced and cynical person; the Seven Ages speech is as much a parody of life as it is Shakespeare's summary of human development across the lifespan.

Yet it is a strange statement that Jaques makes, especially the part about the eyebrow. Critics have various explanations about the meaning of the mistress's eyebrow, but the one we favor is that singing a woeful ballad to an eyebrow was simply satirical, suggesting that the young Lover was behaving in a silly fashion—like our young lovers, **Orlando** and **Rosalind**, in *As You Like It.*[16] **Pericles** is one more exemplar, and when you read his play you will see that he is the Lover who looks for love in all the wrong places, often barely escaping with his life. The tragic archetype for the age of the Lover, of course, is **Romeo**, who *didn't* make it out alive.

As for Shakespeare, we know little about him at this stage of life, except that when he was about eighteen he impregnated a twenty-six-year-old woman, Anne Hathaway. Will, himself, was young enough that he had to get his father's permission to marry Anne.

The Soldier

We don't know whether Shakespeare ever saw military service, although that could account for one of his "lost years" periods. If, however, another stage was as important in the Bard's mind as "the Lover," it would be the Fourth Age of Man, "the Soldier." His history plays are full of soldierly exploits and bloody tragedy perpetrated by the likes of **Macbeth**, **Bolingbroke**, **Joan of Arc**,

Talbot, **Othello**, **Richard III**, and many others. And, of course, **Prince Hal** finally becomes Shakespeare's Soldier archetype and takes up England's crown as **Henry V**.

Many of Shakespeare's plays celebrated the nobility and their exploits, but being a noble was not all collecting taxes and luxurious living. The noblemen of that era were expected to be warriors who would risk their lives in combat, both in civil wars (as in the Wars of the Roses) and in wars against the dreaded Spanish and the hated French (as in the Hundred Years War). Many "men of note" died in those battles, along with their serfs and yeomen, who were also required to serve the call to military service by their overlords.

As for Jaques, he seems to see the good Soldier as a glory-seeker, disregarding danger, "jealous in honor, sudden and quick in quarrel, / Seeking the bubble reputation / Even in the cannon's mouth."

The Justice

"In fair round belly with good capon lined, / With eyes severe and beard of formal cut, / Full of wise saws and modern instances." Again, in his Seven Ages speech, Jaques gives us a cartoonish version of this stage, exemplified by a prosperous middle-aged man. It is a stage at which a gentleman—which Will and his father both aspired to be—would have become successful enough to have a comfortable life and status in the community. Shakespeare himself achieved that level of success by 1600, with a series of successful plays and an interest in a well-attended theater company. His entrance into the Justice stage, then, as a man of about thirty-six, is also the time that he wrote Jaques's Seven Ages of Man for *As You Like It*. And it was near the end of this period in our playwright's life that he moved back to Stratford, in about 1611, at about age forty-seven. Life back home in Stratford, however, wasn't exactly retirement, for he wrote or collaborated on four more plays.[17]

To our minds, the exemplars of this Fifth Age in Shakespeare's work are three characters. The first is **Portia**, as a woman who role-plays a lawyer (as well as a lover) in *The Merchant of Venice*. The second we also find in *The Merchant of Venice*: the long-suffering **Shylock**, who is something of the dark underside of Portia. He is the angry Jew who originally demanded "a pound of flesh" for a lifetime of prejudice and an unpaid bill. The third exemplar is the **Duke of Vienna** in *Measure for Measure*, Shakespeare's morality play that puts to the test the question of whether human nature is essentially good or evil. The duke pretends to leave town but returns in disguise to see how things will go in his absence. Not so well, it turns out. His lieutenant, Angelo, now in charge, enforces the law strictly and harshly—except when it comes to his own desires for a woman he wants to seduce.

The Lean and Slippered Pantaloon

This phase we might now call the early years of retirement—which, of course, was not widely enjoyed by Elizabethans. Our William was wealthy enough to have a retirement in Stratford, but it only lasted about three years. Then, at age fifty-two, he died after being taken ill with a fever. Life expectancy was not so long in those times, so Shakespeare would have been seen as having lived a rather normal lifespan, even though he did not make it far into the final two Ages of Man.

By the yardstick of the times, the Bard beat the odds for longevity, since the average lifespan was only a little over forty years. That number can be misleading, however, since childhood mortality was extremely high, and it hit Will's family hard, as you may remember, when his son, Hamnet, died at age eleven. Nevertheless, those who made it to adulthood, especially if they were not among the urban poor, could hope to live into their sixties or seventies.[18]

In the plays, he inserted several characters representative of this Sixth Age: **Timon of Athens** squanders his wealth entertaining fair-weather friends, leaving us with the apparent moral that friends who must be bought are no friends at all. Then, in *As You Like It*, we have **Duke Frederick**, who retires from his usurped dukedom to become a religious hermit, and **Prospero**, of *The Tempest*, who, after being deposed by his brother, amuses himself with wizardry.

Second Childishness

Hamlet tells Polonius that old age (Jaques's seventh Age of Man) is one of physical and mental decline:

> Old men have grey beards, that
> their faces are wrinkled, their eyes purging thick amber
> and plum-tree gum, and that they have a plentiful
> lack of wit, together with the most weak hams . . .
>
> —*Hamlet*, 2.2.214

Presumably, this was a raving insult that Hamlet hurled at the doddering old **Polonius**, and it echoes Jaques's description of the seventh Age of Man: "mere oblivion, sans teeth, sans eyes, sans taste, sans everything." The poster boy for Second Childishness, however, is **King Lear**, who famously slips into decline and dementia, pulling his family down with him.[19]

ERIKSON ADDS AN EIGHTH STAGE—AND EIGHT LIFE "CRISES"

Psychology's best analog of Jaques is the pioneering psychoanalyst, Erik Erikson, whose eight stages of human development both mirror and magnify Jaques's Seven Ages of Man[20] (figure 2.1). To be sure, Erikson devised a comprehensive theory of human development, while Jaques's monologue is quite brief. Both are "stage" theories that map onto each other. (See table 2.1.) We will see that the similarities show how Shakespeare again anticipated important ideas that reappeared four hundred years later in psychology.

Importantly, Erikson taught that every stage of life confronts us with a "crisis" that we must solve—for better or worse—as we advance to the next stage. Erikson's point is that a person who resolves these crises in a healthy manner will move on happily to the next stage, while those who deal poorly with a life-stage crisis will face the next stage with anxiety, confusion, isolation, and feelings of personal inferiority. Two of these crises have become household terms in American culture: the "identity crisis" and the "midlife crisis."

And how does this connect with Shakespeare? We will argue that many of the Bard's plays—perhaps his *best* plays—deal with an Eriksonian

Figure 2.1 Developmental psychologist Erik Erikson. *Wikimedia Commons.*

developmental crisis. Table 2.1 shows the life crisis identified by Erikson at each of his eight stages of psychosocial development. With these in mind, then, let's take a closer look at Erikson's stages to see how they compare with Jaques's Seven Ages of Man.

Table 2.1 Comparison of Jaques's Ages and Erikson's Stages

Jaques's Seven Ages of Man	Erikson's Eight Stages of Psychosocial Development (and the crisis for each stage)
The Infant	Infancy: trust vs. mistrust
	Early Childhood: autonomy vs. self-doubt
	Preschool: initiative vs. guilt
The Schoolboy	School-Age/Youth: competence vs. inferiority
The Lover	Adolescence: identity vs. role confusion
The Soldier	Young Adulthood: intimacy vs. isolation
The Justice	Midlife: generativity vs. stagnation
Slippered Pantaloon	Old Age: integrity vs. despair
Second Childishness	

Infancy

Unlike Jaques, who characterized infancy merely by "mewling and puking," Erikson saw it as a much more complex period when we establish our foundational view of life. Is the infant's world a welcoming place, filled with love and support? Or is life a constant struggle in a world filled with powerful and threatening people? Therefore, said Erikson, the infant faced a developmental "crisis" that he called *trust vs. mistrust.* At this stage, each of us must deal with these questions, and the conclusions we reach can send us down different developmental paths.

Early Childhood

As table 2.1 shows, Erikson proposed two more stages that Jaques either ignored or subsumed under the age of The Infant. Erikson's approach, then, is a reflection of the increased importance that psychology has placed on child development since Shakespeare's time. It also reflects a different view of children's minds.

In our time, we have come to accept the idea that children think differently from adults because connections in the brain have not yet matured. For example, for a very young child, objects that are out of sight quickly fall out of mind. Sharing is difficult, too, not because they haven't *learned* to share, but because they have not fully developed the capacity for understanding that another's perspective can be different from their own. In contrast, children in the Elizabethan era were assumed to be empty vessels whose minds must be filled with appropriate facts and rules.

The modern mind takes a more *maturational* view of development. A child in Erikson's Early Childhood stage, from about two to five years, gradually becomes a more sophisticated observer and actor, developing an increasing ability to interact with people and objects. If things go well, the child develops a sense of independence and increasing confidence that Erikson called *autonomy.* If, on the other hand, the child emerges from this stage with a sense of defeat and lack of confidence, the result is *self-doubt*, which lays a poor foundation for the next stage of life.

Preschool

By about age five, Erikson argued, a child should increasingly be showing initiative, as in choosing what to wear, how to play, or what to eat. Developmental problems may arise when adults attempt to overcontrol the child, producing resentment and anger or feelings of personal inadequacy. Erikson called this the crisis of *initiative vs. guilt.* Most of Shakespeare's few child characters seem to fall into this stage, including Fleance, Banquo's son in *Macbeth*; and young Martius in *Coriolanus*, along with the melancholy child Mamillius in *The Winter's Tale.*

School-Age Youth

Moving on to Jaques's Schoolboy age, Shakespeare and Erikson are back in synchrony, for the schoolboy spans the same period that Erikson describes as the *school-age youth*—the years up to what we would now call adolescence. For Erikson, the school-age youth must deal with the crisis of *confidence vs. inferiority*, while in Shakespeare's writings, there are few examples in this category, except young Arthur who leaps to his death from the castle wall (in *King John*) and the princes murdered in the Tower (in *Richard III*).

Adolescence

None of Shakespeare's characters go through adolescence—because adolescence is a modern invention involving a period of schooling that keeps teens in a limbo status for a few, often rebellious, years. Our concept of adolescence, as a sort of holding pattern between youth and adulthood, emerged in the twentieth century, when we enacted laws curtailing child labor and keeping children in school, out of the workforce, until they were eighteen.[21] Shakespeare's equivalence is The Lover—which makes sense in terms of the raging hormones and other physical changes that individuals undergo during this period.

Aside from the physical changes, the culture of the times influences people in their teenage years. In the Elizabethan era, the normal life span was shorter, on average, than it is today, and English children went from childhood rather abruptly to marriage and other adult roles as soon as they were deemed old enough—at the end of grammar school, or when physically strong enough to take on adult vocations. This could be around the time of puberty, which could be as young as twelve or thirteen in those days. After all, there was work to be done, and, to borrow Benedick's phrase, "The world must be peopled!" (*Much Ado*, 2.3.223) No wonder, then, that Shakespeare called this the stage of "The Lover."

Remarkably, however, each of the Bard's adolescent "lovers" faces some form of Erikson's famous adolescent "identity crisis," which is the problem of finding a sense of direction for one's life—discovering what one "wants to be" both inward, in terms of self-concept, and outward, in terms of what it takes to succeed in the world. Again, the adolescent/lover faces two possible trajectories, which Erikson called the crisis of *identity vs. role confusion*. It's at this stage that we first encounter Prince Hal, as he vacillates in the transition from wild youth to soldier to king, exemplifying the identity crisis, writ large.[22]

While Prince Hal's resolution of the crisis was successful, things did not go so well for his counterpart, Prince Hamlet, in Denmark, who shuffled off the mortal coil in mid-identity crisis, still an angry and melancholy young man. And in another example of arrested development, the Bard gives us Romeo and Juliet, who find themselves in love yet caught in a long-standing family feud that prevents them from establishing adult roles. The result, for this tragic couple, is suicide. Then, in *The Taming of the Shrew*, we find both Kate and Petruchio, experiencing their own developmental issues but with a less deadly ending than befell the couple in Verona. Much the same obtains with Beatrice and Benedick in *Much Ado About Nothing*. Readers will have to decide whether Shakespeare anticipates the Eriksonian identity crisis in other plays involving lovers, such as *Troilus and Cressida*, *Cymbeline*, and *Love's Labour's Lost*.

Young Adulthood

Jaques's "soldier" Age is Erikson's Young Adulthood, where the challenge is learning to develop close relationships—emotional, moral, and sexual—with other adults. Erikson calls the crisis at this stage *intimacy vs. isolation*.

Shakespearean characters in this category would include Prince Hal, as he transitions from rebellious youth to the quintessential soldier, along with a troupe of antiheroes such as Coriolanus, Titus Andronicus, Othello, and Richard III. It seems that most of the latter are clear failures at the task of establishing close relationships. Again in these dramas, the plots do not deal with the developmental process of young adulthood so much as they reveal characters who had already failed in the face of their earlier developmental tasks.

As for Portia, in *The Merchant of Venice*, she does undergo a striking developmental change, but a change unlike any other in Shakespeare. Prior to her appearance disguised as a faux judge, Portia is a modest yet willful young woman, trapped by her father's stipulation that she must marry the suitor who chooses correctly among gold, silver, and lead boxes. Fortunately, young Bassanio (with whom she is in love) is clever enough to make the right choice, realizing that "all that glisters is not gold" (*Merchant*, 2.7.70). But then the plot thickens. Bassanio's friend Antonio unexpectedly owes a debt that he cannot pay to the Jewish lender, Shylock, a man embittered by years of prejudice and mistreatment at the hands of the Christians of Venice. Shylock demands his recompense according to their loan agreement: a pound of Antonio's flesh. The matter goes to trial, where Portia appears, disguised as a young man of the law. Surprisingly, she agrees with Shylock that the agreement is legal, but she finds a clever way to negate the contract. As in most other comedies of the period, all the young people become married and presumably live happily ever after. Portia emerges as the hero of the play and as a clever, competent young woman—only technically, because of her gender, not The Soldier. And Shakespeare gave her wisdom beyond her years, for she skillfully plays the role of The Justice, which is the next age of "Man." And having done so, she shows us success in meeting Erikson's challenge of intimacy vs. isolation.

Midlife

Shakespeare's Age of the Justice is Erikson's Midlife stage. And now the developmental problem for healthy adults is finding a way to make a satisfying contribution to family, work, society, or future generations in whatever professions or social roles they have chosen. This is the crisis of *generativity vs. stagnation.* Those who do not succeed face a life of merely treading the psychological waters, going nowhere. The playwright gives us an example of a midlife character in the story of the duke in *Measure for Measure*. We find a clearer example of a midlife crisis, however, in Coriolanus, who feeling unappreciated for his battlefield deed in his home country, switches allegiances from the Romans to the enemy Volscians.

On another hand, Shakespeare sometimes used the midlife crisis as comedy—much as writers do today—making fun of the foolish things people do when they see their youth receding and their lives without a sense of purpose. *Love's Labour's Lost*, *A Midsummer Night's Dream*, *All's Well That Ends Well*, *As You Like It*, and *The Merry Wives of Windsor* all feature elements of the midlife crisis.

Old Age

Erikson wraps the final two of Jaques's Ages of Man into one that he called simply "Old Age." We can debate whether, by separating the Slippered Pantaloon from Second Childishness, Shakespeare was making a finer distinction than Erikson did. Or was Shakespeare merely displaying an Elizabethan prejudice against his elders?

For Erikson, the challenge of Old Age involves increasing awareness of one's physical decline and mortality, yet being able to look back on one's life without many regrets and with a sense of a life well-lived. He called this the crisis of *integrity vs. despair.* Life comes to a happy and healthy finale for Prospero, in *The Tempest.* But for Timon, the man from Athens, old age ends in tragedy and despair, much as it does for King Lear. Other examples include the doddering Polonius in *Hamlet*, perhaps King Duncan in *Macbeth*, John of Gaunt in *Richard II*, Nestor in *Troilus and Cressida*, and Henry IV in his dying days.

THE PATH OF PSYCHOLOGICAL DEVELOPMENT

The storyline in most of the plays unfolds over a short period of days or weeks; as for the Bard's beloved *Midsummer Night's Dream*, it occurs merely overnight. Consequently, few of his characters age on stage—which limits the playwright's opportunity to address lifespan changes in all but a few of his works. Those few, happily, include Prince Hal's coming-of-age story (*Henry IV, Parts 1* and *2*) and the unhappy life story of Henry V's successor, Henry VI, told in his eponymous trilogy. For the most part, Shakespeare takes what psychologists would call a *cross-sectional* approach, showing us different characters at different life stages, as he does in the events surrounding King Lear's struggle with dementia and Benedick's antics as he meets his love match in Beatrice (*Much Ado About Nothing*). This leaves us with mere snapshots of the several Ages of Man, rather than a portrayal of the developmental process itself.[23]

One exceptional figure deserves to be singled out: King Henry VI (figure 2.2)—not just because he was an especially weak king, but because he, among

Figure 2.2 King Henry VI of England portrait painted circa 1540. *Wikimedia Commons.*

all Shakespeare's major characters, is the only one whose story stretches across the entire lifespan. You may object that he died in the Sixth Age, not the Seventh, and not so much in the comfort of slippered pantaloons but as a prisoner in the Tower of London. But that was *his* lifespan.

In brief, here's how his life story plays out over the *Henry VI* triad: He was only nine months old when he became king upon the death of his father, the legendary Henry V. At the funeral, Shakespeare gives us a cryptic hint of things to come for the young, titular monarch. In an exchange of unpleasantries with Gloucester, the Bishop of Winchester credits the dead king's success to the prayers of the Church:

> WINCHESTER: He was a king blest of the King of kings;
> Unto the French the dreadful Judgment Day
> So dreadful will not be as was his sight.
> The battles of the Lord of Hosts he fought;
> The Church's prayers made him so prosperous.
>
> —*Henry VI, Part 1*, 1.1.28

Whereupon Gloucester responds with a slur on the bishop's motives and a derisive mention of the infant-king:[24]

> GLOUCESTER: The Church? Where is it? Had not churchmen prayed,
> His thread of life had not so soon decayed.
> None do you like but an effeminate prince
> Whom like a schoolboy you may overawe.
>
> —*Henry VI, Part 1*, 1.1.33

In *Part 1*, young King Henry VI is oft mentioned but seldom seen or heard, even after his coronation at age eight, marking the end of his infancy. During his Schoolboy years, a council of the most prominent and powerful nobles continued to rule England until Henry was eighteen and deemed fit to rule—and still we see little of Henry on stage. Finally, in act 4, Shakespeare pushes him out as a young man, presumably in the Age of the "Lover"—although we see none of that activity in the play. Rather, the Bard's focus is on the politics involved in the loss of English territories in France.[25]

In *Part 2*, Henry is fully an adult, but seemingly no more competent than he was as a child, unable to make peace among his squabbling nobles. This discord accords with history, which also tells us that around age thirty-three Henry began having bouts of mental illness so severe that he was, at times, unable to speak and barely able to move. These episodes of apparent mania and melancholy worried the nobles enough that they reestablished a protectorate that again served as the de facto government of England. When Henry was in the depths of an apparent depressive phase, the Yorkists captured the forty-year-old king and imprisoned him in the Tower, where he wrote this sad poem:

> Kingdoms are but cares,
> State is devoid of stay,
> Riches are ready snares,
> And hasten to decay
> Pleasure is a privy prick
> Which vice doth still provoke;

Pomps, imprompt; and fame, a flame;
Power, a smoldering smoke.
Who meanth to remove the rock
Owst of the slimy mud
Shall mire himself, and hardly scape
The swelling of the flood.[26]

The melancholy nature of the verse, along with Henry's history of mental illness—including reports that he was laughing and singing inappropriately during the Second Battle of St. Albans—all suggest his was a case of bipolar disorder (manic-depressive illness). Finally, in 1471, imprisoned in the Tower of London for a second time, Henry died (stabbed by a Yorkist, says Shakespeare) at the age of fifty—in his Sixth Age of Man.

DEVELOPMENTAL CRISIS PLAYS, SIMPLE HISTORIES, AND SITCOMS

Now, if you will humor us, please imagine these hypotheticals: How much more titillating might some of Shakespeare's works have been had he woven a bit more developmental psychology into the storyline where none of the characters seem to have much depth of personality! The more obscure plays such as *Pericles*, *Coriolanus*, and the *Henry VI* histories come to mind. Or consider what the rather plotless *Henry VIII* or the *Henry VI* trilogy could have been if the storyline had turned on developmental challenges faced by those monarchs—challenges like those seen much more clearly in Hamlet, Prince Hal, and King Lear. Let us tread a little further into this thicket.

It seems that nearly half of the plays pose a serious test or Eriksonian "developmental crisis" for at least one of the leading characters. Prince Hal must deal not only with Hotspur but with the even greater challenge of becoming king of England. Hamlet must decide how to respond to the murder of his father. Macbeth must weigh the risks and rewards of regicide. We may disagree on exactly which ones we might call *developmental crisis plays*, but the point is that they are substantial in number. In a few of them—the plays telling the story of Prince Hal becoming Henry V, for example—the hero resolves a life crisis successfully. In the tragedies, however, the hero does not succeed. We believe that the Bard must have been thinking, on some level, in terms of developmental challenges as he wrote most of his mature histories and tragedies, for in them he seems to have deliberately wrapped the storyline around developmental issues. By our count, there are nineteen of these developmental crisis plays, although, again, one may quibble on the exact number. (See table 2.2.)

A second group forms a cluster that we might call *situational comedies and tragedies*—the Tudor versions of our "sitcoms" and "soap operas." All are pure fiction and meant to entertain, but they contain no serious exploration of developmental challenges. This list also contains the plays featuring comic midlife crises.

A final group of six comprises what we will call the *simple history plays* because they mainly record Shakespeare's version of historical events, employing rather two-dimensional characters, with relatively little psychological depth. These were among his least successful works (in the opinion of modern critics, although the *Henry VI* trilogy was a great commercial success at the time). Taking a psychological perspective, we suggest that nearly half his works fit the description of *developmental crisis plays*.[27]

Two things are certain, however. One is that people *do* develop and change over a lifespan. This allows us to predict with modest accuracy the stresses

Table 2.2 Three Groups of Shakespeare's Plays

Developmental Crisis Plays	Situational Comedies and the Other Tragedies	Simple History Plays
Antony and Cleopatra	*All's Well That Ends Well*	*Edward III*
Coriolanus	*As You Like It*	*Henry VI Part 1*
Hamlet	*A Comedy of Errors*	*Henry VI Part 2*
Henry IV, Part 1	*Cymbeline*	*Henry VI Part 3*
Henry IV, Part 2	*Love's Labour's Lost*	*Henry VIII*
Henry V	*The Merry Wives of Windsor*	*King John*
Julius Caesar	*A Midsummer Night's Dream*	
King Lear	*Much Ado About Nothing*	
Macbeth	*Pericles*	
Measure for Measure	*Taming of the Shrew*	
Merchant of Venice	*Titus Andronicus*	
Othello	*Twelfth Night*	
Richard II	*Two Gentlemen of Verona*	
Richard III	*The Two Noble Kinsmen*	
Romeo and Juliet		
The Tempest		
Timon of Athens		
Troilus and Cressida		
The Winter's Tale		

many people will face at various ages and stages. The other certainty is that, as people grow and develop, the differences among them in personality, intellect, and the trajectories of their lives become increasingly diverse. For this reason, predictions based on theories of life stages will never perfectly capture everyone. Moreover, developmental predictions are likely to be wrong about the outliers—who can be among the most interesting people among us.

The importance of our classification scheme is twofold. First, the identification of the developmental crisis plays in such numbers supports the idea that Shakespeare must have been thinking in terms of developmental challenges when he wrote these works. The second point is that five of the six *simple history plays* were written early in Shakespeare's career, perhaps before he had mastered the techniques of leading us inside the characters' minds. *Henry VIII* is an outlier, a collaboration with John Fletcher, and more of a pageant than a story. One difficulty is that *King John* was presented a year *later* than *Romeo and Juliet*. If our hypothesis is correct, he may have started writing *King John* a bit earlier.

While this classification is frankly conjectural, so is Erikson's rubric for the eight stages of psychosocial development. Although widely used and based on life in modern Western society, Erikson's stages are his attempt to make sense of the patterns he saw in people's lives. But we would emphasize, there are no objective and universal stages of development that apply to all people, in all places, and at all times.[28]

WHAT SHAKESPEARE DID AND DID NOT KNOW ABOUT DEVELOPMENT

Will Shakespeare was not a scientist but rather an entertainer, an actor, a playwright, and a poet. He was also a curious observer of the human condition and a fellow who seemed to know at least a little about nearly everything. Unfortunately, the big discoveries that would come from the sciences of psychology, biology, medicine, and genetics lay in the future. Certainly, Vesalius had made advances in human anatomy, but medicine mostly remained anchored in the ancient writings of Galen and Hippocrates. Essentially nothing of the mechanisms of heredity was understood beyond what livestock breeders had known for centuries. And Mendel didn't start fiddling with his peas until the mid-1800s. All of this means that Shakespeare had little understanding of the fundamental processes underlying human development.

Psychologists in training now learn that development involves the interaction of three basic processes: heredity, environment, and *maturation*. We looked at the first two of these in our discussion of *The Tempest*, where Prospero first paired the terms "nature" and "nurture." "Maturation," however, is a slipperier concept that can take on two somewhat different meanings. The more common meaning refers to *maturity* as the desirable mental state reached in adulthood. We think of a "mature" adult as wise, competent, and restrained. But in developmental biology and psychology, *maturation* also refers to the process by which the interaction of nature and nurture unfolds in an individual over time. In this sense, maturation refers to the timetable that removes our "baby teeth" and gives us permanent ones, and to the process that initiates puberty, deepens voices, and prunes connections in the brain, leading eventually to adulthood and, too soon, to the physical and mental decline of old age. Shakespeare knew of the physical and mental manifestations of maturity and something of its timetables, but he knew nothing of the processes behind them.

That, of course, didn't hurt the drama or the poetry. After all, black bile and blood hold more metaphorical and dramatic interest than do DNA or neural circuits. Still, one wonders what references Shakespeare might have embedded in his works had he known about the double helix, hormones, or the uneven maturation of the adolescent brain.[29] And what if he had known that the hereditary basis of nobility differed not a whit from that of the commoner's lot—that no one was "born to be a king or queen"? What dramas would he have written had his world known that women were the mental equals of men? (Or perhaps he intuited as much, living as he did during the reign of the first Elizabeth.)

What else he would not have known was the outsized role of culture in shaping our understanding of development. He *did know* something of cultural differences—that the French, Italians, Spanish, and Greeks all seemed to think and act differently from the English. He knew that the Jews often maintained a cultural distance, both by choice and prejudice. But he did not foresee the social changes that, over the next few centuries, would change his own culture, lengthening lifespans, reducing child mortality, and forcing us to reconceive the notions of childhood and old age. Our culture now recognizes that adults can be "generative" much longer in their lives. Likewise, psychologists have developed deep insights into the mental and social processes seen in children. Underlying all this, of course, are remarkable advances in sanitation, medicine, and public health—again, none of which could have been forecast by a playwright of Tudor times.

What can we learn from Shakespeare about human development? At times, he is a pioneering feminist, showing us that women should not be thought of as inferior creatures of arrested development. We can also thank him for stories like Hamlet's, showing that rumination on one's troubles leads to an unsatisfactory resolution of life's challenges. And, reading between the lines and between the texts of the *Henry V* and *Henry VI* histories, where England's greatest hero begets one of its greatest failures—we find that heredity does not always mean "like father, like son." This we already know, of course. The real lessons about human development come from the stories. Where else can we find such vivid descriptions that convey the coming-of-age transition than in the Henrys tetralogy or the sad tale of *Romeo and Juliet*? Where else is there such a tragic drama highlighting dementia than *King Lear*? In that sense, then, Shakespeare can bring the idea of psychological development to mind and to the stage, where it can be experienced larger than life.

WHERE ARE THE WOMEN IN THE SEVEN AGES OF *MAN*?

We celebrate Shakespeare for his strong female characters, created in a time and place that was not known for its egalitarian treatment of women. And don't forget that, in Shakespeare's world—the English Renaissance—several strong women appeared, as if from nowhere: Queen Elizabeth was one of them, as was her sister, "Bloody Mary," and yet another Mary, "Queen of Scots," who was beheaded by her cousin Elizabeth. Yet the strong women are nowhere in any of Jaques's Seven Ages! They *are* mentioned in the second line of Jaques's speech: "All the world's a stage, / And all the men and women merely players." But then they disappear. On the internet, one can find many versions of "Seven Ages of Woman," going as far back as 1827 when poet Agnes Strickland noticed Shakespeare's rather glaring omission.[30] (See figure 2.3.) Most of these are gender-equity, consciousness-raising versions of Jaques's Seven Ages, but stated in more modern terms.

With all these ideas in mind, we decided to look at the Bard's works and attempt to craft our own Seven Ages of Woman, based on Shakespeare's female characters. That is, we intended to see if there were clusters of female characters whose development roughly paralleled the ones that Jaques outlined for men. Table 2.3 summarizes our take, but we encourage further refinement of these suggestions.

What did we learn in this exercise? First, we found that Shakespeare's male-oriented ages do have female counterparts because women, even (or especially!) in a male-dominated society, must deal with related issues—but in ways that the culture inevitably constrains. Moreover, women's biology, broadly speaking, takes them through pubescence and aging in ways that parallel the stages men experience, albeit at a different rate.[31]

Figure 2.3 *The Seven Ages of Woman* by Hans Baldung (1534). *Wikimedia Commons.*

Second, we noticed that Shakespeare's female characters in the third stage seem to split into two subgroups that we are calling The Lover and the Assertive Woman. The first involves women following a rather traditional find-a-man-and-get-married path, while the assertive women are much less conventional, to the point of defying cultural norms.[32]

Table 2.3 Comparison of Jaques' Ages and Our Proposed Shakespearean Ages of Women

Jaques's Seven Ages of Man	Proposed Shakespearean Ages of Woman
The Infant	**The Infant** There are no female infants in Shakespeare's works
The Schoolboy	**The Young Maid** Lavinia (*Titus Andronicus*), Cordelia (*Lear*)
The Lover	**The Lover and the Assertive Woman** Examples: Rosalind (*As You Like It*), Helena (*All's Well*), Beatrice (*Much Ado*), Portia (*Merchant of Venice*), Imogen (*Cymbeline*), Cressida (*Troilus and Cressida*), Ophelia (*Hamlet*)
The Soldier	**The Wife, Mistress, or Daughter** Kate (*Shrew*), Desdemona (*Othello*), Mmes. Ford and Page (*Merry Wives*), Goneril and Regan (*Lear*)
The Justice	**The Ruler** Lady Macbeth, Cleopatra (*Antony and Cleopatra*), Calpurnia (*Julius Caesar*), Titania (*MSN Dream*)
Slippered Pantaloon	**The Manipulative Dowager** Volumina (*Coriolanus*), Cymbeline's (unnamed) queen
Second Childishness	**The Crone** The Weird Sisters, of course (*Macbeth*)

And third, we realized that certain Shakespearean women are difficult to place into our Seven Ages of Woman. Juliet, for example, is both very young and very in love, placing her in both the Young Maid and Lover categories. Then we have Isabella, a nun-in-training, who falls afoul of Angelo's sexual advances in *Measure for Measure*. And what of Antiochus's daughter, caught in an incestuous relationship with her father? Is she also a Young Maid/Lover hybrid? The point is this: Shakespeare's women do not fall quite so neatly into "age" categories as do his men. Our take is that his characterizations of women are often more complex than those of his leading male characters.

SHAKESPEARE'S OWN DEVELOPMENT AS A WRITER

Many scholars have commented on how the man from Stratford matured as a playwright, beginning with comparatively simple plays, such as *The Comedy of Errors*, *Two Gentlemen of Verona*, and *The Taming of the Shrew*. These were successful, but it was the still-unsophisticated *Henry VI* trilogy that established his reputation. Then, in mid-career, he produced such towering works as *Hamlet*, *Henry V*, *Othello*, and *Macbeth*. And finally, in the final phase of his literary career, his palette changed, as he turned to "romances"—tragicomedies that often had a dark twist—among them *Cymbeline*, *Pericles*, *The Winter's Tale*, and *The Tempest*. (See figure 2.4.)

Since the *Henry VI* histories were among the Bard's earlier works, they offer us an especially interesting perspective on young Will Shakespeare's development. Indeed, many critics consider these to be his lesser efforts, perhaps

Figure 2.4 Portrait of William Shakespeare (2019). *Wikimedia Commons.*

because he crammed in too many battle scenes to cover up his immaturity at character development. "In the *Henry VI* plays, the playwright had not yet honed his skills," says Shakespeare scholar Charles Boyce. "Henry VI can merely speak of his woeful ineffectuality while the world sweeps him away." Boyce continues:

> As Henry's character develops through the plays, we can observe the young playwright learning how to devise a suitable tragic figure whose very virtues are his undoing. The germ of some of Shakespeare's great characters is here: a man who is good finds himself in a situation where his limitations generate an evil that crushes him.[33]

One of those "great characters" is Jack Falstaff, seen initially as Sir John *Fastolfe* in *Henry VI, Part 1*, where his cowardice leads to General Talbot's capture.[34] Later, as Will Shakespeare honed his play-crafting chops, fully formed characters began to emerge in his mature plays, such as *Hamlet*, *Macbeth*, *Romeo and Juliet*, and *Othello*. Perhaps the most outsized and memorable is the fully developed, rotund Falstaff, first as a jolly, loquacious old carouser, and later as the coward, and finally as the developmental failure whom his former drinking buddy, Prince Hal—now King Henry V—rejects with the words, "I know thee not, old man." The great tragedy of Falstaff, then, is his developmental failure. He never grows up; the jolly knight never squarely faced the life crises that would allow him to become the soldier, justice, or slippered pantaloon.

Part II

THE PERSON VS. THE SITUATION

Shakespeare's greatest and most unlikely hero may be King Henry V, who spent his early years as Prince Hal, in a life of sloth and debauchery. But something precipitated a profound transformation. Was there a spark of responsibility, somewhere deep in his personality, urging Hal to renounce the sins of his youth? Or was it the external circumstances, in which the crown of England was being thrust upon him, that produced his conversion? That is to say, was it the *person* or the *situation* that altered Prince Hal's trajectory? For psychologists who study personality, factors within the *person*—traits and character—weigh more heavily, while social psychologists emphasize external factors—the *power of the situation*. And what was the Bard's view? We'll see in the next three chapters.

On the other side of the same coin, we might ask what happens when the social and moral constraints on an individual are removed. What happens when unexpected events challenge our principles—when circumstances tempt us off the path of righteousness?

One of us (Zimbardo) sought an answer to these questions in a mock prison peopled by volunteers who were randomly assigned to be prisoners or guards but given little guidance on how to perform their roles. But, wouldn't you know, Mr. Shakespeare seems to have posed the same questions four hundred years earlier in *Measure for Measure*—also staged in a prison!

In the next three chapters, then, we will look closely at each of these narratives to see what they can tell us about the Bard's psychology and about the answers to the underlying questions: Which exerts the most influence on our own behavior? Is it the *person* or the *situation*? And how will we respond when the customary constraints are removed and temptation conflicts with our moral principles?

· 3 ·

HENRYS, HUMORS, AND THE PSYCHOLOGY OF PERSONALITY

Richard II–Henry V

England and France (1399–1415)[1]—Henry Bolingbroke would have lost his head had his treasonous plot failed to overthrow his cousin, King Richard II. Instead, he captured Richard and seized the crown of England for himself, becoming King Henry IV. The bloody consequences of this brazen act are the threads binding together a string of Shakespeare's history plays—*Richard II*, *Henry IV (Parts 1* and *2)*, and *Henry V*.[2] Together these works became a sort of serial Elizabethan soap opera that ran on the London stage, beginning in 1595, with new episodes appearing over some five years.

Bolingbroke's usurpation of the throne and his high-handed treatment of the remaining nobles sparked rebellions, which flamed into the Wars of the Roses—and served Shakespeare as the backdrop for these plays. The real problem for King Henry IV and his successor, however, was *redemption*: removing the cloud of illegitimacy that formed when Bolingbroke overthrew the weak Richard II, thereby challenging the "divine right of kings." Many saw this as a challenge to God himself, from whom the authority of the king derived. Thus, Henry IV spent his entire reign attempting to legitimize his claim to the throne. As Shakespeare put it, "Uneasy lies the head that wears a crown" (*Henry IV, Part 2*, 3.1.1). You can see the relationships among the rival factions, all from the House of Plantagenet, in figure 3.1.

In a nutshell,[3] that was the historical context behind the four plays. What packed the theater, though, was not so much the history but Shakespeare's coming-of-age story of Henry IV's son, the rebellious Prince Hal, who spends his leisure in the taverns and stews (brothels) in the company of drunkards, thieves, prostitutes, and other reprobates. Among these was a particularly close friend, the dissolute, rotund old knight, Sir John Falstaff, perhaps the Bard's

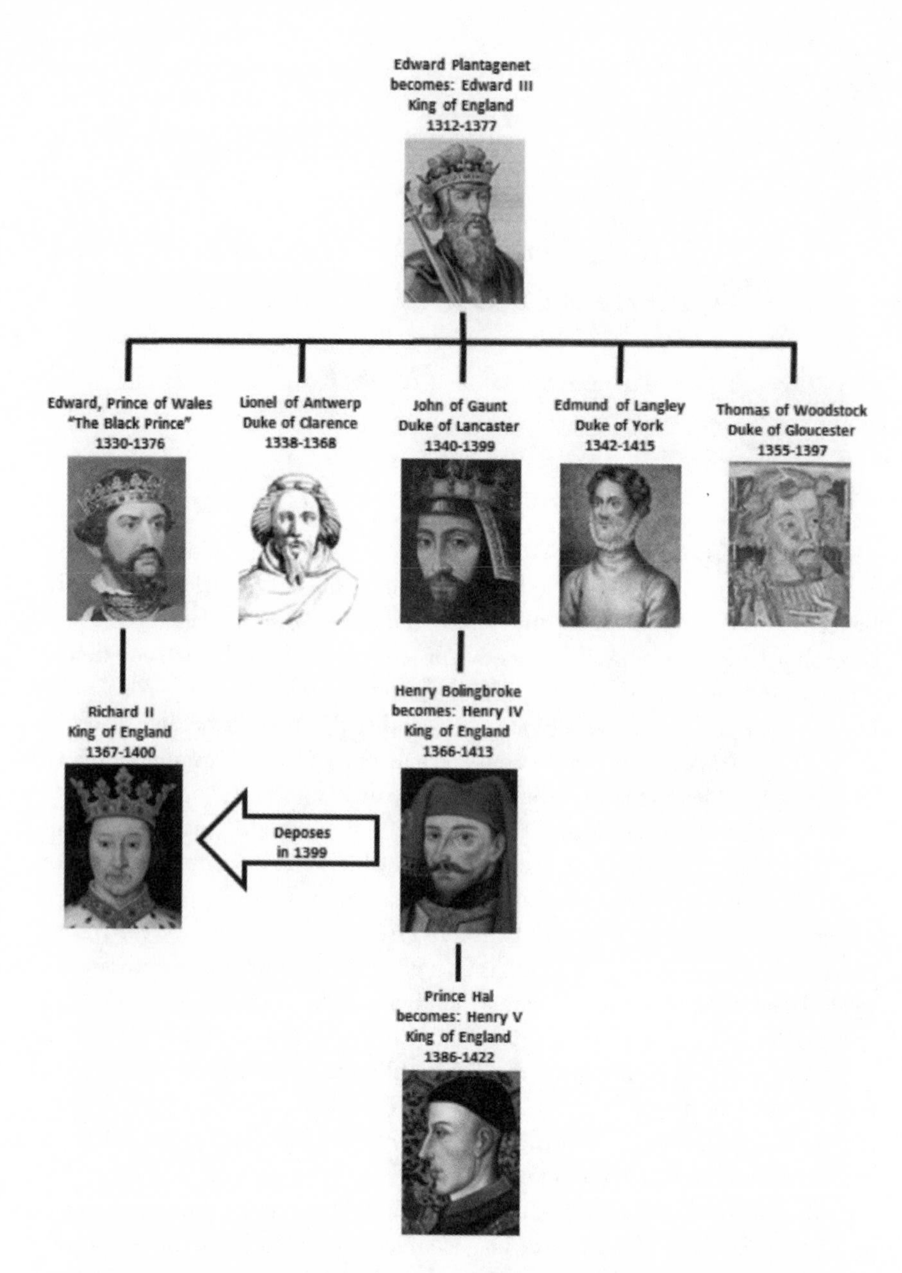

Figure 3.1 Monarchs and other prominent nobles in the House of Plantagenet, from Edward III to Henry V. Animosity over succession to the English throne originated among the sons of Edward III, especially between John of Gaunt and Edward, "The Black Prince." *Chart by the authors, with images from Wikimedia Commons.*

most endearing and charismatic personality. (See figure 3.2.) He epitomizes everything that a good warrior-knight should not be: a vain and boastful sot, a thief, a whoremonger, and, worst of all, a coward. Happily, Shakespeare also endowed Falstaff with the mitigating qualities of wit, humor, and humanity—so audiences loved him even more than the flamboyant Prince Hal.

Figure 3.2 *Falstaff*—painting by Eduard von Grützner (1906). *Wikimedia Commons.*

When Hal eventually defeats his arch-rival, Harry "Hotspur" Percy, in single combat and ascends the throne as King Henry V, he knows he must renounce the degenerate ways of his youth, along with his low-brow friends—including the best friend of all, the unrepentant Falstaff. So, when King Henry (formerly Prince Hal) proclaims, "I know thee not, old man" (*Henry IV, Part 2*, 5.5.47), all eyes in the theater are damp. Suddenly, the once jolly knight becomes possibly Shakespeare's most tragic figure—formerly the personality who stole the show from the two kings named Henry.

Unlike most other great Shakespearean roles, Falstaff has few lengthy speeches. Instead, the Bard gives him bursts of witty banter, as we see in this exchange when Falstaff enters a tavern to carouse with Hal:

> PRINCE HAL: That villainous abominable misleader of youth, Falstaff, that old white-bearded Satan.
> FALSTAFF: My lord, the man I know.
> PRINCE HAL: I know thou dost.
> FALSTAFF: But to say I know more harm in him than in myself were to say more than I know. That he is old, the more the pity; his white hairs do witness it. But that he is, saving your reverence, a whoremaster, that I utterly deny. If sack[4] and sugar be a fault, God help the wicked. If to be old and merry be a sin, then many an old host that I know is damned. If to be fat be to be hated, then Pharaoh's lean kine[5] are to be loved. No, my good lord, banish Peto, banish Bardolph, banish Poins, but for sweet Jack Falstaff, kind Jack Falstaff, true Jack Falstaff, valiant Jack Falstaff, and therefore more valiant being, as he is old Jack Falstaff, banish not him thy Harry's company, banish not him thy Harry's company. Banish plump Jack, and banish all the world.
>
> —*Henry IV, Part 1*, 2.4.479

To be clear, Shakespeare spun the Henrys IV–V triad as history plays, but his Jack Falstaff is not a historical figure. Rather, the fictional fat knight is a pastiche of several individuals that landed Will in a bit of trouble. It's complicated but interesting—so let us explain.

Our playwright lifted the idea of Prince Hal's wild youth from an earlier play, *The Famous Victories of Henry V* (author unknown). That play named a Sir John Oldcastle as one of Hal's rowdy companions. Shakespeare borrowed both the name and the idea of a sidekick for Hal, but nothing of Oldcastle's character. He then endowed his version of Oldcastle with a personality fabricated out of whole cloth. Thus, the old rascal emerged from Shakespeare's brain to become one of literature's most beloved tragi-comic personalities, debuting in *Henry IV, Part 1*, as Prince Hal's witty companion who bore the name Oldcastle.[6]

The problem, possibly unknown to Shakespeare at the time, was that the historical Prince Hal did have a friend named Sir John Oldcastle who married into a wealthy family, the Cobhams, whose lineage still existed in Shakespeare's day. The real Oldcastle had none of our Falstaff's celebrated characteristics, and the stuffy Cobhams were not amused to see one of their lineage cast in such a risible light. We do not know whether they followed up with threats of a lawsuit or merely grumblings of displeasure, but we do know that Shakespeare quickly changed the character's name to Falstaff.[7] Further, the ending of the play adds this disclaimer:

> EPILOGUE: One word more, I beseech you: if you be not too
> much cloyed with fat meat, our humble author will
> continue the story, with Sir John in it, and make
> you merry with fair Katherine of France, where, for
> anything I know, Falstaff shall die of a sweat, unless
> already he be killed with your hard opinions; for
> Oldcastle died a martyr, and this is not the man.
>
> —*Henry IV, Part 2*, EPI.28

But why did he choose the name "Falstaff"? Again, it's complicated.

Asimov's Guide to Shakespeare tells us that a Sir John Fastolfe was a brave warrior who fought many battles alongside King Henry V in France.[8] At one point, however, he was wrongfully accused of cowardice (and eventually exonerated). In Fastolfe's name, our Bard seems to have recognized a pun on "Falstaffe," which was a slang term that implied impotence—a bawdy allusion that would have appealed to his audiences.[9] With a slight change of spelling, everything fell into place, and Falstaff it was.[10]

So, that's the story of Falstaff's name. But on to the main point of our chapter: What does Falstaff's character tell us about Shakespeare's understanding of human personality? (And what, if anything, can we learn about the personality of Will Shakespeare himself?)

In *The Friendly Shakespeare*, Norrie Epstein writes of Falstaff: "A Freudian would say he represents the pleasure principle; to a moralist, he is the seven deadly sins in human form."[11] From another vantage point, Professor Harold Bloom has suggested that Falstaff represents the unfettered free spirit of the individual, while his foil, Prince Hal, embodies the oppressive, Machiavellian state. In support, Bloom quotes novelist Anthony Burgess:[12]

> The Falstaffian spirit is a great sustainer of civilization. It disappears when the state is too powerful and when people worry too much about their souls. . . . There is little of Falstaff's substance in the world now, and, as the power of the state expands, what is left will be liquidated.

These, of course, are all modern interpretations. But what we *really* want to know is how Shakespeare and his Elizabethan audiences understood Falstaff.

In the late 1590s, when the Bard gave birth to Falstaff, the word *personality* did not exist in the sense of an individual's traits or characteristics. Instead, people used the terms "disposition" and "temperament," which they conceived as one's balance of the four "humors": blood, yellow bile, black bile, and phlegm.[13] (And by the way, the word *humor* did not mean "funny" until nearly a century after Shakespeare's shuffle from the mortal coil.)

The humor theory is ancient, dating back at least to the time of the Greek physicians Galen and Hippocrates, in the fourth century BCE. It spread across the Middle East and into Europe, where the humors were considered medical gospel in the Middle Ages, into the Renaissance, and up to the rise of modern medicine in the late nineteenth century. Table 3.1 summarizes the supposed effects of the humors on one's temperament when one or another was out of balance in a person's body.

Curiously, we still use the figurative language of humors even though we have long since discarded the underlying concepts. We say, "Just humor me!" Or we may label others whose temperaments are deliberate or sluggish as "phlegmatic"—having a phlegm-like temperament. We sometimes hear warm, happy people called "sanguine," derived from the term's original meaning as an

Table 3.1 The Four Humors: The Sources and Temperaments

The Four Humors	Source	Temperament
Blood	Heart	Sanguine (cheerful)
Yellow bile (choler)	Liver	Choleric (angry, aggressive)
Black bile (melancholer)	Spleen	Melancholy (sad, depressed)
Phlegm	Brain	Phlegmatic (sluggish)

abundance of warm blood. Occasionally we even hear of anger metaphorically equated with "yellow bile" (thought to be produced in the liver)—so an angry person may be called "bilious." Likewise, depression may be referred to as "melancholy," a reference to "black bile," which was similarly known as "melancholer."[14]

The humors had other qualities associated with them. Yellow bile was regarded as "hot," and so associated with summer, manhood, and high temperatures. Phlegm was its opposite: cold and wet, also a symbol of decline. Likewise, each of the humors was linked to one of the four elementary substances: air, earth, fire, and water, as shown in figure 3.3.

During Renaissance times, the language of the humors was not just medical jargon but deeply embedded in the popular culture, much as people now use the terms "neurotic" or "depressed." Theatergoers were so familiar with the humors that playwrights commonly wrote characters with exaggerated humoric characteristics as a shorthand way of connecting with their audiences.[15] Indeed, Ben Jonson, Shakespeare's friendly rival, famously took the idea of humors to extremes in his comedy *Every Man in His Humor* (in which Shakespeare himself appeared on stage).[16]

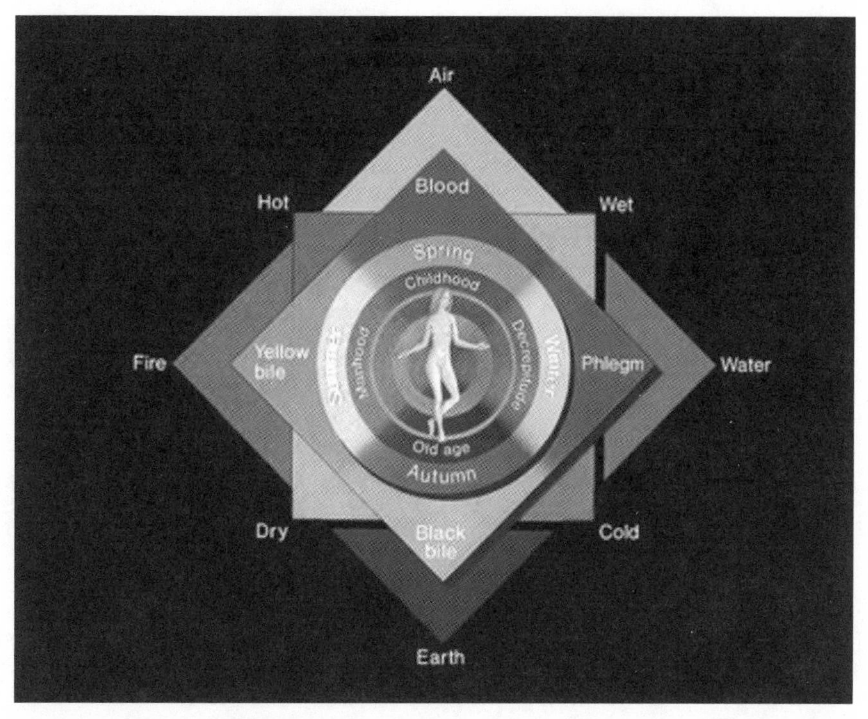

Figure 3.3 The four humors and their associated qualities. *Wikimedia Commons.*

Because people were so familiar with the language of the humors, audiences readily understood that a player was referring to black bile when he spoke obliquely of "sable-coloured melancholy" and "the black-oppressing humor." They also understood Beatrice when she responded to Benedick's protestations that he would never fall in love:

> BEATRICE: I thank God and my cold blood, I am of your humour for that: I had rather hear my dog bark at a crow than a man swear he loves me.
>
> —*Much Ado About Nothing*, 1.1.126

Again in *Troilus and Cressida*, the public hears references to humors peppered throughout this description of the warrior-lummox Ajax:

> ALEXANDER: . . . He is as valiant as the lion,
> churlish as the bear, slow as the elephant: a man
> into whom nature hath so crowded humours that his
> valour is crushed into folly, his folly sauced with
> discretion: there is no man hath a virtue that he
> hath not a glimpse of, nor any man an attaint but he
> carries some stain of it: he is melancholy without
> cause, and merry against the hair . . .
>
> —*Troilus and Cressida*, 1.2.23

Falstaff and Prince Hal, too, are personas endowed with humor-based personalities, as are their foils, Hal's father and Hotspur. According to cultural historian Nelly Ekström, the two parts of the *Henry IV* play center on these four characters, each modeled after a different humor and each with about the same number of lines. Says Ekström:

> So the play itself is very close to the ideal humoral balance. King Henry IV himself is melancholic, Prince Hal sanguine, Sir Harry Hotspur choleric and the knight Sir John Falstaff is phlegmatic. Jack Falstaff is fat, lazy, cowardly, dishonest and sentimental, but the audience loved him despite all his faults.[17]

Indeed, the phlegmatic Falstaff himself, when explaining to a judge his failure to obey a court summons, drops the name of the humor theory's founding father, the Greek physician Galen:[18]

> Falstaff: It hath it original from much grief, from study, and perturbation of the brain. I have read the cause of his effects in Galen. It is a kind of deafness.
>
> —*Henry IV, Part 2*, 1.2.118

But Falstaff is much more than an overstuffed bag of humors. While his personality *is* both sanguine and phlegmatic, it is also comic and self-deprecating. He loves the clever use of language, he loves the life of the stews and pubs, and he loves Prince Hal. He takes nothing seriously, except perhaps sack and sex. And he knows how to light up a room to brighten everyone's spirits. Yet deep inside we briefly glimpse fear of rejection and death. Shakespeare's portrait of Falstaff is one of a complex and nuanced figure that the humors alone cannot fully capture.

SHAKESPEARE "INVENTS" THE HUMAN PERSONALITY

When he came to London, young Will found a vibrant theatrical world. Permanent theaters with resident companies were a new phenomenon in the city. Skilled playwrights, such as Jonson, Marlowe, and Kyd, were writing plays that attracted packed houses. Many of their characters, however, were relatively undeveloped and two-dimensional. Aside from the humors, they might represent a religious figure, royalty, a Greek god, or one of the Seven Deadly Sins. That was the context in which Shakespeare found his theatrical niche—bringing more fully developed personalities to life on stage. We see this in Brutus's metaphor for mental conflict, as he pondered the assassination of Julius Caesar:

> Brutus: Since Cassius first did whet me against Caesar,
> I have not slept.
> Between the acting of a dreadful thing
> And the first motion, all the interim is
> Like a phantasma, or a hideous dream:
> The Genius and the mortal instruments
> Are then in council; and the state of man,
> Like to a little kingdom, suffers then
> The nature of an insurrection.
>
> —*Julius Caesar*, 2.2.63

With such words, Shakespeare brought human mental processes to the fore, opening a window on the private thoughts, feelings, and motives of his characters. Just as important, he wove these inner experiences into the narratives of their lives. Elizabethan audiences had seen kings and queens, gods and goddesses, traitors, magicians, and warriors, but they had rarely heard characters make their psyches public on the stage, as they did with Shakespeare's soliloquies.

To be sure, this concept was not altogether original with Shakespeare. There were earlier literary models, particularly the characters in Chaucer. Remember, too, that Homer gave us the brooding Achilles. Such great writers of the past undoubtedly inspired Shakespeare's work.[19] Harold Bloom, however, gives full credit to Shakespeare for filling out his subjects' personalities with portrayals of their innermost thoughts, feelings, and desires. Declares Bloom, "Personality, in our sense, is a Shakespearean invention."[20]

Whether we fully agree with Professor Bloom that Shakespeare "invented" personality, we must agree that the personalities peopling the plays went beyond those seen on the stage before. It's not that Shakespeare gave up the humors. Rather, he added new dimensions and nuances to them. Falstaff and Hamlet are among the most complex. Hamlet's dominant humoral characteristic is melancholy, but he is also clearly intellectual, consumed with questions about the meaning of life and death, relationships, and government. And he knows about swordplay, as well as theatrical plays, as when he arranges to reveal his father's murderer with *The Mousetrap*, the play-within-a-play.[21] Then there is Shylock, the Jewish moneylender (*The Merchant of Venice*)—yet another melancholy character whose other dimensions include a dark sense of humor and the absurd, plus a choleric need for revenge. Behind the humors, Shakespeare lets us see Shylock as a complex personality with understandable motives, rooted in antisemitism.[22] Here's his famous description of a desire for revenge on those who have persecuted him:

> SHYLOCK: I am a Jew. Hath not a Jew eyes? Hath not a Jew hands, organs, dimensions, senses, affections, passions; fed with the same food, hurt with the same weapons, subject to the same diseases, healed by the same means, warmed and cooled by the same winter and summer as a Christian is? If you prick us do we not bleed? If you tickle us do we not laugh? If you poison us do we not die? And if you wrong us shall we not revenge? If we are like you in the rest, we will resemble you in that. If a Jew wrong a Christian, what is his humility? Revenge. If a Christian wrong a Jew, what should his sufferance be by Christian example? Why, revenge.
>
> —*The Merchant of Venice*, 3.1.57

Shakespeare seems to have been at his most "humorous" when he developed choleric characters, the likes of Hotspur, Coriolanus, Pericles, Titus Andronicus, Edward III, and King John. These are largely portrayals of people given to anger and aggression, but with relatively little depth. Even Mr. Macbeth, that most choleric of personalities, does not have the depth of Ms. Macbeth, whose personality is rendered more complex by cunning, ambition, cruelty, and signs of mental illness.

OTHER INFLUENCES ON PERSONALITY

In the Elizabethan mindset, the humors were not the sole determiners of disposition and temperament. People also believed that the gods, chance, Fortune, and even lines in their palms influenced people's psyches.[23] The stars, too, had their roles. So, we shouldn't be surprised to hear that Shakespeare inserted more than one hundred astrological references in his plays—as when Lear observes, "It is the stars. The stars above us govern our conditions" (*King Lear*, 4.3.38).

Astrology served another purpose, as a link in the chain connecting mortals with the humors and the cosmology of the Middle Ages. Each astrological sign had its connections with a dominant humor and with the heavens.[24] Above the layered world of air, earth, fire, and water lay the realms of the planets and stars, moving on concentric crystalline spheres—yet somehow influencing our lives on the earth below. These celestial bodies influenced not just "our conditions" in the external world but our internal dispositions as well. What a gift it was for a playwright! If you were developing a melancholy character, for example, you might describe her as born under Capricorn with Saturn ascendant, so that audiences would know her immediately as prone to shyness, withdrawal, and depression. You can come up with other such combinations by examining table 3.2.

The ancient world knew of seven "planets"—Jupiter, Saturn, Mars, Venus, and Mercury, plus the Sun and the Moon (which were also considered planets). Some were benign; others could be ominous or unfortunate.[25] In *The Winter's*

Table 3.2 Temperaments, Astrology, and the Humors

Temperaments	Celestial Body	Constellations	Humors
Sanguine	Jupiter	Gemini, Aquarius, Libra	Blood
Choleric	Mars	Aries, Leo, Sagittarius	Yellow bile
Melancholic	Saturn	Taurus, Virgo, Capricorn	Black bile
Phlegmatic	Moon, Venus	Cancer, Scorpio, Pisces	Phlegm

Tale, for example, a pickpocket blames his lot on being born ("littered") under the planet Mercury:

> AUTOLYCHUS: My father named me Autolychus, who being, as I am, littered under Mercury, was likewise a snapper-up of unconsidered trifles.
>
> —*The Winter's Tale*, 4.3.1[26]

Audiences of the day would have smiled, knowing that the god Mercury is the patron of thieves.

If you were born under Mars, named for the god of war, that fact might bode well for your success as a soldier—or it could signal the unfortunate tendency of an uncontrolled temper. In *All's Well That Ends Well*, however, the braggart Parolles's birth under Mars gives Helena an opportunity to skewer him for his cowardice:

> HELENA: The wars hath so kept you under, that you must needs be born under Mars.
> PAROLLES: When he was predominant.
> HELENA: When he was retrograde, I think rather.
> PAROLLES: Why think you so?
> HELENA: You go so much backward when you fight.
>
> —*All's Well that Ends Well*, 1.1.201

To get the joke you must know (as Shakespeare's audiences did) that the motions of the planets sometimes appeared to be backward, or *retrograde*. That is, the planets occasionally seemed to shift direction and move in a path opposed to the other planets and the stars.[27] So, Helena was suggesting that Parolles was a coward because his motion in the face of danger was "retrograde."

Comets were especially ominous, though they influenced events more than personalities. A comet could foreshadow wars and pestilence, but it might also bring bad news for princes and other powerful people. (Common people did not rate having their misfortunes portended by spectacular celestial events.) So, Shakespeare has the Duke of Bedford tell us that a comet foretold the death of his brother, the beloved King Henry V:[28]

> BEDFORD: Hung be the heavens with black, yield day to night!
> Comets, importing change of times and states,
> Brandish your crystal tresses in the sky,

And with them scourge the bad revolting stars
That have consented unto Henry's death!
King Henry the Fifth, too famous to live long!
England ne'er lost a king of so much worth.

—*Henry VI, Part 1*, 1.1.1

Eclipses, like comets, were also bad omens, more often foretelling social upheaval than affecting mortals' actions or personalities. Coincidentally, both a lunar and a solar eclipse were visible in England in 1605. Shakespeare seems to have incorporated them into *King Lear*, written shortly after Queen Elizabeth's death:

GLOUCESTER: These late eclipses in the sun and moon portend
no good to us: though the wisdom of nature can
reason it thus and thus, yet nature finds itself
scourged by the sequent effects: love cools,
friendship falls off, brothers divide: in
cities, mutinies; in countries, discord; in
palaces, treason; and the bond cracked 'twixt son
and father.

—*King Lear*, 1.2.109

For plays that he set in the distant past or a romantic never-land, Shakespeare might call upon the Greek and Roman gods to addle someone's mind. At the behest of his pen, the gods steered people into madness, as Apollo did to Cassandra in *Troilus and Cressida*. And of course, he held Cupid accountable for love. Similarly, he used the gods (who were known to act on whim) to change the plot's direction, as in *Cymbeline*, when Jupiter appears in a dream to persuade Posthumus to scrub a foolish suicide mission. And on occasion, he employed the gods to give his audience a psychology lesson in what we would now call *negative reinforcement* (loosely, the rewarding feeling when an unpleasant stimulus is removed):

JUPITER: Whom best I love I cross; to make my gift,
The more delayed, delighted.

—*Cymbeline*, 5.4.95

While our playwright's works frequently attributed people's temperaments to heavenly objects or the gods, we know that he had a skeptical streak, which showed in *Julius Caesar* when Cassius says, "The fault, dear Brutus, is not in our stars, / But in ourselves" (1.2.142). And again, in *King Lear*, we hear the bastard son of Gloucester lament:

> EDMUND: This is the excellent foppery of the world, that when we are sick in fortune (often the surfeits of our own behavior) we make guilty of our disasters the sun, the moon, and stars, as if we were villains on necessity; fools by heavenly compulsion; knaves, thieves, and treachers by spherical predominance; drunkards, liars, and adulterers by an enforced obedience of planetary influence; and all that we are evil in, by a divine thrusting on. An admirable evasion of whoremaster man, to lay his goatish disposition on the charge of a star! My father compounded with my mother under the Dragon's tail, and my nativity was under Ursa Major, so that it follows I am rough and lecherous. Fut, I should have been that I am, had the maidenliest star in the firmament twinkled on my bastardizing. . . .
>
> —*King Lear*, 1.2.125

Most likely, Shakespeare viewed the gods, the stars, and the humors simply as metaphors—as romantic fictions that have always been employed by writers to personify human desires.[29] He seems to have viewed Christianity in much the same way—although we can be less certain of that. Playwrights in those days usually avoided anything that might appear critical of official Church doctrine and result in heresy charges stemming from disapproval of the government-appointed censor, the Master of the Revels.[30]

LEGACY OF THE HUMOR THEORY

Today, most of us have abandoned the idea that the humors, the stars, planets, and divine forces influence personality and behavior. Yet we still find traces of the old humor theory in the brain sciences, where they survive under the names of *hormones* and *neurotransmitters*—the "body fluids" that *really do* influence how we think and act. The fifty-some hormones (adrenaline, cortisone, thyroxine, insulin, etc.) coursing through our bloodstreams carry messages between the brain and the muscles and internal organs.

Similarly, some one hundred neurotransmitters transport messages between nerve cells within the brain. Too little of the neurotransmitter serotonin, for example, may tip the mental balance toward depression. Likewise, a deficiency of acetylcholine correlates with dementia. But a flood of the hormone adrenaline may exaggerate both our thrills and our fears, while an abundance of testosterone can encourage risk-taking behaviors. Psychoactive drugs, such as cocaine, morphine, LSD, caffeine, nicotine, and alcohol, can also change our personalities by mimicking or altering the effects of our hormones and neurotransmitters.[31]

The humors have left a legacy in the psychology of personality, too. But there we speak of *traits*, which are analogs of the *dispositions* once associated with the humors: sanguine, choleric, melancholy, and phlegmatic. In psychology, traits are the characteristics that everyone possesses to different degrees and that account for the consistency of an individual's personality from one situation to another.

In the short history of psychology, many lists of fundamental traits have been proffered. But recently, a consensus has formed around a concept called the Five-Factor Model, or simply, the Big Five. This view sees personality as a description of a person in terms of five major traits: *openness to experience*, *conscientiousness*, *extraversion/introversion*, *agreeableness*, and *neuroticism*.[32] (See figure 3.4.) Think of these five traits as *dimensions*, each ranging from the named trait to its polar opposite. A person's score on a "test" of the Big Five traits, then, forms a snapshot of that person's personality. Here's a brief description of each of the five trait dimensions:

- ***Openness to experience*** lies at one pole of this dimension (which also includes *curiosity* and *independence*). *Closed-mindedness* anchors the opposite pole.
- ***Conscientiousness*** could also be called *dependability*, *cautiousness*, or *perseverance*. Its polar opposite is *impulsiveness*, *carelessness*, or *irresponsibility*.
- ***Extraversion*** may also assume the names of *assertiveness*, *sociability*, or *self-confidence*, while its opposite extreme is *introversion* and *contemplation*.
- ***Agreeableness*** is also known as *conformity*, *likeability*, or *warmth*; its antithesis is *coldness* or *negativity*.
- ***Neuroticism***, sometimes called *anxiety* or *emotionality*, falls at one pole of this dimension, while *emotional stability* or *emotional control* occupy the opposite pole.

Trait theorists consider the five-factor model to be one of psychology's major achievements because these same five factors emerge (mathematically) from scores on many personality "tests."[33] Moreover, research shows that these same five factors apply to people in cultures all over the world.[34]

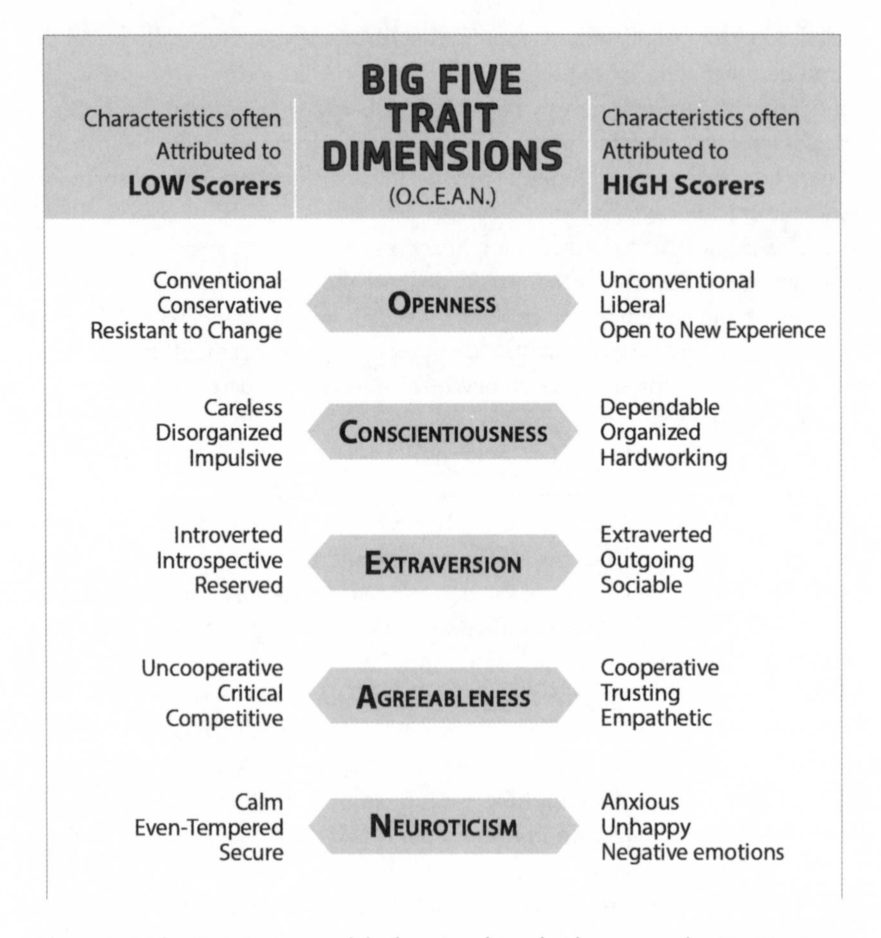

Figure 3.4 The Five-Factor model of personality, also known as the Big Five traits.

We will illustrate the five trait dimensions with a few Shakespearean characters. If you have read or seen the plays, you may think of other examples:

- Prince Hal, with his willingness to change his lifestyle from rascal to ruler, would probably score high on *openness to experience*, as would Prospero, Pericles, and Rosalind. Low scorers might include Coriolanus, Lear, and Holofernes (in *Love's Labour's Lost*).
- Had Prince Hal repeatedly taken a Big Five personality assessment as he matured and assumed the role of Henry V, his score would likely

have moved upward on *conscientiousness*. At the opposite end of this dimension (closed-mindedness), we would find Achilles, Lady Macbeth, and Angelo (the villain in *Measure for Measure*).

- The exuberant Falstaff is exhibit A of *extraversion*, along with Beatrice and Benedick, Petruchio (*Taming of the Shrew*), and the Dromio twins (*A Comedy of Errors*). Near the introverted extreme, we would certainly find the brooding Hamlet.
- As for *agreeableness*, it's Falstaff again—along with Miranda (*The Tempest*) and, perhaps, Puck (*A Midsummer Night's Dream*)—while Kate (*Taming of the Shrew*) and Shylock would fall near the opposite (negativity) pole.
- Many of Shakespeare's most volatile and tragic figures leap to mind as examples of *neuroticism*: Lear, Timon (of Athens), Othello, Hamlet, and Richard II, along with Malvolio (from *Twelfth Night*) and nearly everyone in *Titus Andronicus*. Should we include Falstaff among them?

Here's the point: *Psychology's focus on traits has put the field in much the same position as Shakespeare found the London theater scene in the early 1590s.* Traits, like the humors, paint a simplified—even simplistic—picture of a person in terms of just a few dimensions. But, say psychologists Dan McAdams and Jennifer Pals, people are more complex than a mosaic of just a few traits.[35] A trait perspective glosses over the myriad individual differences that make each of us unique. Imagine how much you would miss in the personalities of, say, Michelle Obama, Barbra Streisand, or Woody Allen if you were to describe them only on the Big Five dimensions.

Indeed, people *do* seem to have certain traits that explain their consistency from one situation to another. Yet people change—although we may only see these changes over time, as a person's life unfolds. We know that Shakespeare understood this, too, for he made personality changes a hallmark of his greatest works. Professor Stephen Greenblatt tells us:

> I think that what is interesting about Shakespeare is he thought through the possibility, not simply that there would be change, mutability, but that it would have the form of a narrative. It would take the shape of a story.[36]

Let's examine this idea more closely, to see whether Shakespeare (with the help of some modern psychologists) can lead us beyond the concept of personality as a collection of humor-like traits.

LIFE STORIES AND THE PSYCHOLOGY OF REDEMPTION

Psychologist Dan McAdams and his colleagues have collected life stories from hundreds of people to show how people weave the events of their lives into some sort of private narrative.[37] Such stories reflect an individual's mental state, says Prof. Barbara Woike of Barnard College.[38] Stories of growth and insight indicate a sense of well-being, while people whose stories are impoverished in meaning and emotion are typically more anxious or defensive. And it is no surprise that depressed people (like Hamlet) tell more melancholy stories.

Most interesting for our purposes are those told by mature and well-adjusted—so-called "generative"—adults, whose stories are much like the ones Shakespeare tells of heroic characters like Henry V. They tell of deliverance from pain, suffering, and adversity. They also tell of helping—or even saving—others. Professor McAdams calls them stories of *redemption*. In his book *The Redemptive Self*, he describes several forms of the redemptive story told by these healthy, generative adults.[39] The tellers of these narratives often describe an early sense that they have been endowed or "blessed" with some sort of talent or privilege, while life has dealt others a much more difficult hand. They grow to feel an obligation or duty to improve the lots of others, but they meet obstacles, conflicts, and failures along the way that help them clarify their goals and focus their efforts. Eventually, their stories tell of overcoming problems and looking back on a life well spent.

McAdams notes that there is a distinctively Christian version of the redemptive story that involves *atonement* for some "sin" committed by oneself or one's family.[40] Shakespeare seems to have wrapped his histories of the Henrys around this form of the redemption narrative. We may see it in the comedies, too—the plays with happy endings—such as *A Midsummer Night's Dream*, *The Tempest*, *As You Like It*, and *All's Well That Ends Well.*

A *failed* struggle for redemption is the stuff of tragedy. Hamlet forfeits his life in seeking to avenge his father's murder; Macbeth and King Richard III believe that others have unfairly blocked their paths to the crown and attempt to seize what should rightfully be theirs—and again both pay for their efforts with their lives. Likewise, the Roman war hero Coriolanus seeks approval of the citizenry, yet he stumbles over his pride and joins forces against the city with his former enemies, the Volscians, who also turn against him. The pattern repeats itself in others who believe they have the best of intentions or justifications but fall tragically short: Othello, Lear, Timon of Athens, Juliet, Friar Lawrence—and don't forget Bolingbroke and Falstaff.

In sum, this development in the psychology of personality, pioneered by McAdams and his colleagues, echoes Shakespeare by showing that a complete human personality can be found in the flowing narrative of a person who grows, develops, and responds to the changing conditions of life, not merely in a collection of traits.

WHAT SHAKESPEARE DID *NOT* SAY ABOUT PERSONALITY

Since the psychology of personality was fundamentally invented by Freud, we feel compelled to reiterate here a point that we made in the Introduction: While Shakespeare invoked the gods, the stars, and the humors to explain human behavior, he never suggested that forbidden desires fester in our unconscious minds and drive our actions and obsessions.[41] On the contrary, his characters typically seem quite conscious of their pleasurable desires and aggressive tendencies—as we see in a Falstaff who revels in debauchery and a Richard III who nearly sneers and twirls his mustache melodramatically. So, while Freudian psychoanalysis may offer one way to interpret Hamlet, Falstaff, the Richards, and the Henrys, Shakespeare's work neither requires nor anticipates a Freudian model of the mind, with its unconscious battles among factions of the id, ego, and superego.

To be clear, psychologists generally revere Freud as a pioneer and an astute observer of the human condition. At the same time, we have largely abandoned Freud's theories of repressed memories and a mind at unconscious war with itself. Nor do we believe that we can read the sources of neurosis in the dreams of a midsummer's night. Rather, psychological science conceives of people as variations on a Darwinian theme—as creatures whose evolved priorities are survival and sending their genes into the next generation.[42] Everything we do derives, one way or another, from those two drives. This includes defending our self-images with what Freud would have called *ego-defense mechanisms* and modern psychologists might term the *self-serving bias.* Previously, we have seen the defense mechanisms of *projection* and *reaction formation* at work in Shakespeare, as when Hamlet's mother says, "The lady doth protest too much, methinks" (3.2.254). We see another example of these self-serving defenses when Falstaff plays dead to avoid being killed in battle. Here Shakespeare makes us privy to Falstaff's comic (and quite conscious) rationalization:

> FALSTAFF: 'Sblood, 'twas time to counterfeit, or that hot termagant
> Scot had paid me, scot and lot too. Counterfeit? I lie, I am
> no counterfeit: to die is to be a counterfeit, for he is but the

> counterfeit of a man, who hath not the life of a man: but to counterfeit dying, when a man thereby liveth, is to be no counterfeit, but the true and perfect image of life indeed. The better part of valor is discretion, in which better part I have saved my life.
>
> —*Henry IV, Part 1*, 5.4.113

To be sure, rationalization *is* one of the Freudian ego defense mechanisms.[43] And we should give the psychoanalysts their due for calling our attention to these self-protective tricks that people play on themselves when they are anxious or fearful, using the ego defense mechanisms of *denial*, *displacement*, *projection*, *reaction formation*, *rationalization*, and others. Yet it wasn't Freud, Jung, or any of their acolytes who first observed these psychological defenses at work. Shakespeare anticipated them three centuries earlier. It's just that he didn't attribute them to conflicts in a Freudian unconscious.[44]

SHAKESPEARE AND THE "PERSON-SITUATION" CONTROVERSY

King Henry V is perhaps the most celebrated of the English monarchs for his near-legendary victories against the despised French, especially at Agincourt. Shakespeare's eponymous play portrays him as a man without peer who emerged just when England needed him the most. Historians call this the "Great Man Theory," a nineteenth-century notion that great leaders (usually men) were born, not made. Here is a second side to that issue: perhaps the great men (and women) *learned* to be great. And wouldn't you know, psychology has yet another answer: the "great men" may have been no different from the rest of us but were responding to the *situation* in which they found themselves.

Consider this: You stop your car when you see a red traffic light, you sing along with the national anthem at the ballpark, and you rush home when you get a message that your child is sick. Social psychologists call this *the power of the situation*, by which they mean that social pressures, expectations, and other circumstances in our environment can be the dominant forces in shaping the ways we think, feel, and act. Indeed, the power of the situation has been demonstrated over and over in psychological experiments. We will consider more of those experiments in chapter 4, where *Measure for Measure* examines the mind and behavior from the "situationist" perspective of social psychology.

Psychologists interested in personality, however, see things from a perspective more akin to the Great Man theory. They point out that different people often behave differently in the same situation, so an individual's personality traits must also play a role in determining behavior—perhaps a dominant role.[45] But we might ask, Which exerts *more* influence on us: our *personal traits* or the *situations* in which we find ourselves? Psychologists call this *the person-situation controversy*, and it continues as an intramural debate within psychology, between personality theorists and social psychologists.

Was the Bard aware of this issue in some form? We will have more to say about this as we go through the remaining chapters in the first half of this book, where we will see just how Shakespeare's plays certainly addressed the person-situation controversy from both angles. Our reading of *Henry V*, as well as the *Henry VI* trilogy, sees the playwright on the side of the *person*. But in *Hamlet* he seems ambivalent, and in *Macbeth* and *Measure for Measure*, it's the *situation* that seems to dominate.

WHAT DO WE KNOW ABOUT SHAKESPEARE'S OWN PERSONALITY?

If you were an Elizabethan looking for good conversation and a pint of ale, you could have done no better than to join Will Shakespeare and his friends, perhaps at a table in the Anchor, the Tabard, or the Mermaid. You might recognize him by the famously bald pate. (It is not certain that any of the artists who rendered the familiar portraits actually knew him in life. Still, there is consensus on certain physical features, such as the extensive forehead [see figure 3.5]).[46] According to the gossipy seventeenth-century biographer John Aubrey, "He was a handsome, well shap't man: very good company, and of a very readie and pleasant smooth witt."[47] Those who knew him well were clearly charmed by him, even the curmudgeonly Ben Jonson, who wrote in posthumous praise of his friend: "I loved the man, and do honour his memory." With another refill of his quill, Jonson added that Will Shakespeare was "of an open and free nature; had an excellent fancy, brave notions, and gentle expressions, wherein he flowed with that facility that sometime it was necessary he should be stopped."[48]

The love and admiration of his friends moved them to assemble and publish a collection of his plays in a volume now called the *First Folio*, many copies of which still exist.[49] In a poetic tribute introducing the *Folio*, Jonson again praised Shakespeare's character and abilities, writing that he was the "Soule of the Age!" and, as if that were not enough, "He was not of an age, but for all time!"[50]

All this suggests that Will Shakespeare was more subdued in his temperament than many of his fellow thespians who were notorious for their flaring tempers, jealousies, and brawls. Christopher Marlowe was killed in a fight,

Figure 3.5 The Droeshout portrait of Wm. Shakespeare, an engraving done by the English engraver Martin Droeshout. This portrait appeared on the title page of the First Folio of the Bard's works. *Wikimedia Commons.*

one of Shakespeare's troupe died from knife wounds inflicted by a fellow cast member in a drunken tavern altercation, and Ben Jonson himself killed another actor in a duel. Because of their rowdy reputation and itinerant lives, men of the theater (they were all males) were looked down upon, even though people of every social class eagerly patronized their productions.[51]

The theaters, as well as the actors, had associations in the public mind with rowdiness, violence, crime, and disease. Accordingly, no theaters were allowed within the city walls of London. Writer Bill Bryson describes the scene:

> It was a banishment they shared with brothels, prisons, gunpowder stores, unconsecrated graveyards, lunatic asylums (the notorious Bedlam stood close by the Theatre), and some enterprises like soapmaking, dyeing, and tanning—and these could be noisome indeed. Glue makers and soapmakers rendered copious volumes of bones and animal fat, filling the air with a cloying smell that could be all but worn, while tanners steeped their products in vats of dog feces to make them supple. No one reached a playhouse without encountering a good deal of odor.[52]

Despite the tawdry reputation of thespians and their theaters in Elizabethan London, Will Shakespeare managed to present himself as a likable fellow and a respectable citizen. But it must also be noted that his own theatrical company, The Lord Chamberlain's Men (later, the King's Men), was itself atypical. Compared to their brawling, hard-drinking contemporaries in Shoreditch and Southwark, members of Shakespeare's troupe had a reputation for being comparatively well-behaved. Most were family men whose lives were comparatively low-key. Yet even in this group, Shakespeare was an outlier in that his family did not live with him in London.

We have one exception to the positive portrait of the Bard. It appears in a snarky little volume titled *A Groats-worth of Witte*, where playwright Robert Greene alludes to young Will as "an upstart Crow, beautified with our feathers."[53] Apparently, Greene (who was known to make snide remarks about his rivals) was jealous of Will's rapid rise in the London theater scene as both an actor and a writer. Greene's comments are the only derogatory references we have about Shakespeare from his contemporaries.

Greene notwithstanding, we can conclude with some certainty that Will Shakespeare was well-liked and respected by most who knew him. In the language of the humors, we would say that he had a ***sanguine*** temperament. But what else do we know about the man? Here are a few other traits suggested by the meager records we have of him.

He was ***funny***. As the sardonic Ben Jonson noted, "His wit was in his own power; would the rule of it had been so too. Many times he fell into those things, could not escape laughter."[54] A century later, Dr. Samuel Johnson (the other Johnson[55]) noted that Shakespeare's natural talent was for comedy: "In his comick scenes, he seems to produce without labour, what no labour can improve. . . . His tragedy seems to be skill, and his comedy to be instinct."[56] John Aubrey added that "his comoedies will remaine witt as long as the English tongue is understood."[57]

And he was ***intelligent***—an assessment based on evidence from his contemporaries and from his writing. There were no intelligence tests in his day, of course, but Shakespeare's work leaves no doubt that he would have occupied a point on the high tail of the IQ curve. The plays show that he was widely read, which indicates that he had both the intellect and interest to study and understand literature—in several languages. He also had a special gift for using language, especially for crafting metaphors, which is an indicator of both intelligence and creativity.[58]

In addition, Shakespeare had a reputation for ***honesty***. We can infer this from his position as the trusted financial officer for his company. More direct evidence of this trait comes from two sources, one being his friend Ben Jonson, who wrote, "He was, indeed, honest, and of an open and free nature." The other source, ironically, comes from the publisher of Robert Greene's screed against the "upstart crow." After meeting Shakespeare in person, Mr. Henry Chettle testified that "divers of worship have reported his uprightness of dealing, which argues his honesty."[59]

In his political views, Will Shakespeare was ***conservative***—or at least that is the side he presented in his writing. It might be more accurate to say that he lived in dangerous times when one could be imprisoned or even executed for seeming to oppose the Crown or portraying someone in the royal family tree in a bad light. All playwrights were careful to observe political correctness in their works, to the point of rewriting history in favor of Elizabeth's or James's ancestors.[60] We see this clearly in Shakespeare's theatrical tribute to Elizabeth's father, Henry VIII, and in his pillorying of Richard III, whose defeat by Elizabeth's grandfather launched the Tudor dynasty.

The politically conservative theme running through the plays suggests both deep respect for the established order and fear of populist mob rule—which almost came to pass in Shakespeare's time in the wake of the infamous Gunpowder Plot to blow up the Parliament building and, with it, the top-drawer of English nobility. Coincidentally, the Plot was hatched in 1605 by one Robert Catesby, son of William Catesby, who lived in Warwickshire—and was a friend of John Shakespeare, Will's father.

Here's the story, in brief. Two members of King James's Parliament caught the Catholic conspirator Guy Fawkes red-handed underneath the Parliament building, ready to light a fuse leading to kegs of gunpowder, enough to make smithereens of the assembled nobles. Under torture, Fawkes revealed the names of his collaborators, many of whom lived in Warwickshire—where Stratford is located. They, in turn, were rounded up, tortured, and eventually suffered public executions, whence they were first disemboweled and then beheaded. Many of their fellow citizens in Warwickshire would have been fearful of reprisals by the authorities, and the Shakespeare family could have logically been among them. Happily, the Crown was satisfied with just executing Fawkes and the traitors named by him.

In Protestant England, one could also get in trouble for the lesser crime of being a practicing Catholic or "recusant."[61] The reason was that the twice-thrice-married Henry VIII had defied the pope by divorcing his first wife, Catherine of Aragon, because she did not bear him an heir. To obtain the divorce, Henry had to crush Catholicism in his country and establish his own Church of England in its stead. Professor Greenblatt argues that Shakespeare was raised in a Catholic family and so may have harbored secret Catholic sympathies. As we have seen, our playwright was cautious, sanitizing his works of any religious sympathies—although he did include clergymen, such as Cardinal Wolsey, whom he portrayed more as scheming politicians than as men of the cloth. The plays also allude to biblical figures and events, not as Bible lessons but to help his audiences grasp the plotline. We see this in the graveyard scene where Hamlet remarks on a gravedigger's casual treatment of freshly unearthed bones with an allusion to the Genesis story of Cain and Abel:

> HAMLET: That skull had a tongue in it, and could sing once.
> How the knave jowls it to the ground, as if 'twere Cain's
> jawbone, that did the first murder!
>
> —*Hamlet*, 5.1.77

So, we can say that Shakespeare was conservative, either by nature or out of fear of the Crown. Yet he was not a prudish conservative. Says A. L. Rowse, "He was the least Puritan of writers."[62] His plays are full of sexual allusions, typically in the form of salacious puns. And then there is *Pericles*, a story revolving around the dark secret of incest. Nor did Shakespeare shy away from violence and gore—the most outrageous example being *Titus Andronicus*, with its amputations and the Sweeney-Todd-like consumption of two brothers baked in a pie.

If the Bard had a fault, it may have involved the peccadillo of ***pride*** or *hubris*. We know that his frugality and business acumen in London allowed him to retire a wealthy man and move home to Stratford—where he immediately bought the largest house in town. This display of wealth may have originated in observations of his father's financial difficulties and the failed attempt to attain a coat of arms—and thus John Shakespeare's failure to attain the status of "gentleman." Upon returning to Stratford, William renewed the arms application, embellished with some fanciful family history. Finally, the heralds at the College of Arms awarded John Shakespeare a handsome crest and shield embellished with a spear and the words "*Non Sans Droict*," which is Latin for "Not Without Right."[63]

So, we have a trait-style portrait of a man who was likable, funny, intelligent, honest, politically conservative, and possibly a bit vain or prideful. What else?

He was ***contemplative*** and therefore, in the jargon of the Big Five traits, most likely ***introverted***, in the classical Jungian sense of focusing on his own thoughts and feelings.[64] It is not a stretch to suggest that Will Shakespeare must have been introspective and comfortable with himself and with the isolation that extensive reading and great writing require—even though he could be outgoing and effusive in social situations.[65]

Also in terms of the other Big Five traits, it seems clear that Mr. Shakespeare would have scored high on ***openness***. This derives in part from Ben Jonson's characterization of him as "of an open and free nature" but also from his wide-ranging knowledge of different topics that we see in his plays: psychology, medicine, law, biology, geography, history, literature, and more. Just as remarkable was his singular ability to write credible parts for women, as we see, for example, in his lines for Juliet, Lady Macbeth, Cleopatra, and Hamlet's mother Gertrude, among others.[66] As Rowse put it, "Unlike any other dramatist in the age, he shared the woman's point of view."[67]

We also know that Shakespeare, the actor, played many roles both on and off-stage. Among them, as his troupe's principal playwright, he was expected to pen at least two new plays every year—and did so for two decades![68] This prodigious output suggests that Shakespeare would have also rated high on the Big Five dimension of ***conscientiousness***.

We can be less certain about the final two "OCEAN" traits of ***agreeableness*** and ***neuroticism***. To gain a reputation as likable, it seems that he would have to have been at least somewhat agreeable. As for neuroticism, we can be less sure. We will return to that topic in the chapter where we talk about mental disorders.

Finally, we can say with confidence that Will Shakespeare was a ***curious*** fellow—curious about many things, from archery and astronomy to botany to history, hunting, heraldry, law, literature, medicine, mythology, military strategy, sports, and exotic foreign places. Most of all, he was fascinated by odd and interesting people—particularly those with unusual "humors." Aubrey tells us that when the theater company visited Stratford, Shakespeare took "the humor of the constable" and later worked it into the story of *A Midsummer Night's Dream*. Aubrey further notes that "Ben Johnson [*sic*] and he did gather humours of men dayly where ever they came." Such anecdotes fit well with our surmise that it was Shakespeare talking when, in *Antony and Cleopatra*, he had Mark Antony say, "Tonight we'll wander through the streets and note the qualities of people" (1.1.56). Surely we can deduce that Shakespeare was intensely curious about human nature. He was a psychologist-before-psychology.

Unfortunately, this trait-based picture of Shakespeare's personality is a highly pixelated portrait. It gives us an image composed of fragments, while the details of his life story are missing. We know very little, for example, about his relationships with his wife and children, how both he and his family dealt with his long absences in London, and what he did with his "lost years." What we would most like to know are not just the events of his life but the narrative that makes sense of the events and traits that shaped him as a thinker and writer. Did he travel? Who were the people who most influenced him as a writer? How did he connect with the larger theatrical community? Was he mostly happy or conflicted about his life? How much richer it would be to know about these things and about his adventures, aspirations, goals, and disappointments—and how he put them together in his mind to make sense of himself.

We *do* have a few facts about Shakespeare's life, and (in our view) three remarkable books that attempt to weave these facts together into a sort of narrative. One is Stephen Greenblatt's *Will in the World*,[69] another is Jonathan Bate's *Soul of the Age*, and the third is the very accessible *Shakespeare: The World as Stage*, by Bill Bryson.[70] We have drawn on all three, as well as several other sources for the following, much-abbreviated summary of the Bard's own life story. In addition, we know a good bit about the social and environmental forces at work in England during this period, forces that surely must have impacted Mr. Shakespeare.

THE LIFE STORY OF WILL SHAKESPEARE IN BRIEF

Born in the spring of 1564 in Stratford on Avon, William Shakespeare was the third of Mary and John Shakespeare's eight children: large families were the norm in those days as an attempt to offset high infant mortality rates. The Shakespeare family configuration changed with the deaths of his two older sisters when they were very young. This left Will as the oldest child of the family—which may be psychologically significant, in light of research showing that high achievement tends to favor the first and oldest child.[71]

The boy Shakespeare grew up in a solidly middle-class family in a small town where his father, at least for a time, was an important and prosperous man. John Shakespeare owned multiple business ventures. He was a glover, a wool merchant, a landowner, and a money lender. He also served in several civic positions, including as constable and member of the town council. However, John ran afoul of a law restricting the wool trade to licensed dealers and was penalized by the courts. That led to a long downward spiral in both his civic and financial fortunes, which must have impacted his family, including the young Will, who may have been observing his father's troubles with some concern.

Coincidentally, during this period, the Tudor monarchs launched an educational revolution by constructing "grammar schools" all over England, one of which was in Stratford. Boys of any social class could enroll, and although we have no records of Will's attendance, it would be surprising if a lad of his background and talents had not been a student there. He would have studied religion, arithmetic, Latin grammar, and classical works in both Latin and Greek. While at school, he would have been required to speak only in Latin; boys who were heard using English were punished. As far as we know, young Will did not continue his formal education after grammar school, but he continued to read widely as he emerged into adulthood.

At eighteen, he married Anne Hathaway, a woman some eight years his senior—probably because of an unplanned pregnancy. He and Anne eventually had three children: Susanna (born six months after the marriage) plus a pair of twins, Judith and Hamnet—the latter, a son who died at age eleven. Other than these facts, we know little about their family lives, together or apart, during his years in London. His son's name, being so similar to that of the melancholy Prince Hamlet, has been the subject of much speculation, but it seems likely that the boy was named after Hamnet Sadler, a neighbor and friend of the family who witnessed Shakespeare's will.[72]

And then there were the "lost years"—the periods between 1578 and 1582 and 1585 and 1592. The first spans the time when he presumably left grammar school and when he married Anne Hathaway in 1582. Aubrey states that Shakespeare may have served as a rural schoolteacher during the second "lost" period, but we have no independent corroboration for Aubrey's claim. During that time, however, he must have been perfecting his acting and writing chops before bursting on the London theatrical scene. In the period between, we have only spare baptismal records for his eldest daughter Susanna and the twins, Hamnet and Judith. Some say that, perhaps, all the "Italian plays" suggest that he may have spent some of this time in Italy. The case is tantalizing, but unfortunately, the evidence is thin.[73]

In the wider world beyond Stratford, the plague regularly scourged the population, although small towns were safer than the crowded cities, where it was easier for the fleas carrying the Black Death to find hosts. Despite that risk, London's new theatrical scene attracted Will Shakespeare sometime around 1590, when he was in his late twenties.[74] It was an exciting time: the heyday of the English Renaissance, with new ideas about literature, science, and government flowing both ways across the Channel between England and Europe. As you may remember, Shakespeare was a contemporary of such luminaries as Galileo, Cervantes, Francis Bacon, and René Descartes, although we don't know whether he interacted with any of them.

His plays are filled with scenes of rowdy urban taverns, castles populated by nobles engaged in courtly intrigues, and tales of faraway cities, the likes of Vienna, Syracuse, and Athens. Yet, amid these, he sprinkles rural allusions to hunting, tanning, butchering, bowls, archery, and wildflowers. Even though he was drawn to London, Shakespeare's works tell us that he was always a country boy in his heart.

We cannot help but wonder how Shakespeare would have described his own life as a play—his personal *story*. Stories are what Shakespeare gave us of Henry V, Falstaff, Rosalind, Hamlet, and so many others. This is exactly what Professor McAdams tells us we need to fill in the complete narrative picture of an individual. We can only wish that Shakespeare had told us more of his own life.

It would be especially interesting to know whether Shakespeare felt he had achieved "redemption," in McAdams's sense. You will remember that redemption can take many forms. In America, people are likely to make their lives meaningful by telling a story of redemption that involves atoning for some "sin" or unearned privilege not shared by others less fortunate than themselves. In a more general sense, says McAdams, "redemption is a deliverance from suffering to a better world."[75] In Shakespeare's case, the evidence is again thin. We will, however, suggest that he may have found redemption in atoning for his father's failures, first in business and later in the failure to qualify for a coat of arms that distinguishes a gentleman. Of course, we cannot know this for certain—nor do we know whether seeking redemption is an English, as well as an American, phenomenon. That awaits research that extends McAdams's work on life stories across the Pond.

· 4 ·

SOCIAL INFLUENCE FROM STRATFORD TO STANFORD

Measure for Measure

Vienna (circa 1604)[1]—Duke Vincentio is a curious man—especially curious about his subjects and how they might comport themselves if he were not around to maintain order. To find out, he stages a quasi-experiment by pretending to have been called out of town, leaving his trusted assistant Angelo in charge.[2] Disguised as a friar, the duke remains in Vienna to watch what happens. Says Vincentio, "Hence shall we see, / If power change purpose" (1.3.37).

Angelo has a reputation for a strict moral character and as a stickler for rules and regulations. True to his form, he closes the brothels almost immediately upon the duke's supposed departure, and to make an example for the people of Vienna, he arrests citizen Claudio for fornication.[3] Technically, Claudio *is* guilty, for Juliet, the woman to whom he is engaged, has become visibly pregnant. Obviously, they had not delayed their intimacy while their marriage was held up by negotiations over the dowry.[4] But Angelo invokes an old and nearly forgotten law providing the death penalty for sex outside of marriage, even though such an extreme punishment has not been used for years.

As the unfortunate Claudio languishes in prison (figure 4.1) awaiting the executioner, his sister Isabella visits Angelo to plead for her brother's life. Unexpectedly, Angelo finds himself attracted to Isabella and says he needs a day to think over the case. The next morning, he proposes a shocking bargain: If she will agree to have sex with him, he will spare Claudio, exchanging her virginity for her brother's life. This development adds another dimension to the story, and we now see Angelo not just as an overzealous official but as an opportunist and a horny hypocrite.[5]

Figure 4.1 A cell in a medieval prison (museum display). *Wikimedia Commons.*

Shakespeare adds one additional complication: Isabella is not merely a virtuous woman; she is a novice nun. Therefore, she is unwilling to commit what she deems a grave sin, even to save her brother's head. What is to be done?

This being a comedy, albeit a very dark comedy, only one person dies: a pirate whose execution supplies a severed head to be substituted for Claudio's, fooling Angelo and temporarily traumatizing Isabella. So Claudio is spared, and the "bed trick"[6] saves Isabella's virtue because Angelo thinks he is romping in the dark with Isabella when it is really Mariana, a woman to whom Angelo had once been betrothed but later abandoned (because she lost her dowry).[7]

Having let the charade play out long enough, the duke finally reveals himself, and a happy ending ensues for nearly all. The duke punishes Angelo by forcing him to marry the woman he had previously rejected.[8] And we are left with a mystery. Twice the duke asks Isabella to marry him, and twice she responds . . . with silence.

Literary critics have had a field day interpreting *Measure for Measure*, especially in light of the political changes that loomed with the death of Queen Elizabeth and King James's ascension to the throne of England.[9] Sketchy records show the play was first performed for King James I at Whitehall Palace on the day after Christmas in 1604.[10] Therefore, many scholars see *Measure for Measure* as Shakespeare's attempt to draw the favor of the new English monarch, who was known to be concerned with piety and his subjects' morals.[11] Whatever was the motive, the play apparently did its job, for Shakespeare's company, which had been a favorite of Queen Elizabeth's, soon came under James's patronage, calling themselves The King's Men.

Full disclosure: one of your authors (Zimbardo) has a special interest in *Measure for Measure* because much of the play follows a script that parallels a well-known psychological study that Zimbardo conducted, known as the Stanford Prison Experiment.[12] Both the play and Zimbardo's research address the question of how people respond when given power over others but with little guidance or accountability. And for both Zimbardo and Shakespeare, what better setting than in a prison in which to explore the issue![13]

Like Duke Vincentio, the young Dr. Zimbardo was a curious man, interested in the effects of power on people's moral behavior. Again like the duke, he wanted to find out how people would behave when given authority with few limits and little guidance. As Zimbardo put it, "I wanted to know who wins, good people or an evil situation."[14] The question is an ancient one, going back more than two millennia, when Glaucon, the cynic in Plato's *Republic*, asserted that people will behave ethically only if they think they are being watched.

Zimbardo's approach to these issues was to construct a mock prison in the basement of the psychology building at Stanford and recruit twenty-four male volunteers for "a study of prison life." Each agreed to being paid $15 a day (an attractive sum for a college student in 1971) for an experiment that was expected to run for two weeks. All were screened with a battery of psychological tests to weed out any who might have deviant personality traits that could bias the research. Zimbardo then randomly assigned those who passed the initial screening to two groups: "prisoners" and "guards." The part of the prison superintendent he took upon himself, so he could be a participant-observer.

Then, on a sunny Sunday morning in August, with the cooperation of the Palo Alto police chief, officers rounded up and arrested Zimbardo's "prisoners," transporting them in squad cars to the simulated cell block at the university. There the police fingerprinted, showered, and deloused them, issuing gowns that would serve as prison uniforms. Meanwhile, the participants who would be their "guards" donned police-style uniforms, along with mirror sunglasses that hid their eyes from view. These manipulations produced a degree of anonymity by reducing the identities of participants in both groups to their garb and their assigned roles. In social-psychology speak, they were "deindividuated"—but with a clear power difference between the two groups. Importantly, the guards received only minimal instruction, simply being told to maintain order and respect for their authority. These procedures left both the guards and the prisoners largely relying on their own expectations of the roles they were assigned. Crucially, Zimbardo instructed everyone that no physical violence would be permitted and that anyone was free to leave the experiment at any point.[15] Even though Dr. Zimbardo was not familiar with *Measure for Measure* at the time, he chose a role much like that of Duke Vincentio by setting up the simulated prison situation and stepping back to watch the scenario unfold.[16]

The prisoners quickly fell into their roles, some docile and some rebellious. Likewise, the guards just as rapidly became authority figures, and like Shakespeare's Angelo, they became zealous in their enforcement of "the rules"—threatening and bullying prisoners for minor infractions. (See figure 4.2.) Seeking to control their sometimes-rebellious charges, the guards held numerous "counts," forcing the prisoners to line up and identify themselves only by their inmate number. The guards demanded push-ups from those committing minor violations and placed persistently defiant ones in isolation cells. They also began humiliating the prisoners by calling them degrading names and requiring them to perform thinly disguised homoerotic acts. Predictably—but only in hindsight—the situation rapidly deteriorated, and Zimbardo was forced to halt the experiment when he realized that it had careened out of control.

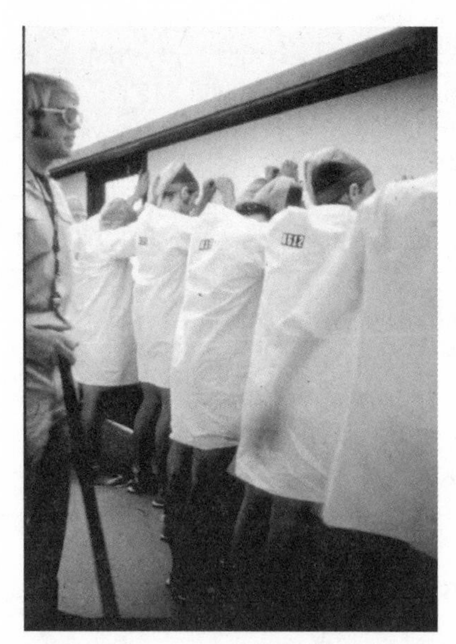

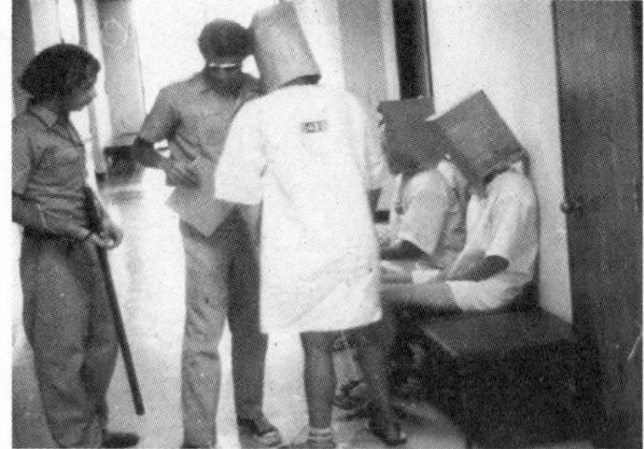

Figure 4.2 Images from *The Stanford Prison Experiment*, a 2015 docudrama about Zimbardo's research on the power of the situation.

Zimbardo's realization came after nearly a week into the simulation, when Christina Maslach, one of Zimbardo's graduate students, stopped by to observe the prison experiment and, like Isabella, was horrified. She called on him to stop the simulation immediately, asserting, "It is terrible what YOU are doing to those boys!"[17] A tense argument followed, but Maslach finally was able to show Zimbardo how deeply he, too, had been swept up in the situation, preventing him from seeing the magnitude of the abuse of prisoners by the guards. When he finally understood that his simulation had indeed gotten out of hand, to the point of influencing his own moral judgment, he shut it down.

The broader social context of the Stanford Prison Experiment was the early 1970s when the news was filled with protests over the Vietnam War and over the government's authoritarian methods of suppressing those protests. Zimbardo had recently moved to Palo Alto, where he sensed a communitarian spirit that contrasted with his previous experience in New York City, where people easily disappeared into a culture of "anonymity." During his time in New York, he had begun to do studies of "deindividuation," showing that when people could not be readily identified, they were willing to act more aggressively toward others—even to inflict high levels of punishment on others for errors in a (supposed) learning experiment.[18]

Not long before, the public had learned about Stanley Milgram's "shock" experiments, showing how ordinary people in urban New Haven, Connecticut, would obey orders to punish others with seemingly dangerous levels of electric shocks for mistakes in remembering lists of items that had been read to them.[19] Zimbardo's prison experiment built on Milgram's work by combining anonymity with authority—plus one more element: ambiguity, by which participants were allowed great latitude in defining their own roles.[20] The interaction of these social pressures interested Zimbardo, who realized that ambiguity, anonymity, and authority differentials all come together in jails and prisons.

Again, the parallel with *Measure for Measure* is striking, for the political context in which Shakespeare wrote the play was, likewise, an uncertain time in English history, when people were anxious about the direction in which their new monarch, King James VI of Scotland, would lead their country as King James I of England.[21] In particular, many worried that the new king might take his religious and moral convictions to extremes—perhaps like an "Angelo" on the throne of England. (It seems that Shakespeare was taking a bit of a political risk with this play—which may account for its surprisingly happy ending.)

Shakespeare named his play's anti-hero, Angelo, after the angels, and that pun would not have been lost on the audience. Too much power would corrupt Angelo's authority, and he would become a fallen Angelo, a devil reminiscent of Lucifer, once God's favorite who fell from grace. Coincidentally, in the wake of the Stanford Prison Experiment, Dr. Zimbardo wrote *The Lucifer Effect*, describing the Stanford Prison Experiment in detail.[22]

Measure for Measure was *not* one of the Bard's more popular works during his lifetime, but it has experienced something of a modern revival, possibly because we feel freer to question authority. As for the Stanford Prison Experiment, it has become a psychological classic that illustrates a foundational principle in social psychology: *the power of the situation.*[23] It is a concept that social psychologists have repeatedly validated, showing that the *situation*—the social context—in which we find ourselves can have more influence on our behavior than do our individual personality traits—perhaps far more often than we realize. We like to think of ourselves as unique individuals, masters of our free will, acting rationally in ways that are consistent with our values. Instead, say social psychologists, we more often respond almost reflexively to the pressures in our circumstances, giving scant thought to the influence of the situation as we go about everyday tasks. We stop at red lights, without giving the social constraints much thought. We follow social norms by standing in lines rather than cutting to the front. We even laugh when prompted by fake laughter tracks on television, even though we despise them.

While our responses to those situations are benign and sometimes even beneficial to society, Zimbardo realized that there could be a darker side to *the power of the situation.* Stanley Milgram's work was a perfect example, showing that people will follow orders to give (apparently) severe electric shocks to a stranger. Multiple experiments have revealed that, in emergencies, bystanders are much less likely to help if there are others present who are not helping.[24] And as psychologist Solomon Asch demonstrated, most of us will cave to the clearly erroneous judgment of a group on such trivial matters as comparing the lengths of lines, even when that judgment contradicts the immediate evidence of our own eyes.[25]

We propose that Shakespeare may have been deliberately wrestling with the power of the situation in *Measure for Measure*. The play also suggests that Shakespeare already knew what Dr. Zimbardo rediscovered in the Stanford Prison Experiment: a lopsided balance of unchecked power could nudge people toward moral (or immoral) extremes. Specifically, the playwright had Isabella describe how the power of an ambiguous situation corrupts the person with "a little brief authority":

ISABELLA: Man, proud man,
Drest in a little brief authority,
Most ignorant of what he's most assured,
His glassy essence, like an angry ape,[26]
Plays such fantastic tricks before high heaven
As make the angels weep.

Measure for Measure, 2.2.139

Steven Pinker used this passage as the epigraph in his book, *The Better Angels of Our Nature*, to describe two distinct viewpoints on behavior that inflicts harm on others.[27] He calls these "the moralist" and "the scientist." The moralist's eyes see evil acts as deliberately performed by an evil-doer: a person with a deeply flawed character. In contrast, the scientist looking at bad behavior tries to understand it in terms of the context in which it occurred and how the individual interpreted that context. Your authors broadly agree with Pinker, but we believe that closer inspection reveals an even finer-grained picture.

At first, we see Shakespeare as taking the moralist position by focusing on the opposing personal traits of Angelo and Isabella. And yet, a close reading suggests that the playwright does not blame one more than the other. Both are flawed but in opposite directions.[28] He gives Angelo and Isabella speeches in which they ruminate and anguish over their moral choices. But what seems to concern Shakespeare most are those who take their morality to the extremes of dogmatism and zealotry, as do both Angelo and Isabella.

In designing his prison experiment, Zimbardo (as the scientist) tried to control for personal traits associated with extremism by screening his volunteers with a battery of tests. Yet personality tests are not perfect probes of one's character. So, on the one hand, a scientist could look at the Stanford Prison Experiment in terms of personality quirks and traits, or on the other hand, at the *power of the situation* to influence behavior. And likewise, we suggest that Shakespeare, unlike Zimbardo, may have deliberately left a measure of ambiguity in *Measure for Measure.*

SHAKESPEARE'S MORALITY ANTICIPATES MORAL FOUNDATIONS THEORY

Measure for Measure is not just a play about the social psychology of authority and power. Critics call it a "problem play" because it deals with discomfiting *moral* issues that we might prefer to keep hidden in the shadows. And yet, it has a happy ending (for most of the characters).[29]

The moral tones of the play should not surprise us since, at the turn of the seventeenth century, there were still troupes of actors wandering from village to village staging the old medieval "morality plays," which were allegories featuring a rather simple plot and two-dimensional characters with names like Virtue, Everyman, Mercy, Justice, Perseverance, and Death. *Measure for Measure* is unashamedly indebted to this tradition, but it is also a morality play of a new sort with which Shakespeare and his fellow playwrights in London were experimenting. These new morality plays featured more sophisticated, fleshed-out characters, the likes of Angelo, Othello, and Miranda who were dealing with more realistic moral conflicts than did the older plays of the genre.

The concept of morality has always had cultural roots, commonly related to religion or that violated cultural norms or customs. We see this in the famous listing of Seven Deadly Sins formulated by early Christian theologians in the fourth century CE, later popularized by Pope Gregory the Great (papacy: 590–604 CE) and revised in the thirteenth century by Thomas Aquinas. Here is Aquinas's list: *wrath*, *avarice*, *sloth*, *pride*, *lust*, *envy*, and *gluttony*.

While morality has always had strong roots in religion,[30] philosophers have also developed moral systems outside of religion: *secular humanism* being an example. Evolutionary biologists have joined the fun, too, searching for *biological* sources of morality that could account for some of our species' most puzzling characteristics, such as the capacity for intense competition (including warfare) yet our being able to come together cooperatively in large groups of unrelated individuals to form cities. In particular, cooperation to the extent we see in those large groups requires a widely accepted system of distinguishing right from wrong; it requires a consensual moral code.

The latest groups of scholars to come to this search party are the evolutionary psychologists and social psychologists who have added *social* and *cultural* perspectives to the study of morality. Among them is Steven Pinker, who writes: "The core of morality is the recognition that others have interests as we do—that they 'feel want, taste grief, need friends,' as Shakespeare put it—and therefore that they have a right to life, liberty, and the pursuit of their interests."[31] And so, once again, we find that William Shakespeare anticipated a topic that would intrigue psychologists centuries later, as he constructed plots around essentially the same moral dimensions that social/evolutionary psychologists have identified in their research. The result, as we can now see from a twenty-first-century perspective, is that just a few moral dimensions seem to be universal, not just peculiar to the England of the Tudors or to modern America, but in our human species across both centuries and cultures. Psychologists call this work Moral Foundations Theory (MFT).

MFT proposes that, as our ancestors dealt with the challenges of living in groups, evolution wrote a "first draft" of a moral sense into their brains by favoring the survival of those with certain moral attitudes. That is to say, morality is not something we humans must entirely *learn* because the basic framework is present in our brains at birth, "wired" into our genetic makeup. But what does this "first draft" contain?

Moral Foundations Theory proposes that natural selection has shaped the human brain at birth with brain circuits, or "modules," that are *prepared to learn* values and behaviors related to common types of social problems that we often lump together under the heading of "morality." Some of those modules deal with empathy, love, and loyalty; others involve anxiety, fear, and aggression (see table 4.1). Everyone is not "wired" the same, of course, nor do we share the same experiences. As a result, there are wide differences in people's concepts of morality that MFT clusters into six categories that make up our moral dimensions or "foundations."[32] MFT labels these dimensions Care/Harm, Fairness, Loyalty, Authority, Purity/Sanctity, and Liberty.[33]

Table 4.1 The Six Bases of Morality According to Moral Foundations Theory

	Care/ Harm	Fairness/ Cheating	Loyalty/ Betrayal	Authority/ Subversion	Sanctity/ Degradation	Liberty/ Oppression
Adaptive Challenges	Protection and care for children, family, self	Benefits in two-way interactions	Forming relation-ships and coalitions	Forming beneficial governing hierarchies	Avoiding contami-nants (both physical and moral)	Limiting powers of dominant people in a group
Examples of Original Triggers	Distress calls from a child or friends	Cheating, deception, cooperation	Threat or challenge to one's group	Behaviors signaling dominance and submission	Waste products; bad sights, tastes, or odors; challenges to one's core beliefs	Bullying, constraining actions of others
Typical Emotions	Compassion, empathy	Anger, gratitude, guilt	Group pride, anger against traitors	Respect, awe, fear	Disgust	Outrage against oppressors
Associated Virtues	Caring, kindness	Fairness, justice	Loyalty, patrio-tism, self-sacrifice	Obedience, deference	Cleanliness, temperance, chastity, piety	Freedom, self-determination, protection of victims

And wouldn't you know? The Bard from Stratford, intuitive psychologist that he was, seems to have anticipated *all six* elements of MFT in plays dealing with lying, treason, murder, prejudice, disloyalty, disobedience of authority, and various forms of sexual immorality, including seduction, rape, infidelity, incest, and fornication. But before we go further down those moral pathways, we should pause to consider how the term *morality* was conceived in Shakespeare's day.

At the turn of the seventeenth century, the terms "moral" and "morality" often carried meanings that seem archaic to us now. To be sure, "morality" carried the meaning of good and bad (or evil), much as it does today, but it was common to use "moral" in referring to a hidden meaning in speech or a text. It could also refer to a lesson, as in "the moral of the story," as we still might say.

Likewise, the list of moral rights and wrongs for English folk in Jacobean times[34] would have been a bit different from our own. For example, people in Shakespeare's audiences were less likely to consider racial or religious discrimination to be a moral issue than are we. But they did consider it sinful to have sex during Lent (although the sin was reportedly often honored in the breach). And they were much more concerned about the evils of witchcraft, so an accused woman could face the death penalty under the prevailing conditions of moral rectitude in Shakespeare's day.

We can see another example of the differences in moral perspectives then and now in the Great Chain of Being, a Tudor-era framework for understanding the concept of *authority*. This concept arranged all things spiritual, human, animal, vegetable, and mineral in a hierarchy, with God at the top, followed by angels, humans, animals, plants, and finally rocks and minerals. There were also sub-hierarchies within the chain, such that kings ranked less than God but higher than their nobles, who were, in turn, higher than rich "gentlemen," who were, in their turn, considered higher than the commoners. Within families, men ranked higher in the Chain than women (who were expected to obey their men). With strong roots in religion, the Great Chain of Being carried a hierarchy of authority and privilege as an unquestioned *moral* principle. Consequently, people presumed to have violated this hierarchy of authority, even by acting or dressing above their station, could be treated harshly, whereas today the idea seems absurd, especially to Americans who have been raised on the ideals of Equality and Individualism, the narrative of the American Dream, and the possibility of upward mobility.

Now, with such cultural similarities and differences in mind, let us begin our search for moral foundations in Shakespeare's work with a look at *Measure for Measure* to see what moral issues it raises.

Critics agree that *Measure for Measure* is about *justice*, and they also agree that Angelo perverts justice in multiple ways. He pronounces an excessive sentence of death on Claudio; he attempts to bargain Claudio's life for Isabella's virginity; and he proposes to violate the civil law's proscription of fornication—the very crime for which he has placed a death sentence on Claudio. Indeed, justice (and injustice) is one of the Bard's recurring moral themes: in *The Merchant of Venice* it is when Shylock demands his pound of flesh; in *Hamlet* it is the prince's quest for his father's murderer; and in *Macbeth,* justice comes to the Thane of Cawdor at the hands of Macduff.[35]

Sexuality, of course, shared the moral spotlight with injustice in *Measure for Measure*, with Angelo plotting to bed Isabella against her will. Likewise, Shakespeare also raises issues of sexual morality with allegations of infidelity in *Othello*, the creepy faux seduction scene in *Cymbeline*, the attempted rape by Proteus in *Two Gentlemen of Verona*, and the incestuous relationship of King Antiochus and his daughter in *Pericles*.

The Bard's audiences would have realized that the moral foundation for sexuality stood on shifting ground, for God's laws were a-changing, depending on who wore the crown. Elizabeth's father, Henry VIII, felt no compunction about serial fornication—but, of course, the nobility had always granted itself more license than it offered its social inferiors. Moreover, under the last few monarchs, the English had been riding a religious see-saw, alternating between Roman Catholicism and the Anglican church. Henry VIII had split with Rome and established his own Anglican church to sanction his divorce from Catherine of Aragon, who had failed to produce a male heir. In the process, Henry criminalized Catholicism, and then his daughter, Queen Mary I (who, like her mother, was a staunch Catholic), restored it. When Elizabeth (who had a different mother) took her turn on the throne, the Anglican church regained its authority, and Catholicism was again declared unlawful. Then, in 1603, when James acquired the crown, uncertainty returned, even though his reign in Scotland (as their James VI) had been marked by his seeking religious compromise.[36]

At this point, then, we can say that *Measure for Measure* spotlights moral issues involving authority, sexuality, and justice as they were understood at the beginning of the Jacobean era. But there is one more moral issue that the playwright almost certainly intended as the crux of the play: moral extremism, including extremes of both civil and religious zealotry.[37] The title of the play itself comes from the Sermon on the Mount in the Gospel of Matthew where the King James Version says:[38]

> Judge not, that ye be not judged. For with what judgment ye judge, ye shall be judged: and with what measure ye mete, it shall be measured to you.
>
> —Matthew 7:1–2

We can be confident that Shakespeare considered certain sexual behaviors, justice, defiance of recognized authority, and zealotry to be moral issues. What other forms of morality appear in his works? Here are six more that we discovered while reading through the canon:[39]

- **Lying:** Shakespeare builds both *Othello* and *Cymbeline* around lies about the sexual infidelity of a woman. Shakespeare also looked askance at the seduction of a woman merely for a man's own gratification, as we also see in *Cymbeline*.
- **Disloyalty:** In his eponymous play, Coriolanus violates another line of the moral code when switching his allegiance to the archenemy of Rome. So do the two famous pairs of lovers, Romeo and Juliet and Troilus and Cressida, who dare to fall in love with enemies of their own families.
- **Treason:** To be sure, this is a special case of defiance of authority—but an extremely important one in early-modern England. We see it when Henry Bolingbroke (later Henry IV) usurps the throne, when Coriolanus gives aid and comfort to an enemy of Rome, when Julius Caesar attempts a coup, when Hamlet suspects that Uncle Claudius has murdered his father, and when Macbeth plunges a dagger into the sleeping King Duncan's heart.
- **Murder:** It's many of the previous suspects again: Macbeth, Hamlet's Uncle Claudius, *et tu Brute*, of course, plus the dastardly Duke of Gloucester in *Richard III* (who probably qualifies in nearly all categories of immorality, if we take Shakespeare's words for it).
- **Cowardice:** The most obvious example is Falstaff's cowardice, revealed by a ruse perpetrated by his friend Prince Hal in *Henry IV, Part 1*.
- **Hypocrisy:** Angelo is once more the winner for proposing to commit the same crime for which he has sentenced another man to death.

By our count, then, the Bard's plays identify at least ten forms of morality, while Moral Foundations Theory lists only six. But we believe we can merge Shakespeare with Moral Foundations Theory by combining a few categories.

We argue, therefore, that nine of Shakespeare's ten dimensions rather closely conform to the six in MFT, as we show in table 4.2.[40] This leaves *zealotry* as a possible new dimension for Moral Foundations theorists to consider—which seems especially apropos in our own climate of political and cultural extremism.

Table 4.2 Comparison of Shakespeare's Morality and Moral Foundations Theory

Shakespeare's Moral Dimensions	Moral Foundations Theory's Dimensions of Morality	Associated Terms in Moral Foundations Theory
Murder*	Care/harm	Kindness, compassion, cruelty, violence
Injustice, hypocrisy, lying*	Fairness/cheating	Justice, trustworthiness, infidelity
Disloyalty, cowardice*	Loyalty/betrayal	Patriotism, self-sacrifice, cohesiveness
Defiance of authority, treason	Authority/subversion	Obedience, deference, respect
Sexual immorality*	Sanctity or purity/ degradation or disgust	Temperance, chastity, piety, taboos, cleanliness
Liberty	Liberty/oppression	Freedom, independence
Moral extremism or zealotry		

*Any of these could also be a violation of law, so the MFT dimension of Authority could also apply. Similarly, other terms in this table could fit well under multiple headings, which suggests that the categories of the MFT list have fuzzy boundaries that could bear better definition.

DID SHAKESPEARE UNDERSTAND COGNITIVE DISSONANCE, TOO?

Indeed he did.[41] In particular, he knew that when people confront moral issues—especially when their behavior conflicts with their own moral standards—they feel an uncomfortable state of mental dissonance that they are highly motivated to resolve, one way or another. In Angelo's words, we hear his struggle with the dissonant notions that he is a person of moral virtue, yet he lusts for Isabella:

> ANGELO: Most dangerous
> Is that temptation that doth goad us on
> To sin in loving virtue: never could the strumpet,

> With all her double vigour, art and nature,
> Once stir my temper; but this virtuous maid
> Subdues me quite.[42]
>
> —*Measure for Measure*, 2.2.198

How does one resolve cognitive dissonance? In Angelo's case, he is struggling with his desire, but he has not yet propositioned Isabella. We hear his anguish in a short soliloquy:

> ANGELO: What's this? What's this? Is this her fault or mine?
> The tempter or the tempted, who sins most, ha?
> Not she: nor doth she tempt; but it is I
> That, lying by the violet in the sun,
> Do as the carrion does, not as the flower,
> Corrupt with virtuous season.
>
> —*Measure for Measure*, 2.2.198

Thus, we see that the Bard also knew that cognitive dissonance can lead either to a contorted *rationalization* to justify one's desires or to a cycle of self-blame, despair, and depression.

With these thoughts in mind, then, let us consider four other plays in which Mr. Shakespeare employed cognitive dissonance, along with his understanding of depression and the ego defense mechanism of rationalization, to mark a turning point in the plot.

King Lear

In his dotage, Lear decides to retire from the throne and divide his kingdom among his three daughters according to how much they profess to love him. Two of the three go overboard with flattery and praise, but the youngest, Cordelia, refuses to play the sycophancy game. This creates dissonance for Lear because Cordelia is his favorite. Rather than admit that she was right and he was foolish, he disowns her and bestows the kingdom to the two evil, obsequious sisters who soon tire of the old man and throw him and his retinue out into the cold, quite literally, where things continue to deteriorate for him as he vents his frustrations and rationalizations to the wind and rain.

Macbeth

The Thane of Cawdor,[43] on his way home from stunning successes in battle, encounters three women boiling a batch of newts' eyes, frogs' toes, and dogs' tongues. Macbeth hears them prophesy that he will become king but leave no heir. They also assure him that he need fear no one "of woman born." Coincidentally, the current monarch, King Duncan, decides to overnight at Cawdor, and when Macbeth arrives with the witches' prophecy in mind, the queen points out to him the opportunity to murder the king and seize the crown. At first Thane Macbeth is hesitant, but she is persuasive, and he soon agrees to her plan to dispatch Duncan and blame the murder on two guards that he will also kill. But even after acquiring the crown he so much desired, Macbeth finds his mind filled with dissonant thoughts. Yet, he cannot change what was done. Regardless of his attempted rationalizations, Macbeth cannot rid his mind of ruminations over Duncan's murder. Distraught and filled with self-doubt, his mind cannot escape the idea that life is without meaning, and so we hear Macbeth wail, "Tomorrow and tomorrow and tomorrow" as he tries to make sense of events, full of "sound and fury" (5.5.20). Finally, justice is done by the sword of Thane Macduff who just happens to have been, from his mother's womb, "untimely ripped"—fulfilling the final part of the Weird Sisters' prophecy because, you see, Macduff's caesarian birth means that he was not "of woman born."

Coriolanus

After a great military victory, Coriolanus returns to a hero's welcome in Rome. Friends persuade him to become a consul,[44] but to do so, he must have the support of the common people for whom he has nothing but contempt. When they refuse their support, Coriolanus becomes furious and publicly rants against the idea of popular rule. His enemies then fan the flames of populism and drive the arrogant Coriolanus into exile, where he plots revenge against Rome and shifts allegiances in support of Rome's mortal enemy, which is treason. He howls and whines about his reasons, but finally, he is the only one persuaded, and he meets death at the hands of his new comrades.

Love's Labour's Lost

It's Shakespeare's comedic version of cognitive dissonance. The king of Navarre and three close friends vow to cloister themselves, away from all women and other worldly temptations, for three years of study and serious discussion. Coincidentally, four attractive young women and their retinue appear on the castle doorstep. It is, of course, love at first sight for all four men, but remembering their vows creates cognitive dissonance, which they try unsuccessfully to sweep away with awkward rationalizations delivered, each one to the others. Eventually the truth comes out, and the four decide not to pretend any longer, reasoning that "young blood doth not obey an old decree" (4.3.233). The lads want to marry the lasses, but the latter are not so easily won because they don't trust men who so readily rationalize away their dissonance. (We save more details for chapter 8, which features *Love's Labour's Lost*.)

In these four plays, the circumstances and, often, foolish pride push the characters to resolve their dissonance by rationalization. But let's examine the role of cognitive dissonance in *Measure for Measure* a little more deeply.

We have seen that the play's focus is on justice, sexual immorality, and moral extremism, but it also introduces the ideal of *mercy*. And that is how the Bard resolves the problems in this "problem play"—by having Duke Vincentio dispense mercy for all who might have committed high crimes or misdemeanors.[45] Yet he leaves us to judge whether we will still hold Angelo personally responsible for his actions. How much of Angelo's hypocrisy and moral extremism should we attribute to personal traits and character flaws and how much to the situation itself? Shakespeare, at first, seems to condemn Angelo, but then he introduces dissonance by making Angelo agonize over his desires and his unbecoming behavior. Angelo was surprised by the rise of his own lust, even more so than he was surprised by the duke suddenly investing so much power in him with little guidance or constraint. (Recall the Stanford Prison Experiment!) Was it Angelo, the *person*, who was to blame? Or was it the *situation*? And so we return to the person-situation controversy that has been festering throughout part II of this book.

RESOLVING THE PERSON VS. SITUATION ISSUE: A PROPOSAL

The Stanford Prison Experiment lent evidence for a situational explanation because Zimbardo first screened his student-volunteers for deviant traits and then randomly assigned the ones who passed the screening to the roles of guards or prisoners. Despite these precautions, some of the guards became moral monsters, and some of the prisoners were visibly shaken by the experience. Clearly, we see situational variables at work.

Personality theorists, however, would respond that Zimbardo acknowledged the power of personality by attempting to eliminate it as a variable in his experiment (where it would have been a *confounding* variable). Or had

Zimbardo's test battery missed something? The personality theorists have convincing research, numbering in the thousands of studies, and their mountain of evidence shows that personality traits correlate with many mental health problems. These include anxiety, depression, and psychopathy, as well as predicting criteria as diverse as academic success, political leanings, divorce, longevity, and job performance.[46]

Could there be some sort of middle ground, then, between person and situation? One clue is that neither the person nor the situation perfectly predicts everyone's response in any given situation or, conversely, anyone's response in every situation.

It is also significant that both Mr. Shakespeare and Dr. Zimbardo staged their "experiments" in a prison—a situation that, by definition, emphasizes the moral dimension of authority when one group was given disproportionate power and control over another. In both cases, it was plain to see "if power changes purpose" (act 1, scene 3). And, sure enough, power did so both at Stratford and Stanford.

We surmise, however, that it would have been different had the power differential been removed from the situation—if the guards had been assigned instead to the role of defense attorneys, for example, where the moral dimension of fairness or justice would have been salient. Likewise in Shakespeare's prison, we can imagine a different (but less interesting) outcome had Angelo instead been an Angelina or if Claudio had a brother who was a priest. Our point is that different situations can emphasize different moral dimensions or moral "foundations."

Moreover, people like to believe that "personality" (especially their own) refers to a thread of consistency across different situations—meaning that there is some unity to our thoughts and behaviors, wishes and dreams, throughout our lives. So, we feel uneasy about the suggestion that the situation often controls us—that each of us might show a different side of our personality when our life circumstances change. Perhaps the real "problem" in the so-called "problem plays" (usually identified as *Measure for Measure*, *All's Well That Ends Well*, and *Troilus and Cressida*) is Shakespeare's pessimistic and unsatisfying hint that the situation can trump personality.[47] Yet we know from the Stanford Prison Experiment that this is more than a mere suggestion.

Now consider this fact: not only do social psychologists differ with those who study personality on the relative influence of traits and experience in shaping behavior, but research in social psychology approaches behavior in a very different way. In social psychological research, the experimenter creates a situation and looks for a response. In fact, much of the research in social psychology relies on clever, sometimes almost theatrical, experiments that put people in unfamiliar, ambiguous situations, as we saw in the Stanford Prison Experiment and in Milgram's shock experiment. These studies, therefore, turn the spotlight on the *situation*. In contrast, studies of personality tend to involve interviews or the filling out of questionnaires in which people are asked merely to imagine themselves in a situation and report their preference or to say how they think they would respond. Also, the pace is slower in personality research. The situation may not be familiar (a cubicle in the psychology building, for example), but stressors and threats are minimal, and participants have time to think about their responses. You can see how this works by examining the following items taken from one version of a Big Five personality inventory.[48] The individuals taking the personality "test" are asked to rate how well items, such as the following, describe them on a five-point scale, ranging from Very Inaccurate to Very Accurate. The full questionnaire contains fifty items. (The trait being measured is in parentheses.)

- I feel comfortable around people. (Extraversion)
- I feel others' emotions. (Agreeableness)
- I am exacting in my work. (Conscientiousness)
- I am not easily bothered by things. (Neuroticism or Emotional Stability)
- I have a vivid imagination. (Openness or Intellect/Imagination)[49]

How different responding to such a questionnaire is from being arrested and deposited in Dr. Zimbardo's prison!

Now, bringing Moral Foundations Theory to bear on the issue, we can see that the six dimensions of morality really identify *six sorts of situations* for which our brains have a "first-draft" answer that is continually modified by experience. Each of our brains is a little different, as is our individual experience, yet it is the "default settings" of the circuitry in our brains that give us the ability to respond quickly, almost instantaneously, to situations that arouse those circuits.

Nobel Prize–winning psychologist Daniel Kahneman tells us that this response system operates mainly on an unconscious level, much like an automatic reflex. He calls it "thinking fast"—or more formally, System 1. For our multiple great-great ancestors, who presumably survived because they were the fittest, the ability to respond quickly and appropriately to a situation that was perceived as good or bad could be a lifesaver. In everyday language, we may refer to System 1 as "intuition" or "instinct." [50] Psychologists also call it "long-term memory," but it is more than that, for System 1 also has an emotional component, meaning that it gives priority to memories laid down in an emotional context, such as fear or pleasure. And as we shall see, System 1 can create mental mischief because its quick responses precede rational thinking by the slower

System 2 that we call consciousness. The result is that when we are responding emotionally, we may *believe* that we are being rational. That is, we often use the "thinking slow" processes in our minds to rationalize or justify whatever the "thinking fast" circuits have already decided.

We will discuss Kahneman's ideas in more detail in the next few chapters when we consider Shakespeare's understanding of reason, emotion, and the unconscious mind. For now, here is the point: our moral responses are most often made quickly and unconsciously in response to our emotional associations with the *situation.*

So, what can we finally conclude about the person-situation issue? Your authors submit that both sides have a slice of the truth. The situation will always be important when you suddenly find yourself in a *situation* that is unclear or possibly threatening, such as the prison Zimbardo constructed at Stanford. On the other hand, when the situation is familiar and nonthreatening, and when you have time for reflection, your interests and preferences, thoughts, ideas, and values—the collection of characteristics we call "personality"—are more likely to modulate your responses, say, in considering a vegan lifestyle, or deciding whether to take a new job.

Likewise, in the theater, a good narrative also draws on both by putting a *person* with some interesting personality quirks in an unexpected and unfamiliar *situation* that demands rapid and difficult choices. And isn't that exactly what the Bard did in his best plays—like *Hamlet*, *Macbeth*, *Romeo and Juliet*, and *Othello*? Just think how humdrum the *Henry IV* plays would be had Prince Hal always been a good boy, how pointless Juliet and Romeo's antics would be if the Capulets and the Montagues were good friends, or how diminished *Macbeth* would be had the thane's wife had no regrets. But leave it to Will Shakespeare to offer yet another solution to the person-situation controversy: "Some are born great, some achieve greatness, and some have greatness thrust upon 'em."[51]

CAN THE PERSON-SITUATION CONTROVERSY TELL US SOMETHING ABOUT SHAKESPEARE HIMSELF?

You will recall that we established some conjectures about Shakespeare's personality in chapter 3. But let's consider him now in the context of the time in which he lived and the place in which he most often worked: London, during the English Renaissance (see figure 4.3). That is to say, the power of *his* situation had to have left its imprint on Will Shakespeare's own life, for the Renaissance was a situation like no other. It was an era in which the New World had just been discovered by Europeans and the influence of the Church had been crippled by devastating plagues. It was a period when courageous thinkers, like Galileo, Copernicus, Leonardo, and Francis Bacon, dared to set aside the doctrines that had dominated discourse over the long span of the Middle Ages. It was an exciting time of questioning, rethinking, and discovery. And those tumultuous forces served up social conflicts that the Bard turned into theater. It could not have been mere coincidence that Wm. Shakespeare was busy with

Figure 4.3 *Portrait of William Shakespeare* by Thomas Sully (1864). This is one of many portraits of Shakespeare, but even though they all share a likeness (as do those of Jesus!), we cannot be certain that any were painted from life. *Wikimedia Commons.*

"the invention of the human" on the London stage, while Cervantes was in Spain doing essentially the same thing on the pages of *Don Quixote.* Much like nature and nurture, the person and the situation *interact.* We cannot assess either in isolation.

We do not claim that Shakespeare discovered either the power of the situation or the human personality. Those discoveries happened long ago—perhaps by Homer or one of the authors of the *Gilgamesh* epic. But our playwright deepened the inquiry by showing us, on stage, how changing situations can alter mind and behavior—and sometimes the very personalities—of characters such as the Thane of Cawdor and the princes Hal and Hamlet. Indeed, the transformation of personality through situational influence occurs so commonly in Shakespeare's plays that we must consider it a favorite in his bag of psychological tricks.

Finally, we suggest that the influence probably goes both ways. So, personality may sometimes change the situation itself—as we saw with the influence of one man's personality on the 2021 insurrection at the US Capitol. And we think it is clear that Shakespeare was, in a much different way, an influence on his time—and perhaps for all time.

· 5 ·

HEROES ANCIENT AND MODERN, MAJOR AND MINOR

Othello

Venice (circa 1570)[1]—Ensign Iago is embittered because General Othello has passed him over for promotion in favor of Cassio—now *Lieutenant* Cassio.[2] Iago vows to bring down the lieutenant and take revenge on the general by ruining both his marriage and his career. The marriage is recent—so recent that Othello and Desdemona have had no time to consummate their bond. But it is an obligatory pleasure that must wait even longer,[3] for the Duke of Venice receives word that the Turks are planning to invade Cyprus, whereupon he orders Othello and Cassio to sail immediately to defend Venetian interests. Desdemona, not ready for a separation quite yet, begs to join Othello and follows him on a later ship, accompanied by Iago and his wife Emilia.

Arriving in Cyprus, they learn that a storm has destroyed the Turkish fleet, so the danger of the Ottoman invasion has passed, and the Venetian company is celebrating the Turks' misfortune. Lieutenant Cassio, however, tipples too much, and Iago goads him into brawling with a Cypriot official. The commotion draws an angry Othello from his bed (where, we presume, consummation was imminent), and he strips Cassio of his promotion.

Iago is, of course, secretly happy with this result but pretends to help Cassio get back in Othello's favor by suggesting he ask Desdemona to plead his case to the general. Iago plans to make the general believe that his wife's support for Cassio is evidence of an affair between them. In a soliloquy, Iago gloats to the audience about his elaborate scheme for gaslighting Othello:

IAGO: For whiles this honest fool
Plies Desdemona to repair his fortunes
And she for him pleads strongly to the Moor,

> I'll pour this pestilence into his ear:
> That she repeals him for her body's lust,
> And by how much she strives to do him good,
> She shall undo her credit with the Moor.
> So will I turn her virtue into pitch,
> And out of her own goodness make the net
> That shall enmesh them all.
>
> —act 2, scene 3

Iago's strategy involves multiple tactics, clever and precisely executed. Crucially, he persuades his own wife Emelia, who is Desdemona's attendant, to obtain a special handkerchief—a gift from Othello to Desdemona—which he manages to plant in Cassio's quarters. He then prompts Othello to inquire of Desdemona about the handkerchief, but she cannot produce it. This plants the seed of doubt, and Othello's suspicions grow as he overhears Cassio's lover, Bianca, castigate him for having given her a secondhand handkerchief. Her description suggests to Othello that it is the very one he had given to Desdemona, and so he infers that his wife and Cassio must be "making the beast with two backs."[4]

Each piece of new "evidence" deepens Othello's jealous rage and resolve for revenge, until he confronts Desdemona and accuses her of unfaithfulness. She protests, but he refuses to consider her version of events and tells her to prepare for death. She pleads for mercy, but Othello smothers her with a pillow (figure 5.1).

Iago's wife Emilia hears Othello proclaim he has killed Desdemona for her infidelity, and she finally understands that her husband is behind all the trouble. She reveals what Iago has done. Othello realizes that he has been duped and that Desdemona was innocent. Angrily, he turns on Iago with his sword but only succeeds in wounding him. Finally, he ends this tale of tragedy by turning the sword on himself.

While Othello is a purely fictional character, the wars between Venice and the Turks were fierce and real.[5] The most recent of those conflicts occurred during Shakespeare's boyhood, so it is quite possible that reports of the battles had captured his imagination.[6] Certainly, in 1604 when the play debuted, the Venetian-Ottoman wars were within recent memory of his audiences.[7] Shakespeare's theatrical company, The King's Men, also received a publicity boost for their

Figure 5.1 *Othello and Desdemona* by the French painter Alexandre-Marie Colin (1829). *Wikimedia Commons.*

production of *Othello* from a widely circulated poem written by King James,[8] memorializing the naval battle of Lepanto in which a coalition of Christian naval forces had crushed the Ottoman fleet some thirty-two years earlier.[9]

The play's references to Othello's background and complexion would also have intrigued audiences, for he was portrayed as "the Moor of Venice"—an unusual pedigree for a tragic-hero figure on the London stage in those days. In Shakespeare's world, the term *Moor* carried multiple overtones, one being a pejorative reference to individuals with dark skin and another invoking the notions of "exotic" and "hot-blooded."

Likewise, imagery of Venice itself would have captivated audiences, for the city was, perhaps, the most cosmopolitan in the world.[10] An immensely powerful trading center, it attracted merchants, sailors, and adventurers from cultures all over the Mediterranean.[11] The ruling families of Venice were White, but they deliberately engaged foreign mercenaries, many of them dark-complected

Moors, to fight their battles. Remarkably, this practice included the officer corps, of which (in Shakespeare's telling) the Moor of Venice was one.

Historically, the relationship between the Turks and the Venetians was a classic *approach-avoidance conflict*—a mixture of attraction and revulsion, need and greed—because the two groups were wary trading partners whose economic intercourse enriched each other.[12] Venice was also a strategic target too tempting to ignore for the Ottoman Turks, who ruled the eastern Mediterranean and the Balkans. Indeed, over the previous few centuries, the Ottomans had attacked far-flung Venetian outposts in their region, sometimes winning and sometimes losing. And so it was, in 1570—in the time Shakespeare had set *Othello*—that the Ottomans were again threatening Venetian hegemony on Cyprus, a sprawling island on the Turkish doorstep. It is the eve of this struggle between the two Mediterranean superpowers that serves as the backdrop for the play.

THE PSYCHOLOGY OF THE HERO IN *OTHELLO*

Aristotle may have been the first to define dramatic figures like Othello as "tragic heroes"—larger-than-life characters with whom the audience could sympathize but who are brought down by a "tragic flaw" or an error in judgment.[13] Othello's flaws include one of gullibility and another of jealousy, as we hear in these famous lines:

> IAGO: Oh, beware, my lord, of jealousy!
> It is the green-eyed monster which doth mock
> The meat it feeds on.
>
> —3.3.195

We psychologists would add that Othello was also a victim of *confirmation bias*, having become so thoroughly convinced that Desdemona was unfaithful to him that he was unwilling to consider any contradictory evidence.[14]

In Shakespeare's plays, we find tragic flaws, of one form or another, in the main characters of *Antony and Cleopatra*, *Troilus and Cressida*, *Coriolanus*, *Hamlet*, *King Lear*, *Macbeth*, *Othello*, *Romeo and Juliet*, *Timon of Athens*, *Titus Andronicus*, and *Julius Caesar*.[15] Not all of these plays have tragic *heroes*, of course, but if we were to make a list of the major Shakespearean heroes, it certainly would include several characters from these same plays.

To that list, we could add three other heroic figures.[16] The first is England's most revered heroic figure, King Henry V, who, even though outnumbered at least two to one, gloriously defeated the French at Agincourt in 1415. While Henry does not qualify as a *tragic* hero because he does not fail or die at the end of his play, he is a hero nonetheless, for his story is one of a great triumph. The second might be the Earl of Richmond, although he has a relatively minor part in the play (*Richard III*), where he slays the dastardly King Richard to establish the Tudor line of English monarchs that would include Elizabeth.[17] And the third apparently unflawed hero is Macduff, a relatively minor character in terms of his lines and time on stage—but who saved Scotland by slaying Macbeth in revenge for the murder of his wife and children. Those, then, are the characters who come to mind when we think of Shakespearean heroes.

But there are heroes of yet another sort in Shakespeare—ones who have been hiding in the plays for hundreds of years. None is a warrior-king or other noble swashbuckler who commits astounding acts of derring-do. Rather, each is a minor or supporting character, yet with a pivotal role in the plot. We will introduce more than a dozen of them to you, beginning with one we have already met in *Othello.* Her name is **Emilia**, the villain's wife and Desdemona's handmaid.

When we first meet Emilia, she seems to be in league with her husband and his nefarious plot against Othello. It is Emilia who picks up the handkerchief in Desdemona's suite and delivers it to Iago, uttering the line, "My wayward husband hath a hundred times / Wooed me to steal it" (3.3.334). Yet, when she finally sees the evil web Iago has woven, she dares to scold Othello, a man far above her station, for doubting his wife's loyalty and fidelity, saying, "If you think other / Remove your thought" (4.2.13). And when she finds that Othello has smothered Desdemona, she hurls a curse at the general with these words: "Thou dost belie her and thou art a devil" (5.2.163).

Nor does Emilia stop there. Even though she knows that Othello could kill her, she threatens to bring justice down upon him, screaming, "I'll make thee known / Though I lost twenty lives" (5.2.198). Then she turns her wrath upon Iago, saying, "Your reports have set the murder on" (5.2.223). Although he repeatedly orders her to be quiet, she nevertheless persists, saying, "I will speak as liberal as the north" (5.2.260).[18] Finally, Iago stabs her, and she pays for her persistence with her life.

We submit that Emilia is a hero, along with many more of her ilk whom we will call "selfless heroes," who (in most cases) fit all elements of the following four-part definition:[19]

- Heroes, in our sense, seek some type of **altruistic goal** or **quest**—to improve the lot of another person or group or humankind in general—as opposed to seeking personal fame and power. Think of the first responders at the World Trade Center on 9/11 or "whistleblowers" who call out the illegal maneuvers of their political bosses.
- Heroes must face **some form of sacrifice or risk**. This could involve an act of physical peril, a social sacrifice, an act of speaking truth to power, or even just the risk of looking foolish—which is to say that no one is scoring just *how heroic* the act is.
- Their acts can **either be passive or active**. While we often think of heroics as a valiant *activity*, a heroic act may, instead, involve passive resistance, as when Rosa Parks refused to move to the "colored" section in the back of the bus.[20]
- Finally, their heroic deeds can be **either a brief, one-time act or a longer-term pattern of heroic behavior**.

Note that Emelia fits these criteria; her role in *Othello* is both heroic and pivotal, even though she is a relatively minor character. Her quest is first to expose the truth of Desdemona's innocence and then to bring Othello to justice. In doing so, she confronts both Othello and Iago at her peril—with the consequence of being murdered.

What other such heroes lurk unnoticed in the Shakespearean canon? As you might expect, several are scattered among the tragedies, where they risk their lives by confronting power with their warnings of danger and folly—typically to no avail, showing us the inevitability of fate. Aside from Emilia, these selfless heroes include **Kent**, **Edgar**, and **the Fool**, all three of whom we find in *King Lear*, risking the wrath of the king by counseling him to use better judgment in distributing his kingdom among his three daughters. To that trio, we certainly should add Lear's youngest daughter, **Cordelia**, for speaking her truth to Lear's power, knowing that doing so might incur her demented father's wrath.[21]

Oddly enough, we also find such selfless heroes in comedies—even more often than in the tragedies. Remember that, in Elizabethan-Jacobean times, comedies were defined as plays with a happy ending, typically with many marriages. There, the heroes may save the day when their counsel is heeded, and circumstances conspire to turn an imminent tragedy into a joyous occasion.

Our first exemplar of this genre is **Camillo**, in *The Winter's Tale*. **Leontes** (king of **Sicilia**) orders Camillo to poison **Polixenes** (king of Bohemia); Leontes believes his wife, the queen of Sicilia, is having an affair with Polixenes. Instead, Camillo warns Polixenes, and the two flee Sicilia together.

The second is **Paulina**, also from *The Winter's Tale*—a Sicilian noblewoman who stands up to the paranoid King Leontes in defense of his wife Hermione's virtue and, after Hermione's apparent death, condemns the king.

We also see some heroic qualities in **Miranda**, Prospero's daughter in *The Tempest*. Although actors often present her as a stereotypical teenager-in-love, Rio Turnbull of the Shakespeare Birthplace Trust writes of Miranda as a woman with the courage to confront her powerful father and speak truth to power:

> Miranda is . . . a brave woman, and is willing to stand up for what she believes in. When Prospero is stirring up a storm in the ocean that might cause its occupants to drown, she sympathizes with the men at sea and implores her father not to do so, saying, **"If by your art, my dearest father, you have put the wild waters in this roar, allay them. . . . O, I have suffered with those that I saw suffer: a brave vessel, who had, no doubt, some noble creature in her, dash'd all to pieces. O, the cry did knock against my very heart."** (emphasis in the original)[22]

And likewise, **Escalus**, a nobleman and advisor to the duke in *Measure for Measure*, urges Angelo to employ less punishment and more mercy, even though Angelo has shown himself to be a moral extremist, possessing absolute power of life and death over the people of Vienna.

Portia's heroics derive from her cleverness—primarily in outsmarting men in a male-dominated world. And so, she neuters the constraints of her father's will that prevent her from choosing a husband. But when she finally chooses Bassanio, she finds that his friend Antonio has defaulted on a loan from Shylock, who has a contract with Antonio stipulating the penalty of "a pound of flesh."[23] Portia points out a legal loophole through which Antonio escapes with his life, as she observes, "The brain may devise laws for the blood, but a hot temper leaps o'er a cold decree" (1.2.12).

Finally, among the comedies (so called in the Tudor era), we nominate **Imogen** (in *Cymbeline*), who is banished by her father, King Cymbeline, for refusing to marry the Queen's cloddish son, Cloten, the man who repeatedly attempts to rape her but also the man of Cymbeline's choice. She also endures accusations in which her reputation is sullied and her relationship with Posthumus, her true love, is threatened by the evil Iachimo, who falsely claims to have enjoyed her willing body. In our view, she qualifies as a hero on all counts listed above except *altruism*, for she only defends her own reputation and her relationship with her beloved Posthumus. Still, understood in the context of medieval times, we would celebrate her strength of character and dub her a hero.

The other possible non-altruistic hero is **Lavinia**, in *Titus Andronicus*—if, indeed, there are any true heroes in that play at all, and if the motive of revenge does not disqualify her. Lavinia suffers multiple rapes and other almost unimaginable assaults for refusing to marry the emperor. Yet, despite the disabilities of having her hands cut off and her tongue ripped out, she manages to reveal her assailants' identities. We leave it for you to decide whether her suffering and ultimate revenge on her tormentors qualify her for hero status.

And there is one more judgment we ask you to make. While witches are seldom nominated as heroes, please consider the **Weird Sisters** in *Macbeth*. This may be a stretch, but like so many other minor heroes that populate Shakespeare's works, this trio also dares to speak truth to power. Consider, too, that it was not easy being a "witch" in those days, for they were often demonized, tortured, and killed as conspirators with the Devil.[24]

We end this list with a revelation that surprised us: over half of Shakespeare's minor heroes are women.[25]

MODERN-DAY HEROES EMERGE FROM THE STANFORD PRISON EXPERIMENT

In this context, we would note one more parallel between *Measure for Measure* and the Stanford Prison Experiment. It concerns the young lass named **Christina**, whose real-life role in the Stanford Prison Experiment was much like that of Escalus in *Measure for Measure* (figure 5.2). As you may recall, Escalus was the underling who stood up to Angelo, telling him that the punishment he proposed was excessive. And so it was with Christina Maslach who said the same to Philip Zimbardo, the acting superintendent of the Stanford jail. Maslach called out Professor Zimbardo, telling him that the brutality of the experiment was excessive, and she did so at some potential risk to her status as a graduate student. Happily, her efforts were successful, whereas Escalus's were not.

The results of the Stanford Prison Experiment were disheartening to Dr. Zimbardo, who had been swept up in the situation. Could any good come out of this work? Were there any lessons to be learned? Zimbardo spent decades telling the story of the experiment to the world, emphasizing how good people, who have no history or intention of hurting or demeaning others, can become caught up in the demands of a situation that leads them down an evil path. Then in 2008, Dr. Zimbardo published a best-selling book, *The Lucifer Effect: Understanding How Good People Turn Evil*—a book that deconstructed and analyzed the prison experiment.[26] Importantly, it was aimed at general readers,

Figure 5.2 Christina Maslach, PhD, is professor emerita of psychology at the University of California, Berkley. *Wikimedia Commons.*

alerting the public to the conditions in which good people can respond badly to bad situations. As Zimbardo likes to say, "Evil is not something done by a few 'bad apples.' Rather, it is usually caused by a 'bad barrel'"—meaning that otherwise good people can become caught up in a bad *situation*, a bad *system*. It's the same idea that social scientists invoke when they speak of *systemic poverty*, *systemic discrimination*, or *systemic racism*.[27]

Despite the initially discouraging results of the Stanford Prison Experiment, Zimbardo realized that there was a body of psychological research identifying the obstacles that can prevent people from intervening heroically in distressing situations. You will remember the famous "shock research" by Stanley Milgram, showing that many Americans will follow the orders of an authority figure to give (seemingly) dangerous electric shocks to another person (a confederate of the experimenter who, in fact, received no shocks).[28] And then there were the "bystander" studies, showing that, in emergency situations, the more people present, the less likely it is that someone will help—because everyone sees others doing nothing.[29]

Even so, there are people who spontaneously perform heroic acts—like the now-famous bystander who, at great personal peril, rescued a man who had fallen off a New York subway platform into the path of an oncoming train. In fact, an overwhelming number of heroic acts are performed by ordinary people, says Zimbardo. He and his colleague Zeno Franco call that "the banality of heroism."[30]

> The banality of heroism concept suggests that we are *all* potential heroes waiting for a moment in life to perform a heroic deed. The decision to act heroically is a choice that many of us will be called upon to make at some point in time. By conceiving of heroism as a universal attribute of human nature, not as a rare feature of the few "heroic elect," heroism becomes something that seems in the range of possibilities for every person, perhaps inspiring more of us to answer that call.

With these ideas in mind, Zimbardo founded the Heroic Imagination Project (HIP) in 2010. Targeting the program at young people, he began training them to avoid the mistake he made by not immediately recognizing the degrading and potentially harmful effects on the participants in his prison experiment. The broader goal was to help students recognize potentially traumatic situations and to take "heroic" actions in line with the definition of "selfless heroes" described earlier.[31]

Now, HIP has spread to more than a score of countries around the world, spreading the word to hundreds of graduates that good people are not immune to being engulfed by *the power of the situation* either as active participants—like the guards in Zimbardo's prison—or as passive bystanders responding to cues from others who are doing nothing to help. In brief, HIP aims to teach young people to respond like the real hero of the Stanford Prison Experiment: Christina Maslach.

WHAT PURPOSE DO THE SELFLESS HEROES LIKE EMILIA SERVE IN SHAKESPEARE?

Note that there are eleven tragedies in the Shakespearean canon,[32] but only in *Titus Andronicus* does the Bard portray the leading characters as thoroughly unsympathetic individuals.[33] More often he gives us a there-could-go-I person with a tragic flaw. For Lear it was pride; for Brutus and Julius Caesar, it was pride combined with a desire for power; and for Macbeth it was gullibility plus all the above.[34] And in each case, the playwright employs one or more of our unsung, selfless heroes to highlight the protagonist's shortcomings and mark an inflection point in his downward spiral.

And finally, we submit one more individual, perhaps most tragic of all: **Cassandra** of *Troilus and Cressida*, the hero who tried to warn of Troy's folly and ultimate destruction but was dismissed as insane. Cassandra, you see, was cursed by the god Apollo. It was a two-part curse: she would accurately see the future, but no one would pay attention to her warnings. And so it may be with quotidian heroes of today—many of whom will be trained by the Heroic Imagination Project. Inevitably, some will sound an alarm and speak truth to power, but no one will listen. It must be the curse of some heroes—who must go unsung.

Part III

INTO THE MIND

In our view, Shakespeare's greatest innovation may have been his opening of a "window" on the mind far wider than any playwright had done before—into the private minds of his characters. Most famously he did so with his use of the *soliloquy*, as when Hamlet ponders the two alternatives of regicide or suicide.

Through this window he also helped his audiences peer into sundry *states of consciousness*—as well as the *un*conscious mind. Mental states involving sleep and dreaming induced by drugs and potions seem to have especially intrigued him, as we shall see in the realms of *A Midsummer Night's Dream*, *Romeo and Juliet*, and *Cymbeline*. But did Shakespeare anticipate Freud's theories of the unconscious? Many literary critics would say so, but we will argue that Shakespeare's own views were less like Freud's and more like the recent, science-based concept of the *dual-process mind*.

And finally, we will see the Bard's fascination with the contents of disturbed minds with characters such as Hamlet, Lear, Lady Macbeth, and King Richard III—characters dealing, whether they knew it or not, with depression, mania, dementia, obsessive-compulsive disorder, PTSD, psychopathy, hysteria, and many other mental aberrations.

· 6 ·

SLEEP, DREAMS, AND DRUGS: WINDOWS INTO THE UNCONSCIOUS

A Midsummer Night's Dream

Athens, in Greco-Roman antiquity—Theseus, the duke of Athens, is preparing for his wedding with Hippolyta, queen of the Amazons, when Egeus, an Athenian citizen, interrupts with a request for the duke to adjudicate a dispute with Hermia, Egeus's daughter: she is resisting the marriage he has arranged for her with Demetrius. You see, Hermia loves Lysander, not Demetrius (who also happens to be the ex-boyfriend of her best friend, Helena). Moreover, Demetrius has already agreed to the marriage with Hermia.

As if those complications were not enough, Hermia's not-so-loving father demands that the duke enforce an ancient law requiring daughters to marry suitors designated by their fathers—or be put to death. The duke, however, offers her another choice: to become a nun in the temple of the goddess Diana.

The headstrong Hermia chooses yet another option: eloping with her lover Lysander to the forest outside Athens's boundaries. Hermia foolishly shares her plan with Helena, who relays the secret to Demetrius. Upset at the loss of his arranged wife-to-be, Demetrius follows the lovers into the woods, and Helena follows Demetrius, hoping that he will change his mind and fall in love with her instead.

Not to worry. A bumbling troupe of fairies in the forest will use magic potions to set everything right.

The leader of that troupe is Oberon, the fairy king, who with his queen, Titania, has set up camp in the forest. There Titania proposes to attend the wedding of Duke Theseus and Queen Hippolyta. Unfortunately, the relationship between Oberon and Titania has become strained because she has refused to give him her Indian changeling to be his page.[1] Oberon decides to teach Titania a lesson by ordering Puck, his fairy-servant, to obtain a special purple flower having a juice that causes people to fall in love with the first creature they see.

Oberon's plan is to administer the juice to Titania, but first he comes across Demetrius who has encountered Helena and is berating her for following him into the forest. Oberon, not such a bad sort after all, wants to smooth things over in Helena's favor, so he tells Puck to sprinkle the flower's love-juice in Demetrius's eyes. Puck, however, mistakenly drops it into the eyes of the sleeping Lysander, and when Helena awakens him from his midsummer sleep, Lysander falls in love with her. Then, when Demetrius finally falls asleep, Oberon puts the juice in his eyes, so upon awakening, Demetrius falls in love with Helena, too.

In yet another subplot, Oberon further complicates the situation by juicing his wife Titania's eyes, causing her to become infatuated with a "mechanical" by the name of Bottom, whose head Puck has transformed into that of an ass.[2] (See figure 6.1.)

Chaos and confusion ensue. Oberon orders Puck to straighten out the relationships among the four lovers—Hermia, Lysander, Demetrius, and Helena—who have all fallen asleep after an exhausting day in the forest. The proper potions are applied to their eyes, so when Theseus discovers them the next morning, he finds that Demetrius has shifted his love back to Helena, and Lysander again loves Hermia. Therefore, he lifts his order against Hermia, and all lovers are happily married—in one of the most complicated plots in the Shakespearean repertoire, and one of the few that were original with Shakespeare.

Figure 6.1 *Scene from A Midsummer Night's Dream. Titania and Bottom*—painting by Edwin Landseer. *Wikimedia Commons.*

A Midsummer Night's Dream endures as one of the Bard's most popular plays. But it is a mental workout for the audience, which must follow nineteen characters (see figure 6.2) for nearly three hours on a labyrinthine path involving four interweaved sub-plots, almost as if . . . well, as in a dream.[3] At times, the players confess that they are not sure whether they are dreaming or awake. As Bottom says, "I have had a most rare vision. I have had a dream, past the wit of man to say what dream it was" (4.1.210). Surprisingly, for all its complexity, this play has the fewest number of scene changes of all the Bard's theatrical works, yet it features numerous instances of sleep and dreaming, along with devious fairies, magical potions, and imaginary beasts.[4]

How would theatergoers, viewing a performance of *A Midsummer Night's Dream* in 1596, have conceived of sleep and dreaming? Today we think of sleep as a restorative process,[5] and so would Elizabethan audiences. As Macbeth says, "Sleep knits up the raveled sleave of care" (2.2.47). But from that point, our perspectives would diverge. Shakespeare's audiences would have understood

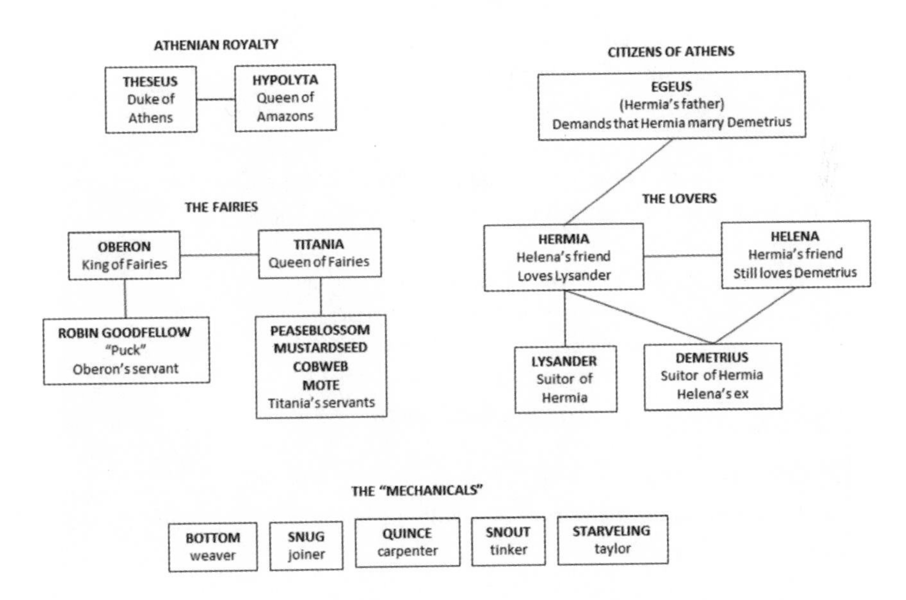

Figure 6.2 Characters and relationships in *A Midsummer Night's Dream*.

sleep in terms of the body's "vital spirits," which were physical products of the soul that coursed through the blood, carrying the force that gives us life.[6] These spirits were understood as a sort of vapor, impossible to touch or see, but nevertheless physical. This concept, widespread in Shakespeare's time, rested on a view of the mind offered by the thirteenth-century philosopher, Albertus Magnus:

> In a normal state of mind and body . . . there exist the spirits which are the vehicle of all the processes of life proceeding from the soul. These spirits originate beneath the heart and are created by the action of bodily heat upon the moisture of food. From the heart the spirits flow first to the liver, where they become natural spirits, governing the involuntary processes (especially digestion); and second, they flow to the brain where they become the animal spirits, governing the functions of the five senses, thinking, and imagination. Now these spirits, being naturally warm and subtle, move outward through the pores and evaporate. When a sufficient amount of these spirits has evaporated, the external organs become cold, driving the spirits inward, and sleep naturally ensues.[7]

Sleep, then, was a process that shut down the external senses, allowing the life forces from the soul to be warmed, moistened, and restored. We see many references to this in Shakespeare. In *Hamlet*, for example, the Player King says:

> My spirits grow dull, and fain I would beguile
> The tedious day with sleep.
>
> —*Hamlet*, 3.2.248

We hear of the body's vital spirits again in *Love's Labour's Lost*, when Berowne objects to the plan of forswearing women in favor of books:

> Why, universal plodding poisons up
> The nimble spirits in the arteries,
> As motion and long-during action tires
> The sinewy vigour of the traveller.
>
> —*Love's Labour's Lost*, 4.3.302[8]

And in *Romeo and Juliet*, Friar Lawrence offers the following words as he gives Juliet a vial containing the potion that will put her into a sleep feigning death:

> FRIAR LAWRENCE: Take thou this vial,
> And this distilled liquor drink thou off:

When presently through all thy veins shall run
A dull and heavy slumber, which shall seize
Each vital spirit: for no Pulse shall keep
His natural progress, but surcease to beat:
No sign of breath shall testify thou livest.
And in this borrowed likeness of shrunk death,
Thou shalt remain full two and forty hours.

—4.11 (Quarto 1)[9]

The notion of sleep was also freighted with associations to death. The ancient Greeks had taught that Hypnos, the god of sleep, was the brother of Thanatos, the god of death. And so it was for Hamlet as he pondered suicide:

. . . . To die, to sleep;
No more; and by a sleep, to say we end
The heart-ache, and the thousand natural shocks
That Flesh is heir to? 'Tis a consummation
Devoutly to be wished. To die, to sleep,
To sleep, perchance to Dream; aye, there's the rub,
For in that sleep of death, what dreams may come,
When we have shuffled off this mortal coil,
Must give us pause.

—*Hamlet*, 3.1.72

Even today we use the same language when a beloved pet must be "put to sleep." And it was likewise for Prospero, addressing the audience as he brings a masque performance to a somber end, reminding us that death is like a sleep that both precedes and follows our existence:

Our revels now are ended. These our actors,
As I foretold you, were all spirits, and
Are melted into air, into thin air:
And like the baseless fabric of this vision,
The cloud-capp'd tow'rs, the gorgeous palaces,
The solemn tem'les, t'e great globe itself,
Yea, all which it inherits, shall dissolve,
And, like this insubstantial pageant faded,
Leave not a rack behind. We are such stuff
As dreams are made on; and our little life
Is rounded with a sleep.

—*The Tempest*, 4.1.163

The fact that the word "sleep" *appears in every one of Shakespeare's plays*, most of his long poems, and several of his sonnets leads us to infer that the Bard had a fascination—even an obsession—with sleep.[10] One scholar has even suggested that his recurrent references to insomnia, nightmares, and other sleep disturbances indicate that Will Shakespeare may have suffered from sleep difficulties himself.[11] Indeed, by our count, sleep, dreams, and sleep disturbances are vital to the storyline in at least one-quarter of his plays.[12]

SHAKESPEAREAN DREAMS

As for dreams, the Elizabethans knew they could reflect our fantasies, as well as the events and worries of the previous day—much as Freud later observed.[13] But they also believed that dreams could have diverse sources, perhaps as messages from God that could portend the future—as when Falstaff referred to the biblical dream about "the *Pharaoh's lean kine*," which forecast a famine in Egypt.[14] Nightmares, however, might be fed to the sleeper by Satan and his minions.

The playwright intended that his audience, watching *A Midsummer Night's Dream*, imagine their theatrical experience to be dream-like, for he had Puck tell them:

> If we shadows have offended,
> Think but this, and all is mended,
> That you have but slumber'd here
> While these visions did appear.
> And this weak and idle theme,
> No more yielding but a dream,
> Gentles, do not reprehend:
> if you pardon, we will mend.
>
> —*A Midsummer Night's Dream*, 5.1.440

For psychologists, dreams show that we humans can experience mental states beyond "normal" consciousness.[15] Indeed, Shakespeare, too, wrote extensively about alternative states of consciousness, including not only sleep and dreaming but alcohol intoxication and drug states induced by magical and medicinal potions and elixirs—and more, as we shall see. But for now, let us make a quick survey of what modern sleep research has uncovered about that hidden third of our lives.

THE SCIENCE OF SLEEP AND DREAMING

One night in 1951, graduate student Eugene Aserinsky fortuitously pasted electrodes on his young son's scalp and around the eyes in preparation for recording brain waves and muscle movements while the boy slept.[16] The next morning, Aserinsky found that the recording showed periodic bursts of fast activity, accompanied by rapid eye movements, as though the lad were awake and alert—even though his eyes were closed! Until that time, scientists had thought that sleep was a time of quiet and rest, but here was proof of intense brain activity, yet in a subject fast asleep. In studies that quickly followed, Aserinsky and his advisor, Nathaniel Kleitman, found that a person awakened during those episodes would often tell of a dream. Their paper in *Science* reported these findings and detonated an explosion of sleep research.[17]

We now know that such episodes of rapid eye movement (now called *REM sleep*) occur about every ninety minutes during the night. (See figure 6.3.) As Aserinsky and Kleitman reported, REM tends to signal vivid, story-like dreams, although other more mundane mental activity does sometimes occur during quiet sleep. Moreover, they found that deprivation of REM sleep, caused by awakening the sleeper during each REM period, produces a strong need for

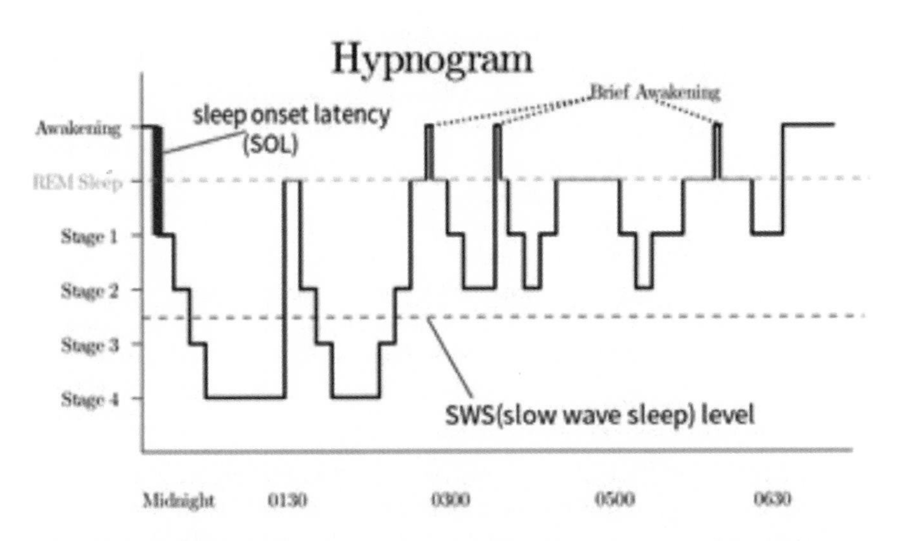

Figure 6.3 This hypnogram shows a typical night's sleep in the pattern of brain waves that vary from fast to slow in cycles of about ninety minutes. Note that, as the night progresses, we spend less time in deep sleep (slow wave) and experience increasingly long periods of rapid eye movement sleep (REM). Most dreams occur during REM. *Wikimedia Commons.*

REM sleep and a "REM rebound" in which the sleeper gets more REM sleep than usual when allowed to sleep uninterrupted.

Strangely, while our eyes are active during REM sleep, we are otherwise paralyzed—presumably by a process that keeps us from acting out our dreams. (Sleep paralysis may also be what makes us feel like we are "running through molasses" in a scary dream.) Shakespeare seems to have hinted about sleep paralysis in *Romeo and Juliet*, where Mercutio metaphorically refers to Queen Mab, an old hag that brings us dreams and sits on sleepers to hold them motionless.[18] Oddly, he does not mention rapid eye movements, even though these have long been obvious in sleeping animals.

Shakespeare does give us one extensive dream report (albeit a secondhand one) telling how Hotspur's wife imagined him dreaming as she watched him sleep:

> In thy faint slumbers I by thee have watch'd,
> And heard thee murmur tales of iron wars;
> Speak terms of manage to thy bounding steed;
> Cry 'Courage! To the field!' And thou hast talk'd
> Of sallies and retires, of trenches, tents,
> Of palisades, frontiers, parapets,
> Of basilisks, of cannon, culverin,
> Of prisoners' ransom and of soldiers slain,
> And all the currents of a heady fight.
> Thy spirit within thee hath been so at war
> And thus hath so bestirred thee in thy sleep,
> That beads of sweat have stood upon thy brow
> Like bubbles in a late-disturbed stream;
> And in thy face strange motions have appear'd,
> Such as we see when men restrain their breath
> On some great sudden hest. O, what portents are these?
> Some heavy business hath my lord in hand,
> And I must know it, else he loves me not.
>
> —*Henry IV, Part 1*, 2.3.39

This observation suggests that the playwright took an interest in, or even solicited, the dream accounts of his acquaintances. And this melds with another of our discoveries about Shakespeare's fascination with sleep. *Not only does every one of his plays mention sleep, but they all speak of dreams or dreamers.*

Aside from agreeing with suggestions from Shakespeare, and indeed throughout all of literature, that sleep rests the body and restores mental acuity, we should note one more remarkable thing sleep research has discovered about the function of sleep and dreams. It now appears that sleep, especially REM sleep, plays a crucial role in forming long-term memories from information placed in temporary storage the previous day.[19]

SLEEP DISORDERS

Shakespeare did realize that sleep is not always restful. Nor are the fleeting, flitting fantasies of the night always like the entertaining imaginations of those in *A Midsummer Night's Dream*. In fact, his references to disturbed sleep outnumber those to normal sleep. Sleep apnea is an example—a common problem in which sufferers stop breathing during sleep, several times per hour.[20] The disorder has two prominent symptoms: chronic daytime sleepiness and sonorous snoring. In its most common form, obstructive sleep apnea, the snoring results from collapse of the airway, while tissues of the nose and throat flap and flutter in the attempt to breathe.[21] Compare these symptoms, then, to the two brief descriptions of the fat knight's daytime sleepiness and snoring reported in *Henry IV, Part 1*:

> FALSTAFF: Now, Hal, what time of day is it, lad?
> PRINCE HENRY: Thou art so fat-witted, with drinking of old sack, and unbuttoning thee after supper, and sleeping upon benches after noon . . .
>
> —1.2.1

And later, we hear:

> POINS: Falstaff!—fast asleep behind the arras, and snorting like a horse.
> PRINCE HENRY: Hark, how hard he fetches breath . . .
>
> —3.3.104

Shakespeare also speaks of *insomnia*, a disorder more common than sleep apnea and likewise thoroughly unpleasant—so unpleasant that, in modern times, it is used as torture. According to Mr. Shakespeare, insomnia made King Henry IV curse the god of sleep in his famous "uneasy lies the head" soliloquy:

> How many thousands of my poorest subjects
> Are at this hour asleep! O sleep, O gentle sleep,
> Nature's soft nurse, how have I frightened thee,

That thou no more will weigh my eyelids down,
And steep my senses in forgetfulness?
Why rather, sleep, liest thou in smoky cribs,
Upon uneasy pallets stretching thee,
And hush'd with buzzing night-flies to thy slumber,
Than in the perfum'd chambers of the great,
Under the canopies of costly state,
And lull'd with sound of sweetest melody?
O thou dull god, why liest thou with the vile
In loathsome beds, and leav'st the kingly couch
A watch-case or a common 'larum-bell?
Wilt thou upon the high and giddy mast
Seal up the ship-boy's eyes, and rock his brains
In cradle of the rude imperious surge,
And in the visitation of the winds,
Who take the ruffian billows by the top,
Curling their monstrous heads, and hanging them
With deafing clamour in the slippery clouds,
That with the hurly death itself awakes?
Canst thou, O partial sleep, give thy repose
To the wet sea-boy in an hour so rude;
And in the calmest and most stillest night,
With all appliances and means to boot,
Deny it to a king? Then, happy low, lie down!
Uneasy lies the head that wears a crown.

—*Henry IV, Part 2*, 3.1.1

So ubiquitous are the references to insomnia in Shakespeare's works that one wonders whether insomnia might have been the Bard's "bed noir" curse.[22] Perhaps most revealing are these lines:[23]

Weary with toil, I haste me to my bed,
The dear repose for limbs with travel tired;
But then begins a journey in my head,
To work my mind, when body's work's expired:
For then my thoughts (from far where I abide)
Intend a zealous pilgrimage to thee,
And keep my drooping eyelids open wide,
Looking on darkness which the blind do see:
Save that my soul's imaginary sight
Presents thy shadow to my sightless view,
Which, like a jewel hung in ghastly night,

Makes black night beauteous and her old face new.
Lo, thus, by day my limbs, by night my mind,
For thee, and for myself, no quiet find.
—Sonnet 27

Finally, of all the sleep disorders mentioned by the Bard, the most famous, perhaps, is Lady Macbeth's *somnambulism* (sleepwalking) scene, where she also cries, "Out damned spot! Out, I say!" referencing the seemingly indelible blood on her hand (5.1.37). She also was haunted by nightmares, which were understood by audiences to arise from stress, physical infirmity, excessive drinking, or perhaps a torment delivered by Satan himself.[24]

SLEEP AND DREAMS: GATEWAY TO THE UNCONSCIOUS

We have seen that dreams were understood in early modern times to occur when sleep shut down the five external senses and left the mental faculty of fantasy uninhibited. People also knew that dreams could be influenced by alcohol, medicines, and herbs or spirits, either divine or evil—and, like us, they understood dreams to exist in an altered state of consciousness.

Three hundred years later, Sigmund Freud called dreams "the royal road to a knowledge of the unconscious activities of the mind."[25] Your authors agree, in principle, although modern psychologists have learned that the unconscious behaves quite differently from the one that Freud conceived. Most psychologists we know no longer paint the unconscious as a seething cauldron of sexual and aggressive desires barely restrained from entering consciousness. We now see the unconscious as all the brain processes that occur outside of consciousness, such as the regulation of body temperature and the coordination of movements. More interesting for our purposes, it is also the storehouse of long-term memories indexed by common patterns and emotions.

Among its talents, this unconscious makes lightning-fast judgments and guides habitual behaviors—the things we can do without much thought, such as walking, driving, and perhaps throwing a ball. Think of the process in which you learned some complex skill, such as swinging a golf club or playing a musical instrument. At first, you had to make every move consciously, and it was exhausting! Later, after the skill became habitual, your movements are controlled by regions of the brain lying outside of consciousness. Other unconscious processes include the learning of language, food preferences, and habitual professional skills, such as those a physician relies on to diagnose disease or a lawyer uses to draft a criminal defense. Social psychologist Jonathan Haidt calls this collection of unconscious abilities "intuition," as contrasted with the processes of our slower and more deliberative rational consciousness.[26]

You will recall that Daniel Kahneman posits a mind that consists of two great "systems." One is a vast and intuitive unconscious that he calls System 1. The other is a smaller, conscious System 2 that is capable of rational thought.[27] Our emotions, however, strongly influence which associations in the unconscious System 1 are pushed into the conscious portion of the mind. Memories tagged with fear and danger generally have priority. It is a mental processing system that served our ancestors well in the struggle to survive and to leave offspring carrying their genes into the next generation.[28]

But which system is in charge of our behavior? Does our "will" reside in the unconscious System 1 or the conscious System 2? Haidt tells us to think of the intuitive unconscious as like an elephant, a formidable creature with priorities of its own—and the largest part of the mind. Meanwhile, our much smaller conscious mind is like the elephant's rider, assuming to be in control—but, in reality, just following the path proposed by the intuitive elephant.[29]

How does all this relate to dreams? The intuitive System 1 can search a lifetime of memories in an instant, looking for recognizable patterns to thrust up into consciousness. That model, then, suggests how our kaleidoscope of dreams forms its ever-changing patterns, unrestrained by the more reasonable and restrictive conscious mind. We can see dreams as our witness to the process by which the unconscious drifts from one pattern to another as it searches for associations in memory. Thus, we suggest that, during sleep, the conscious and rational processes of System 2 are largely shut down, while dreams reflect the associative memory processes that usually occur out of awareness in the intuitive mind of System 1.

Pioneering sleep researcher, William Dement, found that by waking a sleeper during REM periods, he could often trace the pathways leading from events of the previous day linked with a cascade of associative memories through the night.[30] And in your own experience, you have probably noticed how dreams may flit from one fantastic idea, one surreal scene, to another, much as the Bard depicted in *A Midsummer Night's Dream*. They usually have a story, but the story often changes abruptly, taking twists and turns through memories and fantasies—which is what makes dreams so difficult to remember the next day. In this respect, dreams may be as close as we will ever get to watching the processes of the unconscious at work, as it searches its inventory. In sum, we can conceive of dreaming as the process of the unconscious mind leafing through associations in memory, looking for patterns that share common features.

While we are not giving Mr. Shakespeare credit for inventing System 1 and System 2, we *are* suggesting that his understanding of the mind was closer to that of Haidt and Kahneman's *dual-process mind* than it was to Freud's tripartite and conflicted mind of the *id*, *ego*, and *superego*—where dreams are mysterious metaphors for our forbidden desires in a mind continually in conflict with itself.[31]

DRUGS AND ALTERED MENTAL STATES

What were the drugs—especially the mind-altering ones—that Shakespeare used to twist his plots? There were several. In tales such as *Romeo and Juliet*, *Cymbeline*, and *A Midsummer Night's Dream*—where a drug, potion, or herb could make one fall in love, fall asleep, or even mimic death.

The Elizabethans possessed a pharmacopeia of psychoactive drugs, some of which were hallucinogenic (mandrake) or soporific (opium), and some that had lethal effects (hemlock, nightshade, yew, or henbane).[32] As for the love potion in *A Midsummer Night's Dream*, such a libidinal concoction as the potion sprinkled in the eyes on a midsummer night probably never existed except in the hopes of love dancing in our imaginations. But what of the potion that made Juliet sleep like the dead? The experts say, if there ever was such a drug, it was probably an elixir derived from hallucinogenic plants, perhaps *Atropa belladonna* (deadly nightshade), *Hyoscyamus niger* (henbane), or *Datura stramonium* (jimsonweed). These were also believed to be among the ingredients in "witches brews," as we have seen in *Macbeth*, where Banquo wonders about the Weird Sisters and their prophecies:

> Were such things here as we do speak about?
> Or have we eaten on the insane root
> That takes the reason prisoner?
>
> *Macbeth*, 1.3.86

The "insane root" likely was mandrake (*Mandragora officinarum*), grown in the Mediterranean region and imported all over the continent and in England (figure 6.4). Its shape, sometimes like the human body, gave rise to the notion that it contained a spirit that would ward off evil. As with certain other plants of the nightshade family, mandrake contains alkaloids that interfere with consciousness—or as we have called it, System 2—producing hallucinations and blocking communication between certain types of neurons.[33]

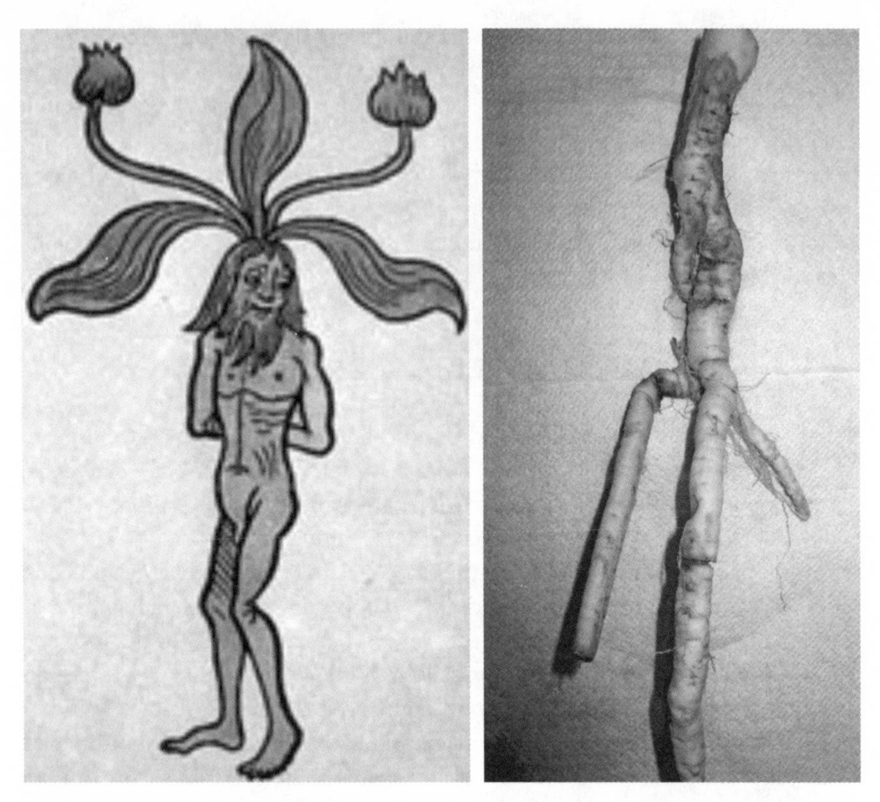

Figure 6.4 Mandrake is a flowering plant common in the Mediterranean region. Because it has strong hallucinogenic properties and its roots can resemble the human figure, it was associated with rituals, magic, and folk medicine in the prescientific age. The plant was reported to scream when pulled from the ground. *Wikimedia Commons.*

The most common of the mind-altering drugs used in Shakespeare's day was, of course, alcohol, usually in the form of wine, beer, and ale and in great quantities.[34] So worrisome was alcohol abuse that Queen Elizabeth issued a ban against beer and ale above a certain strength.

Falstaff, Prince Hal's drinking buddy, obviously differed from the queen about strong drink. His favorite was *sack*, a fortified wine similar to sherry. Says the sotted knight, in praise of sack and its effect on the wits:

> FALSTAFF: A good sherris sack hath a two-fold operation in it. It ascends me into the brain; dries me there all the foolish and dull and curdy vapours which environ it; makes it apprehensive,

> quick, forgetive, full of nimble fiery and delectable shapes, which, delivered o'er to the voice, the tongue, which is the birth, becomes excellent wit. The second property of your excellent sherris is, the warming of the blood; which, before cold and settled, left the liver white and pale, which is the badge of pusillanimity and cowardice; but the sherris warms it and makes it course from the inwards to the parts extreme: it illumineth the face, which as a beacon gives warning to all the rest of this little kingdom, man, to arm; and then the vital commoners and inland petty spirits muster me all to their captain, the heart, who, great and puffed up with this retinue, doth any deed of courage; and this valour comes of sherris. . . . If I had a thousand sons, the first humane principle I would teach them should be, to forswear thin potations and to addict themselves to sack.
>
> —*Henry IV, Part 2*, 4.2.89

Then there was tobacco, recently brought back from the Americas, that quickly became a popular recreational stimulant. As for opium, its use was mostly as a medicine, possibly derived from poppies grown in Spain. It was also one of the main ingredients in an anesthetic called *dwale*, employed during operations performed by barber-surgeons.

To understand the place of medicinal drugs in early modern England, it is important to note some conceptual differences between Renaissance medicine and medical science in our own time. Treatment was still very much under the influence of Galen's theory of the humors. But medicine lay also in the sway of alchemy. And so doctors, led by the renowned Swiss physician Paracelsus, were experimenting with alchemical-derived treatments, in addition to the plants and herbs that were more widely used in antiquity.[35] Among those substances were some very toxic ones, including mercury and arsenic. Paracelsus taught that "poison is in everything, and no thing is without poison. The dosage makes it either a poison or a remedy."[36] Accordingly (and perhaps with the dangers of poisons in mind), the influential Paracelsus advocated treatments that, today, we might call "homeopathic": treating conditions thought to arise from poisons with other poisons in extremely small amounts.[37]

This idea of treating "like with like" (poison with poison) caused much debate in the early-modern medical community because Galen had advocated treating with *opposites*: for example, using bloodletting to release excesses of the humors. This confusion also spilled over into the public mind, where people were worried about the use of poisons to treat disease. We see this conflict in

Romeo and Juliet, where Friar Lawrence ruminates on the opposing forces in life as metaphors for the opposite effects of drugs used as medicines and as poisons:

> FRIAR LAWRENCE: For naught so vile that on the Earth doth live
> But to the Earth some special good doth give;
> Nor aught so good but, strained from that fair use,
> Revolts from true birth, stumbling on abuse.
> Virtue itself turns vice, being misapplied,
> And vice sometime by action dignified.
> Within the infant rind of this weak flower
> Poison hath residence and medicine power.
>
> —*Romeo and Juliet*, 2.3.1

We cannot help but wonder whether the Bard himself (figure 6.5) partook of any mind-altering potions. We have only correlational evidence merely suggesting that he could have. Clay pipe fragments found on his property in Stratford have traces of cocaine, marijuana, and myristic acid (a plant-derived hallucinogen). Nothing ties these artifacts directly to Mr. Shakespeare except their proximity. It was unlikely—but perhaps not impossible—that those fragments were from a pipe Shakespeare used because cocaine had not yet been imported from the New World. However, the pipe fragments do suggest that the use of mind-altering substances was not uncommon in Merrie Old England.

MEDITATION, HYPNOSIS, AND FLOW

Sleep, dreaming, and drug states . . . what other mental states did Mr. Shakespeare recognize? The term *meditation* appears eleven times in the plays and once in the sonnets. Yet there never is a hint that it was meant in the Zen-like sense of clearing the mind or banishing what we might call rational (or System 2) thought. Rather, in each case, "meditation" carries the meaning of "thinking on" or pondering something. Here is a typical example:[38]

> CATESBY: My lord: he doth entreat your grace;
> To visit him to-morrow or next day:
> He is within, with two right reverend fathers,

Figure 6.5 A portrait of Wm Shakespeare painted by ninety artists on sections of canvas on April 23, 2006. In celebration of the Bard's birthday, the canvases were assembled and hung on the wall of the Royal Shakespeare Theater in Stratford-upon-Avon, where the finished portrait was unveiled by Judi Dench and Patrick Stewart. *Wikimedia Commons.*

> Divinely bent to meditation;
> And no worldly suit would he be moved,
> To draw him from his holy exercise.
>
> —*Richard III*, 3.7.60

This passage, of course, refers to *religious* meditation, yet it implies a form of thought that is more like "studying" than "clearing the mind."

What of *hypnosis?* The term comes from the Greek word ὕπνος (*hypnos*), meaning "sleep," and psychologists are divided on whether it is really a distinct state of consciousness or a form of mind-relaxing meditation—or merely an imagined state in highly suggestible persons. There are historical instances that may indicate the use of hypnosis, in some form, by healers in many cultures, but the connection to Shakespeare is tenuous, in our opinion. Nevertheless, hypnotherapist Frank J. Machovec has argued that the Bard knew something of hypnosis because, says Machovec, *The Tempest* is filled with "direct quotes illustrating suggested relaxation and sleep, guided fantasy, behavioral control, and auditory, visual, and tactile hallucinations specific to the suggestibility and unique mental state of the subject."[39] It is an intriguing idea, but we believe it needs further investigation before we can conclude that Shakespeare was referring to hypnosis, rather than to sleep or magic.

And now it is our turn to make a tenuous connection—between Shakespeare and a mental process called "flow." The term was coined by psychologist Mihaly Csikszentmihalyi[40] for the state in which we lose track of time—or time seems to stand still. The flow state commonly occurs when we are fully engaged in an activity that utterly fascinates us, when we become so absorbed that we may ignore the need for sleep, forget to eat, and disregard anything happening around us save the activity that consumes our interest and attention. Csikszentmihalyi suggests that many of the world's greatest achievements in music, art, science, business, and countless other pursuits arise from intrinsically motivated people in a "flow state," pursuing ideas or goals in which they are deeply interested.[41]

We think Mr. Shakespeare would not have been surprised by the concept, for he wrote about *time* even more often than he did of sleep and dreaming! Typically his commentaries on time had a melancholy tone, as he addressed the ravages it produces, bringing on old age and its infirmities (as in *King Lear*). But in the following passage from *As You Like It*, we see clearly that he was aware that time seems to pass at different rates for people engaged in different pursuits:

ROSALIND (as Ganymede): I pray you, what is 't o'clock?
ORLANDO: You should ask me what time o' day. There's no clock in the forest.
ROSALIND: Then there is no true lover in the forest; else sighing every minute and roaming every hour would detect the lazy foot of time as well as a clock.
ORLANDO: And why not the swift foot of time? Had not that been as proper?
ROSALIND: By no means, sir. Time travels in divers paces with divers persons. I'll tell you who time ambles withal, who time trots withal, who time gallops withal, and who he stands still withal.
ORLANDO: I prithee, who doth he trot withal?
ROSALIND: Marry, he trots hard with a young maid between the contract of her marriage and the day it is solemnized. If the interim be but a se'nnight [a week], time's pace is so hard that it seems the length of seven year.
ORLANDO: Who ambles time withal?
ROSALIND: With a priest that lacks Latin and a rich man that hath not the gout, for the one sleeps easily because he cannot study, and the other lives merrily because he feels no pain—the one lacking the burden of lean and wasteful learning, the other knowing no burden of heavy tedious penury. These time ambles withal.
ORLANDO: Who doth he gallop withal?
ROSALIND: With a thief to the gallows, for though he go as softly as foot can fall, he thinks himself too soon there.
ORLANDO: Who stays it still withal?
ROSALIND: With lawyers in the vacation, for they sleep between term and term, and then they perceive not how time moves.

—*As You Like It*, 3.2.303–39

And again, in *Romeo and Juliet*, when the couple has spent the night together after their wedding, and Romeo thinks he sees the light of dawn approaching, Juliet suggests that lovers easily lose track of time:

JULIET: Wilt thou be gone? It is not yet near day:
It was the nightingale, and not the lark,
That pierced the fearful hollow of thine ear;

Nightly she sings on yon pomegranate-tree:
Believe me, love, it was the nightingale.

ROMEO: It was the lark, the herald of the morn,
No nightingale: look, love, what envious streaks
Do lace the severing clouds in yonder east:
Night's candles are burnt out, and jocund day
Stands tiptoe on the misty mountain tops.
I must be gone and live, or stay and die.

JULIET: Yon light is not daylight, I know it, I:
It is some meteor that the sun exhales,
To be to thee this night a torch-bearer,
And light thee on thy way to Mantua:
Therefore stay yet; thou need'st not to be gone.

—*Romeo and Juliet*, 3.5.1–16

So, can we conclude that Shakespeare anticipated Csikszentmihalyi's concept of *flow*? Yes, and we can assume that anyone as productive as the Bard must have experienced flow firsthand.

Finally, let us consider one more process in which we might glimpse the unconscious at work . . . in the mind of volunteers in the psychology laboratory. We are referring to a phenomenon known as *priming*.

Imagine, if you will, that you have volunteered to be a participant in a psychological experiment. When you arrive at the lab, the researcher shows you a list of seemingly random words, such as these:

assassin, octopus, avocado, mystery, sheriff, climate

Then, sometime later, she asks you to fill in the blanks to complete the words on this list:

C h_ _ _ _ n k, O _ t _ _ u s, _ o g _ y _ _ _, _ l _ m _ t e

Chances are you will be quicker to fill in the letters for *octopus* and *climate* because she *primed* you with those words on the earlier list.

Now, let us consider a similar use of priming by our playwright in the scene from *The Merchant of Venice* where Bassanio must correctly choose one of three metal caskets (lidded containers) if he is to win Portia's hand in marriage—according to the demands of her father's will. One is crafted of gold, another of silver, and a third of lead. Portia calls for a song while Bassanio ponders his choice:

> Tell me where is fancy bred,
> Or in the heart, or in the head?
> How begot, how nourished?
> Reply, reply.
> It is engender'd in the eyes,
> With gazing fed; and fancy dies
> In the cradle where it lies.
> Let us all ring fancy's knell
> I'll begin it, Ding, dong, bell.
> —*The Merchant of Venice*, 3.2.65

Undoubtedly you noted that the song is about fancy things and whether they are perceived "in the heart, or in the head." The song says if they are "engender'd in the eyes," then "fancy dies / In the cradle where it lies." In other words, appearances can be deceiving. But did you notice that the first three lines rhyme with "lead"?[42] And so do the words "engender'd" and "fed" in the second stanza.

Did Shakespeare know about priming a response—with rhyming? We suggest that he did, indeed!

· 7 ·

MENTAL ILLNESS AND OTHER ILL HUMORS

Richard III

Bosworth, England, 1485[1]—The Middle Ages came to a bloody end in England when King Richard III's forces clashed with Henry Tudor in the English Midlands. There, at Bosworth Field, Richard uttered his final war cry, "A horse! a horse! my kingdom for a horse!"—at least, in Shakespeare's version of the story (5.4.9).

Richard had rolled the dice, hoping to make quick work of Henry by a charge deep into the opposing forces. But when his horse became stuck in the mud, Richard found himself on foot among mounted and armored foes, led by Henry Tudor, who quickly dispatched him.[2] And thus was Richard III the last English monarch to die in battle. His death also concluded the Wars of the Roses between the Houses of York and Lancaster and began the Tudor dynasty.[3] (See figure 7.1.)

The story begins when Richard's brother King Edward IV ascends the throne but soon falls ill and dies. Uncle Richard is appointed Lord Protector of Edward's son, who becomes Edward V when he is twelve years old. In Shakespeare's telling, Richard is no "protector" but rather an ambitious scoundrel who sees only two obstacles on his path to the throne of England: his older brother George (Duke of Clarence) and the young king Edward V. All he must do is dispose of them. And so he does, first by paying two thugs to drown the duke in a barrel of wine.[4] Next, he arranges for the young King Edward V to be brought to the Tower of London, supposedly for his own protection, where Edward and his younger brother are mysteriously murdered.[5]

As Shakespeare's version of the story continues, people begin suspecting the veracity of Richard's portrayal of himself, first as Lord Protector and later as the only legitimate heir to the throne. However, rumors circulate that the "Princes in the Tower" are bastard children and so not royal heirs. Just to make

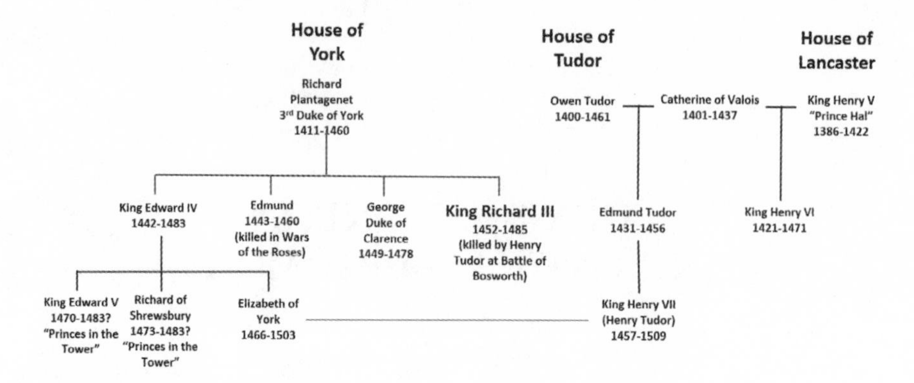

Figure 7.1 King Richard III and the Houses of York, Lancaster, and Tudor.

sure of his claim on the crown, Richard orders the serial beheadings of three relatives of Edward's widow, plus the executions of William of Hastings and the Duke of Buckingham. As additional insurance, he poisons his wife, which frees him to woo his niece, Elizabeth (Edward's daughter), drawing another line linking himself to the crown. Thus, Shakespeare's portrait is of a ruthless man not bothered by empathy, conscience, or compunction. His Richard is the very model of a major medieval miscreant, and when he is killed, the victor, Henry of Richmond, exclaims, "The day is ours; the bloody dog is dead" (5.5.2).

Richard's supporters argue that, in comparison with other nobles of the time, he was no more or less ruthless than his contemporaries.[6] Some critics, however, argue that the Bard was hard on Richard's reputation because he was following flawed sources. But it was probably more complicated by politics. Queen Elizabeth was a Tudor, so it made sense for the playwright to portray Richard, who was slain by Elizabeth's grandfather, as a villain and a scoundrel.[7]

Today, if we accept Shakespeare's version of history, we might label Richard an *antisocial personality* or a *psychopath*—terms that did not exist in Tudor times. Neither was anyone called "neurotic," "psychotic," "schizophrenic," or "paranoid," although people did know about mental illness and often used the terms *mad*, *lunatic*, or *insane*.[8] ("Crazy," however, meant "decrepit" or "weak.")[9]

In fact, London was the site of the famous St. Mary Bethlehem mental hospital—or as it was called in common speech, "Bedlam." There the city kept its most disturbed and disruptive citizens—many of whom we would now call "psychotic."[10] There, for a shilling or two, curious Londoners could gawk at the unfortunate inmates, as we see in figure 7.2. This is what they might have observed, in the words of Shakespeare's contemporary, Robert Burton:

> But see the *Madman* rage down right
> With furious looks, a ghastly sight.
> Naked in chains bound doth he lie,
> A roars amain he knows not why![11]
> Observe him: for as in a glass,
> Thine angry portraiture it was.
> His picture keeps still in thy presence;
> 'Twixt him and thee, there's no difference.[12]

We have no direct evidence that Will Shakespeare ever visited Bedlam, but given his curiosity, his fascination with eccentric behavior, and the fact that the hospital was quite near his troupe's Theater,[13] we assume it likely that he did. Surprisingly, Bethlem Hospital was an improvement over some earlier treatment facilities where inmates might be subjected to beatings and even torture.[14]

Figure 7.2 A scene in Bedlam: The hospital encouraged family visits and tours by the general public (cost: a few shillings in Shakespeare's time). Engraving by William Hogarth: one of a series entitled *Rake's Progress—Scene in Bedlam. Wikimedia Commons.*

THE CHANGING PERSPECTIVE ON "MADNESS"

The times of the Tudors marked a transition in the understanding of mental problems and their treatment from a religion-based view to a secular, medical perspective. In one respect this was a step back—to an understanding of mental disorders in terms of the old humor theory first developed by Hippocrates and his fellow Greek physicians. Yet it was more humane than the medieval view of mental affliction as a symptom of witchcraft and demon possession—when treatment might consist of torture intended to drive out the devilish spirits.[15]

The transition was not abrupt, however, for the conception of "madness" had started to change during the late Middle Ages, when the practice of exorcism began to fall out of favor and be replaced by treatments based on the humors. Some three hundred years before Shakespeare, one Bartholomeus Anglicus, a monk and professor of theology in Paris, named and described sundry mental disorders, along with their causes, in humoral terms. Here is his description of melancholia, written in 1240 (and translated into more modern spelling in an endnote):

> Melancholy is a humor, boystous and thicke, and is bredde of troubled drastes of blode. . . . Of this humor having maistry in the body these ben the sygnes and tokens. Fyrste the color of the skynne chaungeth into blacke or bloo. Soure savor, sharp—and erthy is felt in the mouth. By the qualite of the humor the patient is feynte—and fereful in hert without cause, and oft sorry. . . . Some dread enmyte of some man: Some love and desire dethe.[16]

By the late 1500s, rejection of the old Church teachings on mental illness was widespread among educated people. For example, in a book published during Shakespeare's early adulthood, Judge Reginald Scot described court cases of alleged witchcraft that he had tried, concluding that the women were delusional and insane. "Not witchcraft but melancholie. Voluntarie confessions untruly made," wrote Scot.[17] Likewise, across the Channel, the Dutch doctor Johann Weyer was claiming that confessions of those accused of witchcraft were signs of nothing but mental distress.[18]

Indeed, mental illness fascinated the people of England, in part because of the notorious Bedlam, but also because the theaters—the popular media of the time—also paraded mad and melancholy characters on their stages for the general public. Playwright Thomas Kyd, for example, wrote *The Spanish Tragedy*, with the subtitle *Heronimo Is Mad Again.* People swarmed to see it, where for a half-penny they could experience two hours of melancholia, madness, murder, revenge, and suicide. Shakespeare responded in kind to Kyd's challenge with his own play, *Titus Andronicus*, featuring even more bloody themes and throwing in a bit of unintentional cannibalism for good measure. With such stuff, the playhouses of London became the dissection theaters for mayhem and mental illness.[19]

English law recognized two distinct groups of persons with mental disturbances.[20] In one group were *idiots*, whom we might call "mentally disabled" or "cognitively impaired."[21] Typically, these were individuals whose mental limitations were evident at a young age, and responsibility for their care lay with the monarch. The other group had a rather fluid membership, collectively known as *lunatics*. Among the latter, medicine recognized sub-categories of those suffering from *melancholia* (known today as *depression*), *mania* (severe agitation), *delirium* (abnormal behavior attributed to fever), *phrenitis* (inflammation of, or damage to, the brain), *dementia*, persons with *amentia* ("psychotic"), and *epilepsy* (seizure disorders). Aside from the idea that the moon might be the cause of madness, these categories seem rather sensible to us (although we would today not think of epilepsy as a mental disorder but rather a physical disease of the brain). There was one more, however, that may strike us as amusing: *lycanthropy* was the condition of imagining oneself to be a wild animal, usually a wolf.

We infer that the Bard was particularly curious about mental disorders, from the complex and often unstable characters he wrote into his plays, his frequent use of mind-altering potions and herbs as plot devices, and the references he makes to "madness," "mania," "melancholy," delirium," "insanity," and "antic disposition" and to the London mental hospital, Bedlam. But what was his *theory*, if you will, of mental illnesses and their causes?

As a Renaissance man, he didn't seem to put much stock in religious doctrine (at least, he largely ignored it in his writing). He does, however, employ spirits (in *Hamlet*) and witches (in *Macbeth*)—as we have noted, mainly as a plot device and probably as a sop to King James, who was a strong believer in such things.[22]

But the Bard knew that circumstances could drive people mad, or at least prompt them to behave in odd or even lethal ways. Think of Othello, who was driven to murder his beloved wife by the innuendos served up by the embittered Iago. Or consider Shylock, an old man consumed by anger and revenge from a lifetime spent in a culture that disparaged him because of his Jewishness. Think, too, of Hamlet, whose rapid downward trajectory grew out of the discovery that his uncle had killed his father and married his mother. And then in *Twelfth Night*, there was Malvolio, an odd fellow whose socially awkward antics grew more bizarre under continual bullying.

As we saw in *The Tempest*, Shakespeare also knew anger and bizarre behavior might have a hereditary component as well. He further acknowledged the role of biology when he had Macbeth refer to his wife's condition as the product of "a mind diseased." But in Tudor times, the brain was largely uncharted territory, and the mechanisms of heredity were cloaked in mystery. People did know that the brain was the house of the mind, but as we saw earlier, they held many mistaken beliefs, such as the idea that memory resided in the ventricles of the brain.

Physicians still clung to the old humor theory, a framework that also helped them understand mental states in which the patient was agitated, depressed, aggressive, demented, delusional, or just in love. Excesses of melanin, or black bile, were responsible for sorrow and depression; hot-headedness resulted from the "hot" humor, yellow bile; sluggishness and mental slowness were attributed to too much phlegm, producing a "phlegmatic" personality; and especially passionate individuals owed their temperament to an abundance of blood. Therapy, then, often called for bloodletting. Because blood was considered the transporter of the other humors around the body, bleeding the patient allowed the excess of one humor or another to escape. Another approach involved emetics and laxatives, believed to be especially effective in eliminating yellow bile. This treatment was often delivered by having the patient ingest small amounts of poisonous herbs or plants, such as hellebore.

Perhaps the most positive legacy of the old theory was the idea that humors—and therefore temperaments—were influenced not only by physical conditions within the body but by diet, exercise, events that we would now call *stressors*, and even by climate. That meant that the humoral approach did not displace the ideas that heredity, physical disease, or lifestyle influenced mental problems.[23] Hence, a physician of the time might prescribe certain herbs and urge the patient to get out more into the fresh air. Wine might be administered to the phlegmatic person, or a bit more red meat would be advised to stimulate aggressiveness in those lacking in choler. In other words, the humor theory encouraged physicians to treat not just the symptoms of a disease but the whole person and his or her lifestyle.[24] It was a framework that encouraged doctors to view patients deemed mad more sympathetically, rather than blaming them for a moral weakness or being an agent of demonic powers.[25] Eventually, the most severely ill were treated in comparatively humane "insane asylums," where they could find rest and a haven sheltered from the world. That, however, was not yet the case in Elizabethan England, with its Bethlem mental hospital.

Closer to home, Shakespeare's son-in-law, Dr. John Hall, was one of those who advocated treating mental illnesses with herbs, as well as enemas and other purgatives. His casebook tells of treating in this fashion a seventeen-year-old girl suffering from melancholia, which tells us that physicians of the time considered at least some forms of madness to have a biological basis and thus to be treatable with standard remedies based on the humors.[26]

Let's hear of mental illness in Shakespeare's own words, from *King Lear*—spoken by Edgar, as he decides to disguise himself as Tom O'Bedlam, feigning a condition that we would now label as *psychotic*:

> I will preserve myself, and am bethought
> To take the basest and most poorest shape
> That ever penury in contempt of man
> Brought near to beast. My face I'll grime with filth,
> Blanket my loins, elf all my hairs in knots,
> And with presented nakedness outface
> The winds and persecutions of the sky.
> The country gives me proof and precedent
> Of Bedlam beggars who with roaring voices
> Strike in their numbed and mortified arms
> Pins, wooden pricks, nails, springs of rosemary,
> And with this horrible object from low farms,
> Poor pelting villages, sheep-cotes and mills
> Sometime with lunatic bans, sometime with prayers
> Enforce their charity. 'Poor Turlygod! Poor Tom!'
>
> —*King Lear*, 2.3.1

While Will Shakespeare was not the only playwright or poet to describe mental issues, he arguably did so better than any other—and prolifically, too. Records show his influence extending far beyond the theater, beyond his time, and well into the 1800s, when legal and medical writings quoted his descriptions of melancholy and other disorders.[27] Says scholar Benjamin Reiss, "From the mid-1840s through about the mid-1860s in the United States, during the first generation of American psychiatry, no figure was cited as an authority on Insanity and mental functioning more frequently than William Shakespeare."[28]

Eventually, however, Shakespeare's authority on mental illness waned. In the late nineteenth century, medicine gradually pushed the old humor theory aside, and psychiatry became enamored with the ideas proposed by Sigmund Freud and his psychoanalytic ilk.[29] For much of the twentieth century, Freudian theory held sway in both psychiatry and literary criticism, leading some to believe, for example, that the downfall of Hamlet, the Melancholy Dane, was driven by incestuous desires for his mother.[30] Today, however, most psychologists consider Freud's psychoanalytic theory of a tripartite mind—id, ego, and superego, all fueled by unconscious conflicts—to be prescientific. Nevertheless, we do still revere Sigmund Freud as a pioneering intellect and a perceptive observer of human behavior, perhaps much like Will Shakespeare.[31] But we will note again that nothing in Shakespeare shows that he anticipated anything like Freud's theory of the unconscious.

The last few decades have also seen an explosion of knowledge in the biology of mental disorders, thanks especially to advances in brain-scanning technology and genetics. Under the influence of neuroscience, we have mapped the basic structure of the brain and traced the influence of hormones and other chemicals on emotions, thoughts, and behavior. For us, the brain is no longer *terra incognita*; we know the basic function of all its parts.

This knowledge has also led to medicines that treat mental concerns, such as anxiety, depression, and schizophrenia. Moreover, the "cognitive revolution" in psychology has given us an armamentarium of effective therapies that treat some mental problems better than medicines—particularly fears and phobias, depression, addictions, and obsessive-compulsive disorder.[32]

Advances in genetics, too, have brought new insights into mental disorders—since some seventy years ago biologists began unlocking the secrets of heredity held in DNA. And now we know that autism, schizophrenia, bipolar disorder, major depression, attention-deficit disorder, obsessive-compulsive disorder (OCD), and others have a genetic component.[33]

Recently, neuroscientists opened yet another window on the interaction of heredity and environment. Using the tools of a new discipline called *epigenetics*, we can see much more clearly the ways in which the environment changes how our genes work and how anxiety, fear, and stressors can leave lingering effects on the brain and behavior. At last, we are seeing some of the mechanisms behind the nature-nurture (heredity-environment) interaction!

A thick volume, known as the *DSM-5*, encapsulates American medicine's understanding of mental illness today. Often called the "bible" of American psychiatry, it describes all the mental afflictions known in one grand classification scheme: schizophrenia, bipolar disorder, antisocial personality, phobias, depression—and many more.[34] For example, a clinician today would evaluate the likes of Hamlet, Othello, or Macbeth against the signs and symptoms of a psychotic break with reality, as detailed in the *DSM-5*. These include *hallucinations*, *delusions*, grossly *disorganized thinking*, *abnormal motor behaviors* not explained by physical disease, and *negative symptoms* of severely diminished emotional and social expression.[35] The condition that Shakespeare called "melancholia," also seen so prominently in Hamlet and Macbeth, may also suggest a psychotic break from reality when severe enough.[36] And what our playwright called "mania" we would probably call *schizophrenia*, reserving the term *mania* for the agitated, abnormally energetic, or irritable phase of *bipolar disorder* (formerly called *manic-depressive illness*).

One of the big changes in modern times is that mental health professionals have carved out for themselves a large chunk of human misery as their exclusive territory.[37] Yet, while psychology, psychiatry, and other mental health professions can tell us the signs and symptoms of mental disorders—and precious little about the causes—it is art and literature that can tell us how mental illness *feels*. Meanwhile, in this fog of our understanding of mental illness, we risk losing sight of Shakespeare's pioneering contributions to psychology and psychiatry. As we will see, with no more sophisticated tools than pen and paper, he probed the mysteries of at least fifteen distinct syndromes that we recognize today—some of which were not yet recognized by physicians of the time.

MENTAL DISORDERS ACCORDING TO SHAKESPEARE

We can read a number of the plays as "case studies" that describe conditions now cataloged in the *DSM-5*. As we have seen, the problems he described were given such labels as *madness*, *mania*, *melancholy*, *insanity*, *lunacy*, and *antic disposition* in Elizabethan and Tudor times.[38] Today we might replace those terms with diagnoses such as *psychosis*, *schizophrenia*, *depression*, *conversion disorder*, *posttraumatic stress disorder*, and even *malingering*. No matter the name, Shakespeare described them more vividly than any textbook. In figure 7.3, you will find a chart of the mental disorders Shakespeare portrayed in the plays, along with representative characters in each category.

To elaborate the point, let's return to Richard—an exemplar of what we now call antisocial personality disorder—and then page through other disorders the Bard described in what we might think of as his own *DMS*, or the *Diagnostic Manual of Shakespeare*.

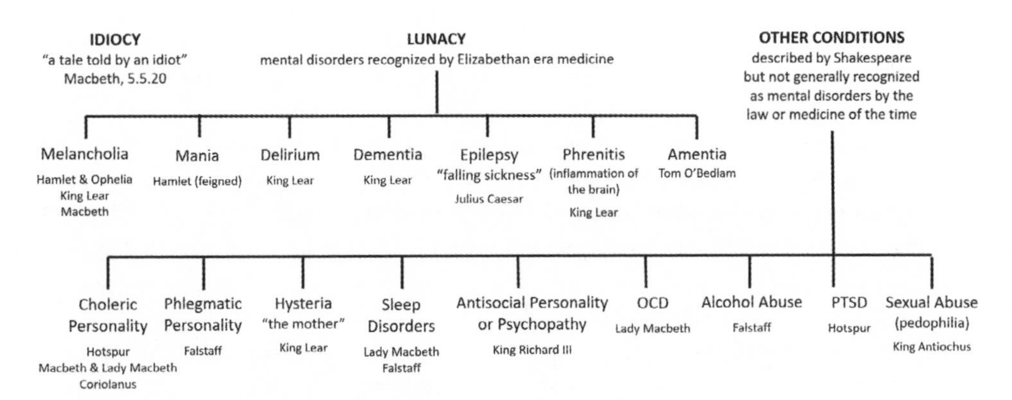

Figure 7.3 Mental disorders in Shakespeare's plays.

Richard III: *An Antisocial Personality*

Consider these five individuals: Richard III, Macbeth, Prince Hal, Achilles, and Julius Caesar. All killed people, probably multiple times. All, of course, were nobles and warriors, so killing was their trade. All were males—so, just to make it a more interesting problem, let's throw in Lady Macbeth along with the revenge-murderer Tamora from *Titus Andronicus*. Now, the questions are these: Which ones are *antisocial personalities* or *psychopaths*? Which, if any, are not?

We have no biopsy, blood test, or brain scan for antisocial personality nor for most other mental problems.[39] As we have seen, the professionals—psychiatrists and clinical psychologists—make diagnoses based mainly on the patient's history of problems and complaints compared against checklists of symptoms found in the *DSM-5*. We don't mean to make the process sound simple and arbitrary, for patients often have multiple issues: Ms. Macbeth, for example, was dealing with hallucinations, delusions, anxiety, obsessive-compulsive disorder, sleep issues, and depression.[40] Others, like Hamlet, were *malingering*—faking a disease for some ulterior motive. And a few cases may be rooted in both an organic brain syndrome and mental illness (like Lear, who apparently has both dementia and delusions).

In cases where antisocial personality is suspected, the clinical checklist includes these features:

- a history of antisocial behavior, accompanied by the absence of guilt or remorse;
- disregard for laws and customs, as well as for the rights, feelings, or welfare of others;
- irresponsibility, impulsivity, irritability, and aggression;
- manipulation of others by threats, intimidation, or brutality, or with lies and personal charm.

Further, we need detailed answers to such questions as, When did the problem behaviors start? How often do they occur? Is there a pattern?

Based on the information Shakespeare gives us, Richard certainly fits the criteria of an antisocial personality (although it is hard to see the characteristics Shakespeare attributed to him in the sixteenth-century painting of Richard in figure 7.4). Based on his behaviors, could we also say that Shakespeare's Richard was a psychopath? The two terms overlap. *Antisocial personality disorder* is a *DSM-5* category, diagnosed primarily by behavioral signs of aggression, manipulation, and blatant disregard for others. In contrast, *psychopathy* is *not* listed in the *DSM-5* but is rather a clinical term focused on personality traits that are associated with antisocial behaviors. For all practical purposes, the two are simply different ways of describing the same individuals.[41]

Figure 7.4 King Richard III of England—painting based on a lost contemporary original of Richard III of England by an unknown artist. *Wikimedia Commons.*

We find it useful to conceive of antisocial personality as a disorder of *empathic understanding*. Richard gives us memories of his own suffering and abuse but seemed to have stored no memories of other people's feelings. Shakespeare takes pains to reveal the *motives* behind Richard's mental state, but he leaves it up to the actor to interpret the *emotions* associated with Richard's memories. How does the playwright reveal Richard's private motives? You will remember that he was the master of the soliloquy, a theatrical device—an "aside"—in which a character seemingly talks to himself (and to the audience) about his private thoughts.

And so we in the audience must infer that his Richard III is an angry man—angry that he is outcast by his peers and shunned by the ladies. Even though he is a wealthy noble, the Duke of Gloucester, he feels he deserves better. We also find him devoid of empathy. In the opening lines, we hear him sniveling about his brother, King Edward IV ("this sun of York"), who has ascended the throne and, to England's great relief, halted the long wars. Clearly, Richard is not among those rejoicing. Why does he get no respect? (Shakespeare calls him "a bunch-back'd toad," 1.3.255). And in the opening lines of the play, he blames his lot on his misshapenness in this famous soliloquy:[42]

> Now is the winter of our discontent
> Made glorious summer by this sun of York.
> . . . . .
> He capers nimbly in a lady's chamber
> To the lascivious pleasing of a lute.
> But I, that am not shaped for sportive tricks,
> Nor made to court an amorous looking-glass;
> I, that am rudely stamp'd, and want love's majesty
> To strut before a wanton ambling nymph;
> I, that am curtail'd of this fair proportion,
> Cheated of feature by dissembling nature,
> Deformed, unfinish'd, sent before my time
> Into this breathing world, scarce half made up,
> And that so lamely and unfashionable
> That dogs bark at me as I halt by them;
> Why, I, in this weak piping time of peace,
> Have no delight to pass away the time,
> Unless to spy my shadow in the sun
> And descant on mine own deformity:
> And therefore, since I cannot prove a lover,
> To entertain these fair well-spoken days,
> I am determined to prove a villain
> And hate the idle pleasures of these days.
> Plots have I laid, inductions dangerous,
> By drunken prophecies, libels and dreams,
> To set my brother Clarence and the king
> In deadly hate the one against the other:
> And if King Edward be as true and just
> As I am subtle, false and treacherous,
> This day should Clarence closely be mew'd up.
>
> —*Richard III*, act I, scene 1

Likewise, Shakespeare's *Titus Andronicus* is a kindred play in many respects, with multiple antisocial personalities and revenge piled on revenge—a work suitable neither for children nor delicate adults.[43]

Titus has just returned with several prisoners captured in a war with the Goths. These include their queen, Tamora, and her three sons—one of which Titus allows to be a human sacrifice to the gods. Tamora vows revenge, which comes to fruition when the Roman emperor falls in love with her and frees the Goth queen and her two remaining sons. The plot is convoluted and melodramatic. It features "14 killings, 9 of them on stage, 6 severed members, 1 rape (or 2 or 3, depending on how you count), 1 live burial, 1 case of insanity, and 1 of cannibalism—an average of 5.2 atrocities per act, or one for every 97 lines."[44]

While the diagnosis of antisocial personality is clear for Richard and for nearly all the characters in *Titus*, does it apply to the likes of Prince Hal, Achilles, and Julius Caesar? Their bloody deeds on the battlefield fall within the limits accepted by their time and culture—so they don't meet the criteria for antisocial personalities. Then, with these thoughts in mind, we leave it to your judgment whether Iago in *Othello* and Iachimo in *Cymbeline* fit the pattern, along with Titus and Richard III.

As for Macbeth and his wife, they are more problematic because, while they murder a king in cold blood, they seem to feel guilt and remorse, emotions that are not evident in either Richard or Titus. Moreover, their case calls for multiple diagnoses, which we shall discover as we turn now to "the Scottish play."[45]

Macbeth: *Hallucinations, Obsessions, Sleep Disorders, and More . . .*

There was a real, historical Macbeth, a Scottish warlord whose full name, in Gaelic, was Mac Bethad mac Findlaích. He was born around 1005 CE into the ranks of the Scottish nobility, and his overlord was, in fact, a King Duncan. Sources tell us that Mac Bethad's forces killed Duncan when the king turned on him and invaded his territory, at which point Mac Bethad became king.[46]

Shakespeare, of course, altered and embellished the story almost beyond recognition. In his retelling, Macbeth and Banquo, another Scottish warlord, have just defeated two invading armies, one from Ireland and the other from Norway. In gratitude, King Duncan awards Macbeth new lands, along with the title Thane of Cawdor, whereupon Macbeth, Duncan, and Banquo plan to celebrate their achievement together at a feast in Inverness.[47] But in the thrall of prophecies by three witches, Macbeth begins to dream of greater things—perhaps becoming king. When he confides these fantasies to his wife, she urges him to kill King Duncan and seize the throne.

That is the point at which the Macbeths' lives begin to fall apart and signs of mental disorder emerge. The result is a yarn that highlights more symptoms of mental illness than any other in the Shakespearean canon. As we watch, the couple develops hallucinations, delusions, obsessions, compulsions, anxiety, sleep disturbances, mania, and finally a deep existential melancholy.[48] In Lady Macbeth's case, the denouement is suicide, and for Macbeth himself, it is at the point of Macduff's sword.[49]

Macbeth confesses his anxiety about the plot when he says to his Lady, "If it were done when 'tis done, then 'twere well it were done quickly" (1.7.1). The scene that follows serves up, perhaps, Shakespeare's best example of a hallucination, as Macbeth steels himself to kill the sleeping king Duncan:[50]

> MACBETH: Is this a dagger which I see before me,
> The handle toward my hand? Come, let me clutch thee.
> I have thee not, and yet I see thee still.
> Art thou not, fatal vision, sensible
> To feeling as to sight? or art thou but
> A dagger of the mind, a false creation,
> Proceeding from the heat-oppressed brain?
> I see thee yet, in form as palpable
> As this which now I draw.
> Thou marshall'st me the way that I was going;
> And such an instrument I was to use.
> Mine eyes are made the fools o' the other senses,
> Or else worth all the rest; I see thee still,
> And on thy blade and dudgeon gouts of blood,
> Which was not so before. There's no such thing:
> It is the bloody business which informs
> Thus to mine eyes.
>
> —*Macbeth*, 2.1.42

The deed is indeed done quickly, and Macbeth will be king. Yet he begins to fear that Banquo or his sons will seize the crown, according to the witches' prophecy. Macbeth's solution: hire thugs to ambush and murder Banquo on his way to the feast at Inverness. There, at the banquet table, Macbeth—but no one else—sees Banquo sitting on an empty stool. He interprets this hallucination as a threat to his destiny as king.[51]

What of Lady Macbeth[52] and her futile attempts to wash away the hallucinated bloodstains of the murdered King Duncan? The famous hand-washing scene opens with a doctor who has been summoned to check on her worrisome sleepwalking habit.

DOCTOR: You see her eyes are open.
GENTLEWOMAN: Ay, but their sense is shut.
DOCTOR: What is it she does now? Look, how she rubs her hands.
GENTLEWOMAN: It is an accustomed action with her, to seem thus washing her hands: I have known her continue in this a quarter of an hour.
LADY MACBETH: Yet here's a spot.
DOCTOR: Hark! She speaks. I will set down what comes from her, to satisfy my remembrance the more strongly.

—*Macbeth*, 5.1.26–36

We would call her handwashing a symptom of *obsessive-compulsive disorder* (OCD), a condition that has two main components, as you may recall. *Obsessions* are persistent and intrusive thoughts, such as the Lady's preoccupation with the blood, which she seemingly cannot get off her hands. In contrast, *compulsions* are behaviors, such as ritual hand-washing, which is common with this disorder.

What makes Lady Macbeth an unusual and especially interesting case is that her OCD symptoms appear during a sleepwalking episode (see figure 7.5). Our current understanding of sleepwalking (or *somnambulism*) is that it is normal if done rarely but pathological if repeated frequently and in response to severe anxiety and stress.[53] Shakespeare seemed to know this, for the play has her somnambulating nightly. Indeed, this is why the gentlewoman has summoned a doctor. As we watch the famous scene unfold, Lady Macbeth walks in her sleep, hallucinating blood on her hands and then washing compulsively. From a literary point of view, the "Out, damned spot!" (5.1.37) monologue is just another form of soliloquy that helps the playwright reveal Lady Macbeth's innermost thoughts and emotions. For our purposes, it also shows Shakespeare's detailed understanding of obsessions, compulsions, and hallucinations.[54]

In *Othello* we again see an example of an obsessional disorder but without the compulsive ritual.[55] You may recall that the story revolves around a manipulative plot by Iago, a disgruntled army officer passed over for promotion, to convince General Othello that his wife has been unfaithful. In a fit of jealous rage, Othello then smothers his wife with a pillow. The *DSM-5* has a diagnosis for Othello's condition, which it labels as obsessional jealousy, and describes it as a pathological "preoccupation with a partner's perceived infidelity."[56] It is the undoing of Othello, his career—and his wife.

Kahneman's Systems 1 and 2 are useful in understanding hallucinations, as well as dreams. Both are products of the unconscious System 1, which continually produces a stream of images, patterns, fears, hopes, feelings, and experiences. And both occur under conditions where the rational System 2 is

Figure 7.5 *Lady Macbeth Sleepwalking*—painting by Eugène Delacroix (circa 1850). *Wikimedia Commons.*

quiescent. As for obsessional disorders, we might frame the disorder in terms of a "feedback loop" between System 1 and System 2. Think of the unbearably loud screeching sound made by feedback between a microphone and a loudspeaker as an analogy for Othello's obsessive thought of Desdemona's unfaithfulness that builds into a compulsive and lethal rage.

Hamlet: *Melancholia, Malingering, and Revenge*

The prince, just home from college, is acting strangely. The king's advisor, Polonius, attributes Hamlet's seeming madness to the melancholy of lovesickness—in those days considered a true disease that afflicts young lovers. Nevertheless, Polonius is puzzled. Something about Hamlet's "antic disposition" doesn't ring true, and Polonius says to himself, "Though this be madness, yet there is method in't."[57] The actor in figure 7.6 shows us both the "mad" and the "method" aspects of Hamlet's personality.

Figure 7.6 Hamlet, the Melancholy Dane—Artist/photographer: Sebastian Bieniek; model: Lars Eidinger as Hamlet. *Wikimedia Commons*.

While Polonius puzzles over whether Hamlet is actually "mad" or merely pretending to be mad, we in the audience know that Hamlet is plotting a way to reveal his father's murderer. The unfolding story of angst and revenge becomes perhaps Shakespeare's most renowned case study of both real and feigned mental disturbance.

Prince Hamlet's initial reaction to his father's apparent murder and his mother's likely complicity fits perfectly the *DSM-5* description of a major depressive disorder—*melancholia* in Elizabethan-speak—with his thoughts of anger, angst, diminished pleasure in life, and suicide. Yet, in a famous scene, he tells his friend Horatio that he plans to fake a mental illness:

> There are more things in heaven and earth, Horatio,
> Than are dreamt of in your philosophy. But come,
> Here, as before, never, so help you mercy,
> How strange or odd soe'er I bear myself
> (As I perchance hereafter shall think meet
> To put an antic disposition on), . . . to note
> That you know aught of me. This not to do,
> So grace and mercy at your most need help you,
> Swear.
>
> —*Hamlet*, 1.5.186

And so he makes Horatio and Marcellus swear that he will not raise questions when Hamlet pretends to have "an antic disposition," no matter "how strange or odd" his behavior. Nevertheless, Hamlet seems not to realize that he is in the throes of one mental illness, even as he is faking another. As we read the play, then, Shakespeare's meaning was that Hamlet is *both* melancholy *and* methodological—both depressed and malingering.[58] Indeed, he is "the melancholy Dane." Yet we see in him one more problem: a morbid obsession with revenge, for which critics have labeled *Hamlet* one of Shakespeare's "revenge tragedies," along with *Titus Andronicus*.[59]

Perhaps Hamlet's character resonates so clearly because Shakespeare knew melancholia himself as a result of his son Hamnet's death just a few years before he wrote the play, or possibly because he was also prone to depression—the melancholy man from the Midlands.[60] How else could we explain why we so often find depression in his works?[61]

In *The Merchant of Venice*, we again see melancholy in a brief description of Antonio, who is weary with worry over his investments:

> In sooth, I know not why I am so sad.
> It wearies me, you say it wearies you,
> But how I caught it, found it, or came by it,
> What stuff 'tis made of, whereof it is born,
> I am to learn.
>
> —*The Merchant of Venice*, 1.1.1

His companion Salarino tries to comfort him, saying that some people are simply more prone to melancholy than others:

> Nature has framed such strange fellows in her time:
> Some that will evermore peep through their eyes
> And laugh like parrots at a bagpiper,
> And other of such vinegar aspect
> That they'll not show their teeth in the way of smile
> Though Nestor swear their jest be laughable.
>
> —*The Merchant of Venice*, 1.1.49

The idea that melancholy may simply be part of a person's makeup comes from the humoral theory, which attributes it to melancholer, or black bile. In other plays, however, Will Shakespeare demonstrates his understanding that melancholy can come from circumstances, as well as heredity—for example, in the tales of Lear and Macbeth. That is how we know that he knew melancholy had components of both nature and nurture.

On another hand, we caution modern readers of the Bard's works that his meaning of melancholy does not map perfectly onto our modern concept of depression. Rather it was a broader notion that applied to most men, in whom a touch of melancholy was considered normal—even desirable—since men were considered more serious and capable of thinking more deeply. Nor was it confined to the sad and cynical, but also embraced lovers, who were thought to behave in a distracted and disturbed way because of a surge in melancholer. We see this in *A Midsummer Night's Dream* when Theseus exclaims:

> Lovers and madmen have such seething brains,
> Such shaping fantasies, that apprehend
> More than cool reason ever comprehends.
> The lunatic, the lover and the poet
> Are of imagination all compact.
>
> —*A Midsummer Night's Dream*, 5.1.2

Two final examples of melancholy better fit our modern meaning. One comes from the murderous Macbeth, who also seems to fall into a deep depression near the end of the play when, in a soliloquy, he describes the emptiness and meaninglessness of life:

> Tomorrow, and tomorrow, and tomorrow,
> Creeps in this petty pace from day to day,
> To the last syllable of recorded time;
> And all our yesterdays have lighted fools
> The way to dusty death. Out, out, brief candle!
> Life's but a walking shadow, a poor player,
> That struts and frets his hour upon the stage,
> And then is heard no more. It is a tale
> Told by an idiot, full of sound and fury,
> Signifying nothing.
>
> —*Macbeth*, 5.5.20

The other example comes from his wife, who has also descended into depression so deeply that Lady Macbeth, like Ophelia, takes her own life.

King Lear: *Depression, Dementia, Delusions, Phrenitis, and Hysteria*

In *Lear*, Shakespeare again serves up a course of melancholia, mixed with anger, delusions, and revenge. The plot centers on a king in his dotage, who decides to divide his realm among his three daughters and step down from the throne.[62]

Foolishly, he decides to partition his holdings according to how much each daughter loves him. Only two daughters play the game by expressing their love in hyperbolic and sycophantic terms. The third, young Cordelia, Lear's favorite, simply says she loves him as a daughter should, whereupon the suddenly enraged king disinherits and banishes her. However, when the remaining two assume Lear's lands, fortune, and power, they turn against him and expel him and his retinue. This is the final blow that drives Lear into madness. The scene shifts anon to a violent storm raging on the heath, where we witness his disorganized and delusional thinking—characteristic of schizophrenia—as he raves incoherently in a fashion clinicians call "word salad":

> Nature's above art in that respect. There's your press-money.
> That fellow handles his bow like a crow-keeper:
> Draw me a clothier's yard. Look, look, a mouse!
> Peace, peace; this piece of toasted cheese will do't. There's my gauntlet;
> I'll prove it on a giant. Bring up the brown bills.
> O, well flown, bird! I' th' clout, i' th' clout! Hewgh!
> Give the word.
>
> —*King Lear*, 4.6.105

His delusions emerge full-blown, as he stands with his Fool ranting in the wind and rain, as shown in figure 7.7. He has delusional symptoms, plus signs of *organic brain syndrome*, or what the Elizabethans called *phrenitis*.[63]

> GLOUCESTER: Is 't not the king?
> LEAR: Ay, every inch a king!
> When I do stare, see how the subject quakes.
> I pardon that man's life. What was thy cause?
> Adultery?
> Thou shalt not die. Die for adultery? No!
> The wren goes to 't and the small gilded fly
> Does lecher in my sight.
> Let copulation thrive: for Gloucester's bastard son
> Was kinder to his father than my daughters
> Got 'tween the lawful sheets.
> To 't, luxury, pell-mell! For I lack soldiers.
>
> —*King Lear*, 4.6.126–35

One daughter describes him as having "the unruly waywardness that infirm and choleric years bring with them" (1.1.341).[64] Now we call it *senile dementia*[65]

Figure 7.7 *King Lear and His Fool*—painting by George Frederick Bensell. *Wikimedia Commons.*

or, in the language of the *DSM-5*, it suggests a *neurocognitive disorder*, the most common of which is *Alzheimer's disease.* The diagnostic features of this condition include cognitive decline, memory deficits, and difficulty with everyday activities and relationships where they had appeared normal before. Sometimes the individual is aware of the problem and sometimes not. Lear describes the terrifying insight of knowing he is mentally unstable, as he swoops into and out of madness.[66]

LEAR: Pray do not mock me;
I am a very foolish fond[67] old man,
Fourscore and upward, not an hour more nor less;
And, to deal plainly,
I fear I am not in my perfect mind.

—*King Lear*, 4.7.68

And only a few lines later, we find this exchange:

LEAR: Am I in France?
KENT: In your own kingdom, sir.

—*King Lear*, 4.7.88–89

The king is alternating between lucidity and madness, mixed with melancholia, delirium, and dementia. Yet there is another issue that also marks his undoing: *hysteria*—a condition that, ironically, marked the making of Sigmund Freud.[68] Hysteria was conceived as a disease involving physical symptoms, such as weakness, paralysis, or loss of sensation, without an obvious physical cause. Its name derives from the Greek word for womb (think: *hysterectomy*) and dates back some 2,400 years to Hippocrates, who taught that the condition's cause was a desiccated uterus wandering around the body in search of moisture. Naturally, the diagnosis became a dumping ground for women's complaints that were thought to be "all in the head." In the play, King Lear terms it "*hysterica passio*" and also "this mother," a puzzling reference to a "wandering womb" that seems to be ascending in *his* body.[69]

LEAR: O, how this mother swells up toward my heart!
Hysterica passio, down, thou climbing sorrow.
Thy element's below.

—*King Lear*, 2.4.62

We now call it *conversion disorder*—but still, no one understands its etiology.[70]

The obvious problem with Hippocrates's theory is that Lear, being a man, lacks a uterus! His symptoms are of choking or smothering, thought by physicians of the time to be a manifestation of melancholia to be found only in women. As we might expect, this seeming misstep by the Bard has caused considerable discussion ever since the play populated the stage in 1606. Certainly, Shakespeare was right in suggesting that a person in great distress may manifest physical symptoms, such as shortness of breath and choking, but now we are likely to think of them in terms of an anxiety or panic attack. But what of his reference to a wandering womb in a man? Our take is that Shakespeare is giving us a Lear who feels old and unmanly (read: "feminine"), and who is metaphorically attributing his powerlessness to "the mother."

How could we frame both Lear's and Hamlet's melancholy/depression? In terms of Kahneman's Systems 1 and 2, we can conceive of both Hamlet's melancholy and Lear's depression as much like an obsession with feelings of worry, dread, and worthlessness. Clinicians sometimes call it a process of *rumination.* And so, like an obsession, we can think of depression as a pathological feedback loop between the conscious and unconscious, but without the compulsive ritual to relieve the symptoms, even temporarily. And as for Lear's delusions, we can conceive of them, in terms of Systems 1 and 2, as much like hallucinations or dreams—fantasies served up by the unconscious System 1 to a confused conscious mind that has difficulty distinguishing fantasy from reality.

Henry IV, Part 1: *Hotspur's Posttraumatic Stress Disorder*

Henry "Hotspur" Percy was undone in a more masculine manner by his braggadocio and, you will remember, by Prince Hal. We should note that the nickname "Hotspur" implies aggression and a quick temper, which was commonly admired in nobles who were eager to fight and sought glory in combat. Shakespeare and his audiences would have thought of Percy as having a *choleric personality.*[71] Today we might think of individuals of Percy's ilk as having a form of *conduct disorder.*

In addition, Hotspur also showed apparent signs of what we now call *posttraumatic stress disorder* or *PTSD.* In World War I the same syndrome was called "shell shock," and in World War II it was "battle fatigue." But PTSD was far from new even in Shakespeare's time, when he gave us a description of it in Hotspur's brief return from mortal combat in the Wars of the Roses. As the young Lord Percy prepares to return to the wars the next morning, his wife anguishes over his worrisome mental state:[72]

> LADY PERCY: O my good lord, why are you thus alone?
> For what offense have I this fortnight been
> A banished woman from my Harry's bed?
> Tell me, sweet lord, what is't that takes from thee
> Thy stomach, pleasure, and thy golden sleep?
> Why dost thou bend thine eyes upon the earth,
> And start so often when thou sit'st alone?
> Why hast thou lost the fresh blood in thy cheeks
> And given my treasures and my rights of thee
> To thick-eyed musing and curst melancholy?
>
> —*Henry IV, Part 1,* 2.3.39

Shakespeare shows us PTSD again when the Macbeths experience "flashbacks" of Banquo's ghost and hallucinated bloodspots on the hands. Such terrifying memories, nightmares, and even hallucinations are common symptoms of this illness.[73] We now know it in terms of stress. Yet, while the word *stress* existed in Shakespeare's time, he never used it in his writings, preferring the terms *distress* and *strain*. People knew, of course, that the distress and strain in life could take a psychological toll, even though they did not couch it in those terms. And for our purposes, the point is that Shakespeare knew the concept of stress and its effects on mental health.[74]

Pericles: *A Secret Disorder*

Pericles: Prince of Tyre is an obscure, seldom-performed play with yet another convoluted and melodramatic plot. Briefly, Pericles goes to Antioch, seeking to marry the king's beautiful daughter. To win her hand, he must answer a riddle. Any suitor who fails will immediately be put to death:

> PERICLES [reading the riddle]: I am no viper, yet I feed
> On mother's flesh which did me breed.
> I sought a husband, in which labour
> I found that kindness in a father:
> He's father, son, and husband mild;
> I mother, wife, and yet his child.
> How they may be, and yet in two,
> As you will live, resolve it you.
>
> —*Pericles*, 1.1.63

Pericles, being a quick study, realizes both the correct answer and a deadly dilemma. The correct answer is that the king and his daughter have an incestuous relationship—yet Pericles instantly knows that revealing the royal relationship also means that he would be killed. Wisely, he flees Antioch, but the journey home is fraught. As we watch the play, we see that Pericles must solve many problems, first at Antioch and later on his journey home. For our psychological purposes, however, it is the king and his daughter who have problems. We might call those problems *sexual abuse* or *pedophilia*—depending on the daughter's age.

No wonder our teachers assigned *Julius Caesar*, rather than *Pericles*, in high school literature classes.

OTHER DISORDERS RECOGNIZED IN SHAKESPEARE'S TIME

Our list grows long, so we will mention only briefly a few other mental health issues that the Bard touched upon. First, we have Falstaff and friends, including Prince Hal, the poster-boys for *alcohol use disorder*, who drink their ale and sack at the Boar's Head Inn and get into a plenitude of trouble.[75] Then, in *The Taming of the Shrew*, we have Petruchio who attempts to "tame" his new and unruly wife, Kate, whom he marries against her will.[76] To do this, Petruchio pretends to be angry, with the intent of showing Kate what it is like to be on the receiving end of defiance, irritability, and anger—for example, yelling at servants to bring mutton for their dinner.[77] Nowadays, a mental health professional might diagnose Kate—and, perhaps, Petruchio—as having *oppositional defiant disorder*. The same diagnosis might apply to Caliban (in *The Tempest*) and the bitter warrior Coriolanus (in the play of his name).

Finally, we should note Shakespeare's mention of Julius Caesar's "falling sickness," now known as *epilepsy*.[78] Plutarch seems to be the original source from which Shakespeare borrowed the story of Caesar's malady and adapted it to the following interchange:

> CASSIUS: But, soft, I pray you; what, did Caesar swoon?
> CASCA: He fell down in the market-place, and foamed at mouth,
> and was speechless.
> BRUTUS: 'Tis very like; he hath the falling sickness.
> CASSIUS: No, Caesar hath it not; but you and I,
> And honest Casca, we have the falling sickness.
> CASCA: I know not what you mean by that, but I am sure Caesar
> fell down. If the tag-rag people did not clap him and hiss him,
> according as he pleased and displeased them, as they use to do
> the players in the theatre, I am no true man.
>
> —*Julius Caesar*, 1.2.262–72

In early-modern England, epilepsy was considered a form of mental disease caused by blockage of the humors in the veins of the head. Today we know that the cause is abnormal electrical activity in the brain. That is, epilepsy is now considered a purely neurological disease that produces seizures and, often, a loss of consciousness—but not otherwise a mental illness.

THERAPY FOR MENTAL ILLNESS IN EARLY-MODERN ENGLAND

In the England of Queen Elizabeth and King James, educated people generally considered the mind and body inseparable, and so they believed that mental illness should naturally be treated as a physical illness. Problems of the mind, physicians thought, involved the humors arising from certain organs. The brain was the organ of thought, while emotions arose from the heart, kidneys, liver, and spleen. By comparison with the medieval notion that one's thoughts and feelings sprung from an intangible soul—sometimes stained by sin, demons, and curses, or chastened by repentance—it was a rather enlightened view, still prescientific, but prescient of what we believe today.[79]

You will recall how, in *Macbeth*, Shakespeare echoed the belief that medicine of the day could treat diseased minds—but not always successfully—as Macbeth pleads with the doctor to cure his wife:

> MACBETH: Cure her of that.
> Canst thou not minister to a mind diseased,
> Pluck from the memory a rooted sorrow,
> Raze out the written troubles of the brain
> And with some sweet oblivious antidote
> Cleanse the stuff'd bosom of that perilous stuff
> Which weighs upon the heart?
> DOCTOR: Therein the patient
> Must minister to himself.
>
> —*Macbeth*, 5.3.49–57

Most victims of madness in those days were cared for in the homes of family members. Others lived the best they could on the streets. Only a few were sent to mental hospitals, such as Bethlem—again, for much the same reason they are today: because families and communities did not know what else to do with them. Only a few who were considered mentally deranged were also deemed dangerous.

You will remember that physicians, including Dr. John Hall, the Bard's son-in-law, believed that mental diseases were treatable with conventional medical remedies of the time, including herbs, laxatives, purgatives, and bloodletting, which they believed could rebalance the humors causing the mental disturbance. Still, Lady Macbeth's doctor admits that he is flummoxed but hopes that she will recover spontaneously:

> DOCTOR: This disease is beyond my practice. Yet I have known those which have walked in their sleep who have died holily in their beds.
>
> —*Macbeth*, 5.1.62

In those days, however, many in the audience still believed mental illness could be caused by witchcraft and by demons that needed to be driven from the person's body. Also an expert in marketing, the Bard was not above pandering to those beliefs. We hear this in *As You Like It*, where Rosalind mocks Orlando, saying:

> ROSALIND: Love is merely a madness and, I tell you, deserves as well a dark house and a whip as madmen do, and the reason why they are not so punished and cured is that the lunacy is so ordinary that the whippers are in love, too. Yet I profess curing it by counsel.
>
> —*As You Like It*, 3.2.407

As we have seen, Shakespeare and his fellow playwrights understood that both mental and physical diseases could be caused by stress, and so they frequently used that notion to contrive plot devices in both comedies and tragedies.[80] Moreover, educated people also realized that treatment for mental illness, particularly melancholia, should involve rest, counseling, and the advice of friends. And if possible, the afflicted also should seek distraction in recreation or business.[81]

To realize just how enlightened was Shakespeare's Rosalind in the treatment of the melancholy of lovers, note the last line in her speech above, where she says, "Yet I profess curing it by counsel." That is, she prefers treating madness with talk, rather than with darkness and whips. In our view, Rosalind is invoking what we might call an Elizabethan version of *cognitive-behavioral therapy*, for she (disguised as Ganymede) proceeds to make an appointment with Orlando for a role-playing session, ostensibly to cure him of his lovesickness.[82]

THE LADY MACBETH EFFECT

Will Shakespeare taught us that madness comes in many interesting and tragic forms, and he told us that the cause may be hidden in nature or nurture—or both. With the characters of Hamlet, Ophelia, Lear, Othello, and the Macbeths, he taught sympathy for the victims of mental illness to some eleven generations of audiences. (Perhaps, for Richard III, not so much.) And we might expect such lessons from a great writer who was so fascinated by mind and behavior. But who would have guessed that the Bard might be able to amaze us, four hundred years hence, with his foresight on an obscure phenomenon now known as *the Lady Macbeth Effect*?

You see, *Macbeth* may hold an answer to a puzzle that neuroscientists and therapists have long argued: whether obsessive-compulsive rituals, such as compulsive handwashing, arise from a moral effort to rid oneself of guilt. Could there even be some sort of "morality circuit" in the brain that becomes deregulated in OCD victims? And could Lady Macbeth, too, have had some sort of method in her madness?

Several studies now affirm that obsessive-compulsive rituals help OCD patients deal with feelings of guilt. But how could scientists possibly test for something so ephemeral as that? Let us see—in an experiment engineered by psychologists from Tel Aviv University.

Orna Ruven and her colleagues guessed that guilt might make people feel "unclean" and so could be absolved by ritual handwashing, especially in people with OCD.[83] To test this, they recruited two groups of twenty-nine people. Each of them agreed to be a subject in the experiment for payment of twenty-five dollars. One group consisted of people diagnosed with OCD, while the other group had no history of the disorder—but, crucially, the two groups were matched on age, gender, and educational level.

The experimenters interviewed and assessed these volunteers, one at a time, for symptoms of OCD and other problems. Next, each was asked to use a computer to write about some unethical act they had committed, along with the emotions they associated with the act. The results were detailed stories of lying, cheating, stealing, and unfaithfulness—along with expressions of regret and remorse—from both groups of subjects. Then the researchers asked half of the individuals in each group to cleanse their hands with a wipe. (They were given the rationale that the lab requested all users of public computers to do so.)

At that point, all were told that the experiment was over—along with another fib: that a graduate student in the lab desperately needed volunteers for a study that was essential for his dissertation. Would the volunteers agree to be a subject in another study—this time for free? This was the crucial element of the study, designed to find out whether the hand-cleansing would help them compensate for the feelings of guilt induced by just thinking and writing about committing an unethical act.

Previous work had established that writing about an immoral act *increases* one's felt obligation to help another person in such situations.[84] Moreover, the opportunity to cleanse the hands after writing about one's failings *decreases* the tendency to help. What remained to be determined by this experiment was whether there would be a difference between the OCD and non-OCD groups.

There was a difference, and it was huge! As expected, those in both groups who were *not* instructed to cleanse their hands were *more* likely to help. But in the OCD group, this Lady Macbeth Effect was far more pronounced. That is, *all but one* in the OCD group who did not clean their hands were willing to help, while *all but one declined* to do so among those who had wiped their hands. The researchers concluded that "washing rituals in OCD are an effective, if only short-lived, way of reducing uncomfortable negative emotions associated with guilt or immorality." It appears that OCD is, as Shakespeare showed us, both a mental disorder and an act of moral cleansing.[85]

Part IV

REASON VS. EMOTION

These two final chapters spotlight the last of the four Big Issues that link Shakespeare with modern psychology—in **bold** below:

- heredity vs. environment
- the person vs. the situation
- the nature of the human mind and its states of consciousness
- **reason vs. emotion**

Here we will first consider *love*—Shakespeare's favorite emotion and one that poets have long known make people either joyous or melancholy and perhaps a little "crazy" (in the colloquial sense of the term). Exhibit A is *Love's Labour's Lost*, in which love makes the infatuated men crazy and the women canny.

Then we turn to reason—for which Hamlet is the ironic poster boy who thinks he is making rational decisions while feigning an "antic disposition." Shakespeare certainly wasn't the first to see this tension between reason and emotion, but no one has explicated it more vividly than did the Bard in his play about murder and revenge in the rotten state of medieval Denmark.

All of this raises these two questions: Are we humans mostly rational beings who sometimes get swept away by emotional events? Or are we, in fact, emotional creatures who fancy ourselves as reasonable, but who rarely resort to reason?

· 8 ·

EMOTION, MOTIVATION, AND ELIZABETHAN LOVE

Love's Labour's Lost

Pamplona, Navarre (circa 1492)[1]—King Ferdinand and his three noble pals, Lords Berowne, Dumaine, and Loganville, vow to withdraw from the world for three years of scholarly study, hoping to bring acclaim to themselves and to Navarre. Quoth King Ferdinand:

> Navarre shall be the wonder of the world;
> Our court shall be a little academe,
> Still and contemplative in living art.
> —*Love's Labour's Lost*, 1.1.1

But no sooner have the men declared their intentions than what should appear but a contingent of single, attractive, and noble ladies led by the comely princess of France. Chivalry demands that the women be treated respectfully, even as the lords' vows forbid allowing the ladies inside the castle walls. The king, however, allows a parley with the women at their camp—a mile distant—at which he falls instantly in love with the princess. Coincidentally, each of the other lords is simultaneously smitten by a different lady in the party. As one might expect, the smittenness is reciprocal. None of the men, however, wants any of his pals to know he is in the thrall of love and about to violate his scholarly oath.

Between the quadruple pairs of lovers, there follows a secret exchange of messages, confounded by the bumbling messenger Costard who delivers the love notes to the wrong recipients, causing everyone's secret infatuation to become public. Berowne, the most voluble of the bunch, offers up a convincing rationalization that love is a legitimate topic of philosophical study. And so, on that technicality, the pact calling for three years of cerebral celibacy is mooted.

Meanwhile, the women, realizing that they are the more rational ones, take charge of the romantic antics, demanding a test of the men's ability to remain true to their vows. They will accept the love-struck men's proposals of marriage—but only after a year and a day.

It's not much of a plot, but critics agree that *Love's Labour's Lost* shows us Shakespeare in his best linguistic fettle, running wild with wordplay in a profusion of puns and topical references.[2] The play is a satire on courtly love and the manners of the then-recent Middle Ages, and it may have been written specifically for a private performance for the Earl of Southampton, who soon would become the Bard's patron. Says Isaac Asimov, "The play was [also] a satire on pedantry, and its complicated verbiage and intrusive Latinity would appeal to the sense of humor of the educated."[3] It may also have been a satirical jab at the French, whose king Henry IV had previously been Henry of Navarre—perhaps since his visage (figure 8.1) does not look much like a young lover.[4] And for our purposes, it is a play about *love*. So, please come with us as we follow Mr. Shakespeare into the Elizabethan-Jacobean mind in search of love and other emotions.

The mindset of Shakespeare's world—like our own—derived from the culture and the times. We have seen how multiple forces were altering the English world, and notions of love, sex, and marriage were changing apace. Wave after wave of the plague was sweeping over England and the Continent, along with the tsunami of Martin Luther's Reformation, the introduction of gunpowder from Asia, the rediscovery of Greek philosophy from texts preserved in the Muslim world, and the birth of science (nurtured by Galileo, born the same year as our Will Shakespeare). At the same time, a Renaissance version of globalization was causing huge social and cultural changes around the world. Voyages of exploration led to vast trade wealth from Asia, the Pacific, and the New World, but the wealth was flowing into new channels—to the urban centers and into the coffers of bankers, merchants, and traders, bypassing the old, landed nobility, with its manors, castles, and farmlands in the countryside.

In the face of these forces, the medieval feudal system was dying, while enterprising young people like William Shakespeare were following the new ideas and the money to seek their fortunes in the urban centers. That migration loosened family ties and weakened parental authority as parents found they had less control over children who had left their family homes behind. And all of this was reflected in the popular theater, especially in plays dealing with traders

Figure 8.1 The real King Henry III of Navarre, who became King Henry IV of France not long before Will Shakespeare wrote *Love's Labour's Lost*: painting by Frans Pourbus the Younger (1553–1610). Shakespeare may have been making fun of Henry in this play. *Wikimedia Commons.*

and moneylenders, like *The Merchant of Venice*; plays featuring the New World, as in *The Tempest*; and plays about rebellious youth like Romeo Montague and Juliet Capulet.

SHAKESPEARE ON LOVE

The Bard did not invent love, but he owns the trademark. He stamped it not only on *Love's Labour's Lost* but on *Romeo and Juliet*, *Antony and Cleopatra*,

Much Ado About Nothing, and his sonnets of love, as we see in figure 8.2. By our count, nearly half of his plays deal with romantic love in some significant way.[5] And we can add the two narrative poems and the 154 sonnets. But what did this writer of romances mean by "love"? The short answer is close to our term *infatuation*. The longer answer recognizes *love* as a slippery word whose many meanings depend on whether we are talking with a physician, a poet, a prince, a lover, or a lunatic.

In the medical terms of the day, love begins with the eyes, which feed stimulation through the veins into the liver causing the generation of blood—hot blood that could make the lover feel as though burning with passion. In *Love's Labour's Lost*, it is this stage of infatuation that fits the men of Navarre who are intoxicated by the attractive visitors from France.[6] It could be a joyous experience—perhaps the best feeling a person could have. Yet for others, love could be an ache, a form of melancholy, a kind of madness, or even more akin to a fatal disease.

Unrequited love could turn passion into melancholy, with the lover falling into a despairing spiral of longing; physically the lover might become wan, weak, pale, and excessively lean.[7] Rejected lovers or those who were separated

18.

SHall I compare thee to a Summers day?
Thou art more louely and more temperate:
Rough windes do ſhake the darling buds of Maie,
And Sommers leaſe hath all too ſhort a date:
Sometime too hot the eye of heauen ſhines,
And often is his gold complexion dimm'd,
And euery faire from faire ſome-time declines,
By chance,or natures changing courſe vntrim'd:
But thy eternall Sommer ſhall not fade,
Nor looſe poſſeſſion of that faire thou ow'ſt,
Nor ſhall death brag thou wandr'ſt in his ſhade,
When in eternall lines to time thou grow'ſt,
So long as men can breath or eyes can ſee,
So long liues this,and this giues life to thee,

Figure 8.2 Sonnet 18 in the 1609 Quarto of Shakespeare's sonnets. *Wikimedia Commons.*

from the objects of their desire might be at risk of mental collapse or even suicide. This melancholy of love was an ailment to which writers and intellectuals were especially prone.

In *As You Like It*, Rosalind riffs on the same theme with a less poetic sentiment: "Love is merely a madness, and . . . deserves as well a dark house and a whip as madmen do" (3.2.407). And in *Hamlet*, Shakespeare offers yet another analysis of the relationship between love and madness through the mouth of Polonius, who notes the prince's melancholy:

> Fell into a sadness, then into a fast,
> Thence to a watch, thence into a weakness,
> Thence to a lightness, and by this declension,
> Into the madness wherein he now raves.
>
> —*Hamlet*, 2.2.140

What Polonius does not realize is that his observation applies not to Hamlet but to his daughter, Ophelia, whose love Hamlet rejects—and so, she will soon commit suicide.[8]

WHAT'S MARRIAGE GOT TO DO WITH IT?

Most of Shakespeare's comedies are about love, and nearly all end in marriage. But that was not the way things necessarily happened for people in early modern England. Marrying for love was not a new notion, yet it was not the overwhelming consideration in most marriages, which were conceived more as contractual arrangements that would secure or improve the family's wealth and status. This did not necessarily mean that the family ignored the wishes of young lovers, but everyone knew that marriage involved the whole family. And, of course, mutual respect and love could grow out of such arrangements.

As we might expect, marriage among the upper classes was more complicated and elaborate since it could involve large fortunes, great swaths of land, and diplomatic alliances. At those upper levels of society, the negotiations had little to do with love. This may explain why, between 1595 and 1620, approximately a third of the old nobility in England lived separately from their spouses.[9]

While nuptials were much simpler among the lower classes, a marriage celebration could also become a party that lasted for days. Customarily, betrothal was announced in church on three consecutive Sundays, when a couple would publicly declare their intent to be married.[10] A month or so later, a priest might perform a ceremony uniting them in wedlock. Couples in a hurry could bypass this ritual by purchasing a marriage license from the church and swearing that they knew of no legal impediment to their marriage. This alternative helped avoid public exposure if the bride were pregnant or if the couple were secretly Roman Catholics.[11] In its simplest form, a pair could even become betrothed merely by a public declaration—which need not even take place in a church—and finalize a "common law" union by having sex together.

The Bard's work shows us just how much gender roles, as well as concepts of love, courtship, and marriage, had changed since the Middle Ages. Remarkably, nearly half of his plays feature a strong woman asserting her will in a male-dominated society, and a third of those plays feature women defying marriage arrangements imposed by their fathers.[12]

Juliet is, of course, the Shakespearean archetype of a daughter going against her parents' wishes. So, by setting *Romeo and Juliet* in fourteenth-century Verona,[13] the playwright shows scorn for the strict medieval marriage ideals. He drove that point home with multiple variations on the theme of stubborn daughters and their unhappy fathers. In *A Midsummer Night's Dream*, for example, Hermia insists that she will marry Lysander, even though her father Egeus demands that she will either marry Demetrius or be put to death.[14] The same theme echoes in *Cymbeline*, where the king's daughter defies her father by refusing to marry Cloten and secretly marrying Posthumus instead. Perhaps the most blatant statement in this headstrong-daughter genre appears in one of his earliest plays, *The Taming of the Shrew*, where Kate defies first her old-fashioned father and then also her misogynistic husband, as they each take a turn at trying to domesticate her. It helped, too, that Queen Elizabeth was a headstrong female role model.

For men in the Elizabethan world, marriage offered advantages and opportunities that were closed to their single counterparts. Marriage was considered, both by custom and law, to be the natural and desirable state for adults (unless they were widowed). Accordingly, boys knew that, if they remained single, they would not be eligible for civic offices, which were open only to married men.

Women's roles were even more restrictive because the country was largely run by a patriarchy, extending from the government down to the family.[15] A woman was expected to follow the orders of her husband without question. Even so, life could be easier for a woman if she were married to a kind and prosperous man. While she could not own property outright or enter into contracts without the consent of her husband, the law did recognize her right to inherit a portion of her husband's property upon his death. And to give a more surprising example, married women could not be "witches," while spinsters were open to the charge of witchcraft and its rather dire penalties. Yet even the power of the patriarchy was changing, if only slightly. Thanks to "Bloody" Mary Tudor and her half-sister, Elizabeth, women could occupy the throne of England—if they had the right ancestry, a good army, and a ruthless streak.[16]

The legal age of consent for marriage also varied by gender, as well as time and place. For girls it was generally fourteen in England and on the Continent, although during the Middle Ages it had been as low as twelve in some regions, perhaps because lifetimes were somewhat shorter and more uncertain, due to the plague. (Shakespeare puts Juliet at thirteen but leaves us guessing about Romeo.)[17] Curiously, the age of consent for boys was sixteen to eighteen, perhaps because people expected that boys matured later (even though they believed men to have greater emotional stability and higher intellectual ability).

Despite the low legal age of consent, the average age of marriage in Shakespeare's time was in the twenties. In this regard, rural Stratford was no different from London at the time when young William married the somewhat older and pregnant Anne Hathaway. Even so, Will and Anne were outliers from the norm in that he was eighteen and she was some eight years his senior.

Among the nobility, the legal age limits for marriage did not apply, again because noble marriages, especially among the royals, could involve alliances, territory, and wealth. One particularly egregious example illustrates how differently the law regarded the nobility: England's King Richard II (who was famously deposed by Henry Bolingbroke *cum* Henry IV) was age twenty-nine and a widower when he was betrothed to six-year-old Isabella, daughter of France's King Charles VI, after a long political negotiation.[18] Another exceptional case, of course, was that of King Henry VIII, who initially married Catherine of Aragon for political reasons but whose divorce and subsequent five marriages arose, not from love, but lust and the desire for a male heir. Ironically, he instead begot the first two females who would ascend the throne of England.

WHAT'S SEX GOT TO DO WITH IT?

The Church and civil law officially frowned upon sex outside of wedlock, but couples acted pretty much as they do today under the influence of their hormones. Henry VIII was the royal archetype, while other hot-blooded young lovers managed to satisfy both their desires and their obligations to the church, the law, and their parents. So commonly did betrothed couples consummate their vows before the formal marriage ceremony that, based on evidence from marriage and birth records, between a fifth and a third of brides presented themselves pregnant at the altar. We are not sure how large that proportion is today.

It is interesting to note how Shakespeare dealt with these ambivalent attitudes about sex. While his characters rather freely spoke of sex, there was never a simulation of sexual intercourse in the plays, although on occasion there is a suggestion that a couple was coupling somewhere off-stage, between scenes. A steamy, in-person performance apparently would have been too strong stuff for the royal censors—yet libidinous language and explicit descriptions of body parts, especially when used for laughs, were approved, perhaps with a wink.

Some of the Bard's sexual references are direct, as when Helena plays the "bed trick" on Bertram.[19] Others merely involve puns, as in Falstaff's comedic scenes that take place in bawdy houses that double as taverns and feature such double-entendre characters as Mistress Quickly and Doll Tearsheet. And you will recall Jaques's speech in *As You Like It*, where he puns on the word "whore," which was pronounced like "hour" in Elizabethan times. (We reproduce it here in an endnote.)[20]

Other sexual references are more oblique, as in *Much Ado About Nothing*, when we hear Benedick rationalizing his change of mind about bachelorhood while delicately but enthusiastically declaring his desire for a bedtime romp with Beatrice:

> BENEDICK: I may chance have some odd quirks and remnants
> of wit broken on me because I have railed so long against
> marriage, but doth not the appetite alter? A man loves the
> meat in his youth that he cannot endure in his age. Shall
> quips and sentences and these paper bullets of the
> brain awe a man from the career of his humor?
> No! The world must be peopled. When I said I would die a
> bachelor, I did not think I should live till I were married.
>
> —*Much Ado About Nothing*, act 2, scene 3

More surprisingly, the playwright was able to hide, in plain view of the censors, still other references to sexual matters, body parts, and excretions. So, we find Malvolio reading a message that he believes to have come from the lady he loves:

MALVOLIO: By my life, this is my lady's hand. These be her very C's, her U's, and her T's, and thus makes she her great P's. It is in contempt of question her hand.

—*Twelfth Night*, 2.5.88

The player would have exaggerated and pantomimed the line to make the meaning more obvious to the audience, but you can get much the same effect by reading it aloud, pronouncing "and" as "N." When you do, you will hear the naïve Malvolio is unwittingly spelling "CUNT" (which, incidentally, was a term used in medieval anatomical texts and not nearly as freighted with taboo as it is today).[21]

THE EVOLUTIONARY PSYCHOLOGY OF LOVE'S LABOURS

In our play, the love-struck men of Navarre take the brunt of Shakespeare's wit—much more than do the more reluctant and rational women from France. This fits with the picture of gender differences as seen through the lens of evolutionary psychology, which views women (on average) as more selective about their mates than are men (again, on average). Why might this be true?

Evolutionary psychology teaches that both sexes are genetically "programmed" to send as many of their traits as possible into future generations—but to do so requires different reproductive strategies for males and females. From this viewpoint, gender differences in mate selection occur because the biological investment in a potential pregnancy and subsequent child-rearing falls far more heavily on women than on men, a fact determined largely by biological imperatives. Childbirth in Elizabethan times also included a substantial risk of death. Moreover, in the time it takes for an impregnated woman to carry one or two developing offspring, a male could theoretically impregnate scores of other mates.[22] No wonder the women visitors to Navarre were a bit wary of a relationship with King Ferdinand and his male friends!

As we saw in chapter 1, from a strictly biological viewpoint, natural selection seems to favor eager men and reluctant women. Taking the long view over our evolutionary history, the best reproductive strategy for males is promiscuity. Sperm is cheap, so males are driven to mate as often as possible with as many different females as they can convince to copulate, even by subterfuge. But since pregnancy is biologically much more costly, females will have better reproductive success if they follow a different strategy. They can produce many fewer offspring than can a male—because they produce only an egg (or two or three) each month, as opposed to the roughly forty million sperm that males leave behind in each sexual encounter. Clearly, it is to women's evolutionary advantage to be more selective about partners than are males. Accordingly, the biology of the ladies from France encourages caution about the sincerity of the men of Navarre and their likelihood of keeping their vows—which, in marriage, normally means being faithful to only this one relationship and willing to take responsibility for any offspring. In this light, then, making the men wait a year and a day for the nuptials was a rather good test.

None of this emotion-driven calculus happens on a conscious level, of course. The eagerness of the men and the reluctance of the women is built into their genes and, consequently, into the "wiring" of their brains, which regulate their emotions and drives—all largely outside awareness. This is again Daniel Kahneman's "System 1 thinking," which he also holds responsible for all the near-instantaneous responses we make when we "act without thinking." It is the brain's System 1 that allows us to drive a car along the highway "on auto-pilot" (while the conscious and rational System 2 attends to the music on the radio).[23] System 1 also dominates our "fight-or-flight" reactions in threatening situations, our aversion or attraction to certain tastes and odors, and our ability to read the emotions on people's faces. For our purposes, however, it is System 1 thinking that is responsible for the "love at first sight" that we witness in *Love's Labour's Lost.*[24]

While our Bard knew nothing of evolutionary psychology, he was quite concerned with the very issues that this new psychological domain explains: infatuation, jealousy, cuckoldry, and infidelity. This tells us that his brain, too, was wired to be sensitive to these evolutionary forces. And so, he was an "intuitive evolutionary psychologist," as were the others who wrote the classic literary masterpieces about revenge, murder, and war driven by the emotional responses involved in sexual relationships. Here are some examples that our teachers probably glossed over in high school literature classes:

- *The Iliad* is an epic poem about a devastating ten-year war fought over a married Greek woman (Helen) who ran away with another man, Prince Paris of Troy.
- *Medea* is an ancient Greek play about a woman who kills her children to spite a husband who left her for another woman.[25]
- Among the hundred stories told in Boccaccio's *Decameron*, the ones that made the book so popular mostly involve trysts, seduction by trickery, and wives being abducted or taking secret lovers.
- In *The Canterbury Tales*, Geoffrey Chaucer tells a series of yarns, many of which follow the same pattern: the Wife of Bath regales her fellow

travelers with ribald stories of her many marriages; "The Merchant's Tale" is a naughty narrative of an old and rich man with a young wife who becomes attracted to someone else; and in "The Miller's Tale," another young wife, with a body "gent and smal" as a "weazel," cuckolds her elderly husband, as she plays two lustful young men against each other.

See the pattern? Now let's look at Shakespeare in the same light:

- In a fit of jealous rage, Othello smothers his wife with a pillow because he (falsely) believes she is having an affair with his lieutenant Cassio.
- In *Cymbeline*, Iachimo conspires to make Posthumus believe that his wife, Imogen, has been unfaithful.
- In *The Winter's Tale*, the King of Sicily (again, falsely) believes that his queen has cheated on him with his friend, the King of Bohemia—and so the child she is carrying is illegitimate (see figure 8.3).

Figure 8.3 Evolutionary psychology tells us that much of art and literature is driven by evolutionary themes involving sexuality, such as suspicions of cuckoldry and jealousy. *The Winter's Tale* is a story of Leontes's obsession with the belief that his wife has cuckolded him, and he is not the father of Hermione's newborn baby girl. In this illustration of act 2, scene 3, he commands Antigonus to take the baby into the wilderness and abandon her to die. The image is an engraving based on a 1793 painting by John Opie. *Wikimedia Commons.*

- In *Much Ado About Nothing*, Don John informs Claudio that his betrothed, Hero, is unfaithful. As proof, he arranges for him to see Borachio (the eponymous drunkard) enter her chamber, where Borachio and a woman (actually Hero's chambermaid) have a sexual liaison (out of the audience's sight—of course). Claudio naively believes only what he sees and vows revenge by humiliating Hero publicly at their wedding.[26]
- And in *The Two Noble Kinsmen*, the title characters, Palimon and Arcite, see a beautiful woman, Emilia, from the window in their cell, where they languish after being captured in a war. Both instantly fall in love with her. To make matters sillier, their love for the same woman drives them to become sworn enemies (though they have been lifelong friends and have no hope of their love for the woman they see ever being requited).

We cannot blame all Shakespearean sexual misconduct on male miscreants. Hamlet's mother, who was (likely) in cahoots with Claudius in the dispatch of Hamlet's father, is certainly no paragon of true love. Nor is the ruthless Tamora (of *Titus Andronicus*), who stops at nothing for revenge on Titus, including the murder of her own baby. Then there is Cleopatra, the Egyptian Queen who becomes enraged when Antony abandons her in favor of a political marriage to a powerful Roman's sister but commits suicide when the wounded Antony dies in her arms. The same "green-eyed monster" of envy and jealousy motivated most of them. According to evolutionary psychology, the take-away message is how extremes of those emotions can interfere with survival and reproduction, even among those who, at first, seem to be the fittest.[27]

THE ELIZABETHAN CONCEPT OF EMOTIONS

We know the Bard best for his treatment of love, but he was not *all* about love, even if we add the related emotions of melancholic longing, lust, envy, and jealousy. He also wrote about rage, fear, pride, mirth, worry, joy, and more. So, how did he and his audiences conceive of emotions?

The Elizabethans knew that the brain was the organ of sensing, thinking, and memory, but they told a different story about emotions, which they considered more "visceral and associated with the humors." Indeed, the word "emotion" does not appear in Shakespeare's works, nor did it even exist in its current sense until some years after his death.[28] In the Bard's time, the responses we now call "emotions" were supposed to be generated by various organs in the body, although it is important to note that different schools of medicine had different opinions on which organs were involved with each emotion.

Likewise, the Bard wrote of the spleen as the source of anger—as even today one might speak of rage as "venting one's spleen."[29] Medicine also believed that melancholy (or *depression*) came from an excess of black bile, again thought by some to be another product of the spleen.[30] Elsewhere, anger and other aggressive passions were said to arise from "hot blood," caused by yellow bile from the gall bladder vented into the blood—which, in turn, could disrupt the function of the liver and exaggerate all human desires and feelings.[31]

Even the brain had its role in emotion—not so much as the place where emotions were felt, but as the source of the humor *phlegm.* You see, the humorism theory conceived of the brain as rather akin to a sponge that could generate and fill with phlegm and then shrink and squeeze it out into the blood. An abundance of phlegm, a thick, viscous, and mysterious substance, could make one "phlegmatic"—that is, calm or even lethargic, with a personality characterized by slowness of thought and action.

Love itself came from the heart, as we still say metaphorically. Yet Falstaff, Claudio, and Malvolio all loved with their livers (from which also came courage and valor, with some help from the stomach).[32] And again, in *The Tempest*, we find Shakespeare invoking the influence of the liver, along with "fire I' th' blood," to explain Ferdinand's love-at-first-sight attraction to Miranda:

> PROSPERO: . . . do not give dalliance
> Too much reign. The strongest oaths are straw
> To the fire I' th' blood . . .
>
> FERDINAND: The white cold virgin snow upon my heart
> Abates the ardor of my liver.
>
> —*The Tempest*, 4.1.56

As we can see, then, the humor theory of feelings, desires, moods, and temperaments was complex and often rather inconsistent.

Yet it all makes a sort of sense. Even today when we feel emotions, we often speak of them as "gut feelings."[33] When we are afraid, we may get a "knot" in the stomach; when we are in love, our hearts seem to beat a little faster; and when we are angry, our faces may turn red, and we may feel an "adrenalin rush." Indeed, emotions do seem to come from the gut, where they can also produce measurable changes in the heart rate, blood pressure, eye dilation, perspiration, and piloerection (goosebumps)—an effect called *frisson*. These are essentially the same responses measured by polygraph operators. After all, the so-called "lie detector" apparatus is nothing more than a detector of a participant's emotional arousal.[34]

All artists, Shakespeare included, strive for an emotional response to their work, and we have ample evidence that at least some of them succeed. Studies show, for example, an association between reported emotions and physiological responses in volunteers listening to poetry and music that they like.[35] Yet, as far as we can tell, no one has measured frisson produced by exposure to the Bard's plays, poems, or sonnets. But who could listen to King Henry V's "Band of Brothers" speech, the depressed ruminations of Hamlet, the wrath of a jealous Othello, or the anguish of Juliet at the death of Romeo without experiencing emotional frisson?

So, the old humoric view that Shakespeare held was not altogether wrong. Emotions *can* come from the feelings generated by the internal organs. Moreover, the subjective feeling component of an emotion may arise from multiple sources at once, including the brain, which senses the body's current state of arousal. The brain's System 1 also stores *memories* of the body's state in similar situations in the past. Neuroscientist Antonio Damasio calls these visceral memories *somatic markers*.[36] So, in response to, say, a near-accident while driving, your brain immediately retrieves a somatic marker (memory) of how your body felt during past encounters with danger, including a racing heart, a "knot" in the stomach, and a cold sweat. At the same time, the brain sends messages to the visceral organs that produce and sustain the very gut responses that System 1 associates with the emotions we call fear, anger, jealousy, disgust, lust, loss, and—yes—also with love.

THE SEVEN DEADLY SINS AGAIN: EMOTIONS OR MOTIVES?

While lust and infidelity caused plenty of trouble in earlier times, there were other desires and temptations that could lead to ruin. It was Pope Gregory (1590–1604) who popularized the famous list of the seven deadliest sins: *wrath*, *avarice*, *sloth*, *pride*, *lust*, *envy*, and *gluttony*.[37] Chaucer later borrowed that list for his "Pardoner's Tale," and Dante constructed his *Purgatorio* around sinners doing penance for these sins.

Oddly, Shakespeare never mentioned the Seven Sins by their collective name—but he described each of them vividly.[38] Here is a short list of characters in his plays who exemplify each one.

- *Wrath*: Personified by the vengeful Titus and Tamora in *Titus Andronicus* and by Coriolanus in his own play.

- *Avarice*: Portrayed by Shylock, the money lender in *The Merchant of Venice*, but also by the antihero of avarice, Timon, in the play named after him.
- *Sloth*: Falstaff is the hands-down paragon of laziness in both parts of *Henry IV.*
- *Pride*: Henry "Hotspur" Percy, the arrogant young buck from York who dies at the hands of Prince Hal, in Part 1 of *Henry IV.*
- *Lust*: The treacherous Iachimo in *Cymbeline*, the libidinous duo of Antony and Cleopatra in their own play, and the brooding Achilles and his lover, Patroclus, in *Troilus and Cressida.*
- *Envy (and Jealousy)*: Iago and Othello, both victims of the "green-eyed monster" in the latter's play.
- *Gluttony*: And again, it is Fat Jack Falstaff!

While it may seem too fine a point, psychology now conceives of *emotions* as feelings that are directed outward at external objects or events, rather than feelings aroused by internal needs (such as hunger or thirst).[39] So, it is our emotions that produce arousal at the sight of a loved one or anxiety about anticipated events, such as an upcoming wedding or a challenging final exam. From this perspective, wrath, pride, lust, and envy qualify as emotions because they are directed outward at people or external conditions in our lives, while avarice, sloth, and gluttony are *motives* because they typically arise from a desire to satisfy a need coming from an internal physical or mental condition.

Most likely, Shakespeare and his audiences would not have cared a whit about this distinction, nor would most psychologists—until recently. In this context, those who studied motivation regarded outward-directed emotions merely as a nuisance that prevented people from using good judgment or, in extreme cases, caused mental illness.[40] Consequently, both emotions and motives were considered problems best left to psychotherapists and their patients—annoyances in everyday life and evils to be rooted out of the unconscious minds of disturbed individuals, to be replaced by more reasonable thoughts. Even the Bard seems to have agreed that emotions could, in the extreme, be a nuisance, as in *Love's Labour's Lost*, or even lethal, as in *Othello* or *Titus Andronicus*. But for him, emotions and motives were also the stuff of great theater—the very essences of human nature, and thereby of his art. And in recent decades, psychologists, too, have discovered that emotions play a surprisingly vital role in our lives . . . as we shall see momentarily.

WHAT DO EMOTIONS DO FOR US?

Shakespeare thought that emotions, especially love, could conflict with reason. But his making fun of Holofernes, in *Love's Labour's Lost*, suggests that he thought pure reason could also extract the joy from life. In fact, many of the plays probe the push and pull between reason and emotion, with emotion often the victor—among these, *Love's Labour's Lost*, *A Winter's Tale*, *Othello*, *Much Ado About Nothing*, and again *Romeo and Juliet*. But is it true that emotion and reason are in perpetual conflict? Consider now the textbook case of Elliot, a true story drawn from the annals of neuroscience.[41]

Once a model employee, Elliot had let the quality of his work slip to the point that he finally lost his job. If anything, said his supervisors, he had become almost too focused on the minor details of his work, yet he had trouble setting priorities. He often latched onto a small task, such as sorting a client's paperwork, and spent the whole afternoon on various classification schemes—never quite getting to the real job he had been assigned. Strangely, Elliot's whole life was unraveling, yet he maintained an attitude of composure. A divorce was followed by a short marriage and another divorce. Several attempts at starting his own business involved glaringly flawed decisions that finally ate up all his savings.

In most respects, Elliot seemed normal. He was obviously smart, and he had a pleasant personality with an engaging sense of humor. Mental exams showed that he understood the political and economic affairs of the day and had unimpaired memories of important events, names, and dates. Elliot also knew who and where he was. In psychiatric parlance, he knew "time, place, and person." Nor did medical examinations reveal anything wrong with his movements, perceptual abilities, language skills, intellect, or ability to learn.

It was complaints of headaches that finally led the family doctor to suspect that Elliot's behavioral troubles might be the result of a brain lesion, and tests proved that suspicion correct. Brain scans showed a mass the size of a small orange that was pressing on the frontal lobes just above Elliot's eyes.

The tumor was removed, but not before it had done extensive damage. The impact, limited to the frontal lobes, bore a remarkable similarity to that of the notorious Phineas Gage nearly 150 years earlier, when an accidental explosion of blasting powder drove a metal stake through the frontal lobe of his brain.[42] Like Gage, Elliot had undergone profound changes as a result of frontal lobe damage. But while the effects in Gage involved remarkable changes in loss of impulse control, the effects in Elliot were more subtle, mainly involving his ability to weigh alternatives and make decisions.[43] His reasoning abilities were intact, but the damage to the circuitry of Elliot's frontal lobes disrupted his ability to use his emotions to establish values and priorities among the objects, events, and people in his life. The result was that Elliot could not value one course of action, or one thing, over another.

For us, Elliot's case demonstrates the broader notion that *emotions allow us to place values on external objects and events and, therefore, to weigh alternatives and make choices and decisions.* One main function of our emotions, then, is to "tag" our memories to help us determine how good or bad an object, a person, or an experience was. In this way, then, emotions associated with memories have survival value. But are emotions perpetually in conflict with reason? Mr. Shakespeare certainly seemed to think so—or, at least, he thought that stories pitting tension between emotion and reason made for good theater, as we will see in the case of Hamlet, the star of our final chapter.

· 9 ·

REASON, INTUITION, AND THE PRINCE OF DENMARK

Hamlet

Elsinore Castle, Denmark (November 1572)[1]—Upon news of his father's death, young Prince Hamlet has made the long trip home[2] from the university at Wittenberg to find that his Uncle Claudius had seized the throne of Denmark . . . and married his mother, Gertrude. By right and custom, the crown should have passed to him.[3] And if that weren't enough. . . .

Further complications come in the filmy form of a Ghost resembling the dead king appearing on the rampart at night (figure 9.1). It tells Hamlet that Claudius has killed his father and demands that the prince avenge the murder. If the Ghost is to be believed, Hamlet realizes that his mother's hasty remarriage makes her complicit in this "murder most foul" (1.5.33).[4]

But the Ghost presents more of a problem than a solution. Is it really the ghost of Hamlet's father? Horatio, the prince's friend from the university, suggests that the apparition may be an evil spirit—even the Devil himself—impersonating Hamlet's father. Horatio warns that the Ghost may be trying to "deprive your sovereignty of reason / And draw you into madness" (1.4.77). If the Ghost is lying and Hamlet follows its orders to avenge his father's murder, then he would be guilty before high heaven of regicide, a mortal sin that could result in his eternal damnation.[5]

Hamlet finds the apparition both disturbing and credible—but how should he respond? "To be, or not to be, that is the question" (3.1.64). Should he "take arms against a sea of troubles," or should he simply end his torment by taking his own life?—regicide or suicide? It is a momentous problem with nothing but negative consequences as far as the prince can see. Picking up on Horatio's idea of "madness," Hamlet decides "to put an antic disposition on," giving him cover while he finds a way to submit the Ghost's claims to a test.

Figure 9.1 Kronborg Castle—which Shakespeare dubbed "Elsinore" in *Hamlet*. *Wikimedia Commons.*

Meanwhile, Rosencrantz and Guildenstern, two of Hamlet's childhood friends, arrive at the castle, bringing with them an acting troupe encountered on their travels.[6] Hamlet quickly realizes that the two are there to spy on him at Uncle Claudius's behest. But he also sees an opportunity to use the troupe of actors to test his uncle's guilt or innocence—to verify the Ghost's revelation with a psychological "experiment." He asks the thespians to stage a play that he calls *The Mousetrap*, in which the plot resembles the Ghost's story of Hamlet's father's murder. Surely, thinks the prince, Claudius will react with some signs of guilt, and so he proclaims, "The play's the thing / Wherein I'll catch the conscience of the king"[7] (2.2.575).

Shortly thereafter, the king, queen, and their courtiers gather to see the production. As Hamlet has predicted, when the play unfolds to the scene where the sleeping player-king is murdered by the king's nephew, King Claudius rises and storms out of the room. At this point, it is not clear to the audience whether the king's response reveals his guilt or—just as logically—his fear that a crazed Hamlet is about to kill him. The prince, however, interprets the king's reaction as proof of his guilt.[8]

The scene shifts, and we see the king kneeling in the chapel. We hear him pray for forgiveness for murdering his brother—but only those of us in the audience hear the confession; Hamlet does not. But he decides not to kill Claudius during prayer because (as audiences immersed in the Christian belief system of Elizabethan times would know) Claudius would go immediately to heaven, rather than hell.

The play upsets Queen Gertrude, too, and she demands to see Hamlet in her chambers. In their ensuing dispute, the Ghost appears to Hamlet (but is invisible to Gertrude), rebuking him for the harsh words to his mother and his failure to take revenge on Claudius. Suddenly, the prince notices the movement of a figure hiding behind the drapes hanging in Gertrude's quarters. Thinking it to be Claudius, Hamlet draws his sword and stabs blindly through the tapestry, killing . . . Polonius, who is spying on the prince on the orders of the king. Coincidentally, Polonius also happens to be the father of Hamlet's fiancée, Ophelia—yet Hamlet shows no remorse. Without apology, he exits, dragging Polonius's corpse.

When King Claudius hears of Polonius's death, he knows that he could have been behind the drapes. Fearing for his life, he arranges to send Hamlet to England in the care of Rosencrantz and Guildenstern, who carry a secret letter bidding the English monarch to execute the prince. Hamlet discovers the letter and substitutes one of his own, designating Rosencrantz and Guildenstern as the victims. He then manages to escape and return to Denmark.

Laertes, Polonius's son, is understandably angry over his father's death; he challenges the prince to a duel in which Laertes and Claudius connive to kill Hamlet in one of two ways: either with a goblet of poisoned wine or with a sword dipped in poison. Their plan goes awry when Gertrude mistakenly drinks the poisoned wine and dies—after which both Laertes and Hamlet, in turn, wield the poisoned blade. Both receive mortal wounds, but Hamlet lives long enough to run the poisoned sword through Claudius and pour the remaining lethal wine down the king's throat. Only Horatio is left alive to mourn and to bid Hamlet to fare well in the next life:

> HORATIO: Good night, sweet prince,
> And flights of angels sing thee to thy rest!
>
> —*Hamlet*, 5.2.397

Shakespeare's Prince Hamlet is an anachronistic character who combines a quasi-historical figure—a prince from the High Middle Ages[9] embedded in the culture that harks back to the Viking age but also seems rather English, perhaps like a Danish version of the Wars of the Roses. Shakespeare's aim, of course, was to make the setting familiar to his audiences.

Critics agree that the Bard's inspiration came from a Danish book of legendary tales told as history but without historical basis, as far as we know.[10] From that source, Shakespeare borrowed the story of a prince named Amleth (sounds like Hamlet, if you say it fast) whose father had died, whereupon his uncle grabbed the throne and married the dead king's widow.[11] Amleth, like Hamlet, even plotted revenge under the cover of a feigned mental illness. If this Amleth ever existed, he lived during Denmark's Viking era, several hundred years before Shakespeare—in about 1050 CE.[12]

Shakespeare obviously wanted audiences to think of the play as occurring during Viking times, when the Danes were more warlike. Why, then, did we date the play (at the beginning of this chapter) as 1572? And why must it be November?

Journalist/astronomer Dan Falk tells us that the clues for the month and year in which Shakespeare sets Hamlet's story point to an astronomical event that occurred within memory of most patrons of the Globe, and we hear these clues in the opening lines of the play when Bernardo, a castle guard, tells Horatio about the Ghost's appearance on the rampart:[13]

> BERNARDO: Last night of all,
> When yond same star that's westward from the pole,
> Had made its course to illume that part of heaven
> Where now it burns, Marcellus and myself,
> The bell then beating one—
>
> —*Hamlet*, 1.1.42

Bernardo is saying that the Ghost appears one hour after midnight—at one o'clock in the morning. Another guard has commented that it is "bitter cold," and another mentions that it is just before "that season comes / Wherein our Savior's birth is celebrated." So, the tale must occur before Advent and most likely on a cold night in late fall.

Which star would strike the eyes of the guards west of the pole at that time and place? "The pole" refers to Polaris, the pole star, says Falk. And if one turns back the astronomical clock to 1572, there was a supernova—an exploding star that was visible in both Denmark and England—in that very spot, in the otherwise unremarkable constellation of stars that form Cassiopeia.[14] It shone brightly, even in the daytime, for almost a year, apparently making an impression on an eight-year-old schoolboy from Stratford-upon-Avon.[15]

What is the relevance of this to psychology? Falk suggests that Shakespeare had a demonstrable interest in science that adds, for us, a new dimension to the scope of his knowledge and worldview.[16]

Moreover, Shakespeare's linking of Prince Hamlet to the pole star was his way of updating a character from medieval Denmark to make the story more immediate and relevant to Elizabethan audiences. It also accounts for the anachronism of Hamlet as a student at the University of Wittenberg in Germany.[17] Wittenberg *did* have a renowned university in Shakespeare's day, but not during medieval times in which Hamlet's story originated. Nor did Elsinore Castle exist in those earlier days.[18]

And why Wittenberg? It was located hundreds of miles from Hamlet's family castle at Elsinore, but the town would have been known to audiences as the site of the famous university at which Martin Luther had taught.[19] The town of Wittenberg also was home to the church on the doors of which the monk Luther had nailed the ninety-five theses that launched the Reformation. This reference would have been especially significant for English audiences because it was their King Henry VIII (Elizabeth I's father) who brought the Protestant Reformation to England—but not for Luther's high-minded reasons.[20]

A little more information about the political differences between England and Denmark will help us understand how Elizabethan audiences would have understood the Danes. In the Viking era and the High Middle Ages, Denmark had not yet become a pacifistic Scandinavian nation but was instead a warlike country with strong territorial ambitions, although not necessarily perceived as a threat to the English during this period.[21] From an English perspective, Shakespeare's audiences would have expected Hamlet to be the logical successor upon his father's death; they didn't realize that the Scandinavian and Germanic peoples often selected their leaders by a vote of the freemen in an assembly known as the *thing*.[22]

In interpreting *Hamlet*, the succession issue is also noteworthy because England, in Shakespeare's day, faced a similar problem. Queen Elizabeth was aging and was without an heir, potentially throwing England into its own "sea of troubles." English claims to the throne normally required either a "blood" connection to the previous monarch or a military victory over the previous ruler. Elizabeth's position was ironically analogous to that of her father, Henry VIII: she needed an appropriate heir.[23]

THE QUINTESSENTIAL RATIONAL MAN—WITH AN EMOTIONAL PROBLEM

It is with *Hamlet* that the Bard probed more deeply into the human mind than he or any other playwright had done before.[24] There he exposed a conflicted fellow, but one who was smart and also convinced that he could control his own intuitions and emotions to solve his problems rationally. Repeatedly, Hamlet proclaims the virtues of reason and the suppression of emotion, as when he says to Horatio, "Give me that man / That is not passion's slave" (act 3, scene 2).[25]

It is no stretch of imagination to see Shakespeare in sympathy with Hamlet, for he warns us repeatedly of extremes of emotion. Yet he also held up to ridicule such hyperrational types as the academic Holofernes (*Love's Labour's Lost*) and the aptly named Pedant (*The Taming of the Shrew*). It was extremes of emotion that seemed to worry him most—when reason becomes the slave of emotion. He underscores this by serving up multiple examples of lovers who are blinded by their emotions, as in *Romeo and Juliet*, *Much Ado About Nothing*, *A Midsummer Night's Dream*, and *Love's Labour's Lost*. Nor was it only love that drove people to emotional excesses, as he demonstrated in *Hamlet*, *Lear*, *Macbeth*, and *Othello*. Such tragic flaws can emerge from any failure to realize that seemingly rational decisions can be driven by emotion and desire. It is fundamentally the same meme that we hear now from psychologists Daniel Kahneman and Jonathan Haidt.[26] Like Haidt's "rider," we believe that we are rational, yet we follow our "elephant's" unconscious desires, emotions, intuitions, ideas, and prejudices uncritically absorbed from our culture; meanwhile, our rational "rider" merely creates a justification.

Professor Neema Parvini says, "Few writers in history have understood this aspect of human thinking better than William Shakespeare, who shows again and again in his plays how emotions and intuitions trump reasoning."[27] Parvini points out that Mark Antony does not win over the crowd with a logically crafted oration but an emotion-laden one; Iago does not persuade Othello to smother Desdemona with flawless reasoning but with innuendo. The quintessential example is Henry V's "Band of Brothers" speech, fairly saturated with emotion. Parvini adds:[28]

> The one exception to this general pattern is Hamlet, who does his very best to make reason his "marshal," trying at all times to put his reasoning up front. He attempts to take intuition out of the equation altogether in pursuit of a fantasy of pure reason. He resists human nature— his own nature . . .

Yet, Hamlet's world comes crashing down around him. Shakespeare also seems to have known that all reason and no emotion could make Hamlet a dull and melancholy fellow—as we hear in the prince's comments to Rosencrantz and Guildenstern:

> HAMLET: I have of late—but wherefore I know not—lost all my
> mirth, forgone all custom of exercises, and indeed it goes so

> heavily with my disposition that this goodly frame, the earth, seems to me a sterile promontory; this most excellent canopy, the air—look you, this brave o'erhanging firmament, this majestical roof fretted with golden fire—why, it appears no other thing to me than a foul and pestilent congregation of vapors. What a piece of work is a man! How noble in reason, how infinite in faculty! In form and moving how express and admirable! In action how like an angel, in apprehension how like a god! The beauty of the world. The paragon of animals. And yet, to me, what is this quintessence of dust? Man delights not me. No, nor woman neither, though by your smiling you seem to say so.
>
> —*Hamlet*, 2.2.316

This is indeed the Melancholy Dane, trying to find a pathway out of depression, while at the same time feigning a Bedlam-kind of mental illness to cover up his desire for revenge. Critics have called him ambivalent and indecisive, a procrastinator, and a dithering slave to reason. Professor Neema Parvini calls Hamlet a victim of "analysis paralysis."[29] But are those analyses fair? We submit that Hamlet must deal with difficult problems that have no clearly good solutions—the sort of problems that could make melancholy ditherers of us all.[30] In our view, Hamlet's tragic flaw is not so much indecisiveness but exaggerated confidence that his reasoning abilities can overcome his emotions. Jonathan Haidt labels it the *rationalist delusion*. Hamlet is caught in a vortex arising from his rumination on the bad choices—which is what melancholic or depressed people tend to do.

HAMLET'S DILEMMA

Hamlet's mother has married the man Hamlet suspects has murdered his father. He risks losing his inheritance, including the throne of Denmark, yet nobody, except perhaps Horatio, seems outraged or even willing to question the strange reality he confronts. Moreover, King Claudius seems to want Hamlet out of the way—perhaps enough to have him executed. In addition, we know that Hamlet's faith in reason has blinded him to his own emotional problems and biases.

The immediate problem on which Hamlet must focus is *not* how to deal with Claudius but rather the Ghost. Is it lying or telling the truth? Claiming to be his father's spirit, the Ghost made Hamlet swear to avenge the murder of his father, but he knows that it could be an evil spirit, perhaps Satan himself, bent on deceiving and damning him. As we have noted, if the prince follows the Ghost's command to avenge his father's death, Hamlet could be judged by God a murderer and so, in his religious framework, he would lose his soul. Here is how Hamlet describes his dilemma:

> The Spirit that I have seen
> May be the devil, and the devil hath power
> T' assume a pleasing shape; yea, and perhaps,
> Out of my weakness and my melancholy,
> As he is very potent with such spirits,
> Abuses me to damn me.
>
> —*Hamlet*, 5.2.575

It is not a simple matter of a revenge killing among tribal factions in Viking Scandinavia. Rather, in this most Christian of Shakespeare's plays, Hamlet sees it as a problem that could decide the eternal fate of his soul.

While critics often see Hamlet as indecisive, psychologist John Orbell has pointed out two other interpretations that emphasize the power of an ambiguous *situation* over Hamlet's seemingly indecisive *personality*. Says Orbell:[31]

> Hamlet is shown to be a case of decision making under uncertainty, where the consequences of error are unthinkable and where a decision cannot be avoided. Far from showing us a personality constitutionally unable (for whatever deep reason) to decide, the play shows us the difficulties attendant on this class of decision problems—a demonstration highlighted by the fact that those difficulties must be faced by an individual better equipped than most for decisive action. The play also shows us the only satisfactory way for resolving such a problem. A de-emphasis on *personality* and a turning of critical attention to the structure of *situations* that normal people confront might contribute not only to an understanding of this particular play and of dramatic tragedy but also to our understanding of decision making in general.[32]

Dr. Orbell's point deserves reemphasis. Perhaps we should see Hamlet not so much as a troubled *person* but as a competent and seemingly rational individual facing an extremely difficult *situation*. In this light, then, Shakespeare is raising the *person-situation* issue—yet again.

What exactly is Hamlet's dilemma? The matrix shown in table 9.1, based on Prof. Orbell's work, shows the only two choices Hamlet sees at first: to believe the Ghost or not.[33] The difficulty is that each of those choices could have a terrible outcome, depending on whether the spirit is truthful or lying.[34] Hamlet suspects that King Claudius is guilty of murder, but he must find plain evidence—for all to see.

Table 9.1 Hamlet's Dilemma: Should He Believe the Ghost and Kill the King—or Not?

	Kill the King	**Don't Kill the King**
If the Ghost is lying (sent by the Devil) . . .	*False positive:* Hamlet is mistaken and is guilty of regicide; he faces eternal damnation.	Hamlet has done the right thing by foiling the Devil's plan.
If the Ghost is telling the truth (his father's spirit) . . .	Hamlet has done the right thing; he has avenged his father's murder.	*False negative:* Hamlet has failed in the son's duty to avenge his father's murder.

You will remember that Hamlet confides to Horatio that he believes the Ghost to be credible but that he needs more information. The first step in his as yet ill-formed plan, then, is deciding to feign a mental illness that can serve as a cover for an attempt to reveal the king as a murderer. And so, he says to his friend Horatio, "I perchance hereafter shall think it meet to put an antic disposition on" (1.5.186).

Throughout act 2, the prince presents this "antic disposition" first to Ophelia, then to Polonius, Rosencrantz, and Guildenstern. Soon he seizes an opportunity to use *The Mousetrap*, the play-within-a-play, as a sort of "lie detector" to confirm the king's guilt. But other psychological forces are also at work.

We know that merely changing one's behavior (e.g., putting on an "antic disposition" or a "happy face") can affect a person's emotions—and sure enough, Shakespeare adds a dose of melancholy to Hamlet's feigned erratic behavior.[35] The result, in act 3, is that Hamlet sees a new alternative: suicide, which he frames as "to be, or not to be." Claudius's response to the play, however, changes the melancholy to anger. Now Hamlet begins to act merely on impulse, erroneously killing his girlfriend's father, thinking it is Claudius. In the terms of psychologist Haidt's dual-process mind, the "rider" has lost control of an enraged "elephant."[36]

If we conceive of Shakespeare as a proto-psychologist, then we can see table 9.1 as the framework for a psychological experiment to be conducted by Hamlet.[37] He has a *hypothesis* (the king is guilty!) that he seeks to test and thereby determine whether the Ghost is lying.[38] The experiment itself involves staging the play-within-a-play to elicit the king's response when he sees a reenactment of the murder similar to the regicide that Hamlet believes Claudius perpetrated. The "data" he plans to collect consist of observations of the king's reaction, to be judged by Hamlet, with the help of his friend Horatio.[39]

No, we are not suggesting that Shakespeare invented the experimental method. Rather, we are saying that he was presenting Hamlet as the most rational of characters, realizing that such people need evidence on which to base their actions. We are also recommending the application of the rubric in table 9.1 to other Shakespearean characters and the decisions they make, perhaps allowing us to see how differently the playwright portrayed his less rational creations. For example, in Lear's case, the king leaps to the conclusion that Cordelia does *not* love him, an inference that creates a tragic *false-negative* error (not believing Cordelia when she was telling the truth).[40] Or, in the case of Othello, the Moor makes a *false-positive* error by believing Iago's suggestion that Desdemona had been unfaithful, which was false. Neither Lear nor Othello considers an alternative or even ponders what might be "the question."

You will remember that in *Pericles* we find a different prince facing a dilemma that is much like Hamlet's, as the prince of Tyre solves the riddle posed by the king of Antioch. Pericles knows that Antiochus will kill him whether he gives the correct answer or not. Unlike Hamlet, Pericles solves the problem by fleeing the country, making us realize, from another perspective, that Hamlet's *blind spot bias* has prevented him from seeing other solutions to his problems.

HOW DID ELIZABETHAN AUDIENCES UNDERSTAND *HAMLET*?

Some critics have suggested that *Hamlet* was Shakespeare's metaphor for a world turned upside down. By the end of the Middle Ages, the Church had lost much of its credibility for its inability to reconquer the Holy Land, to stem the plague, and to clean up its own corruption. Meanwhile, as we have seen, Copernicus and Galileo were challenging the Earth-centered universe, Martin Luther was launching the Reformation, the New World was being explored, and Arabic numerals were being introduced to England.[41]

The Globe's less cerebral patrons, however, would have thought of Prince Hamlet mainly in terms of the four humors. Because nearly everyone was familiar with the humor theory, playwrights commonly associated their melancholy characters, like Hamlet, with black bile, or melanin, from which the term *melancholia* was derived.[42] On stage, those characters were usually rendered as "lean and pale, moving slowly, sad and brooding, perhaps suspiciously looking around for enemies."[43] Melanin was also the humor associated with creativity and genius, and so it was assumed to be prevalent among writers and poets, including, perhaps, Shakespeare himself.[44] Scholars, too, were thought to be touched with melancholy—and so, Hamlet is, fittingly, a student.

According to Paul Jorgenson, the physicians and moral philosophers of the Renaissance customarily diagnosed mental illness as having one of two sources: the body or the mind.[45] If the cause of melancholy was deemed to be in the body, it might be treated by bloodletting, to release the excess black bile, followed by a strict dietary regimen. Such is not the case, however, with Hamlet—at least as perceived by his Uncle Claudius. If the cause were a "rooted sorrow"[46] in the mind, it would be more likely treated with rest, the companionship of "a faithful friend," or religious counseling. Initially, Claudius does seem to favor the latter approach, explaining to Rosencrantz and Guildenstern why they have been asked to come to Elsinore:

> CLAUDIUS: . . . by your companies
> To draw him on to pleasures, and to gather
> So much as from occasion you may glean,
> Whether aught, to us unknown, afflicts him thus,
> That, open'd, lies within our remedy.
>
> —*Hamlet*, 2.2.1

King Claudius at first seems confused about the prince's strange actions and disrespectful tongue.[47] But by the third act, he has decided on melancholy, saying:

> CLAUDIUS: There's something in his soul
> O'er which his melancholy sits on brood,
> And I do doubt the hatch and the disclose
> Will be some danger.
>
> —*Hamlet*, 3.1.176

The audience soon realizes, however, that Hamlet is not only morose but angry. He does not suffer fools, such as Polonius, with equanimity, and he is even more ill-disposed to those who he suspects are complicit in his father's death. In modern terms, Shakespeare gives us a Hamlet that is depressed and angry but also one who would score high on *neuroticism.*

Shifting our perspective back to the framework proposed by Jaques in his Seven Ages of Man,[48] we can see Hamlet as a college boy, perhaps a scholar, beyond the schoolboy stage, who matures quickly when he returns home to Elsinore to discover his father's apparent murder, his mother's likely betrayal, and the possible loss of his inheritance. He also mourns the loss of a father, whom he admired. At this stage in life, Hamlet would otherwise have been "the lover," betrothed to Ophelia, who is also plagued with melancholy.[49] But perhaps he now fits better the description of the cunning "soldier" who lays a battle plan, beginning with *The Mousetrap* and culminating in a swordfight that is not merely lethal to a few courtiers but brings down the whole kingdom. This, then, is the *developmental crisis* that we have found to be a trademark of Shakespeare's histories and tragedies. And the tragedy is that he does not handle it well; he does not overcome the evil forces impinging on him.[50] He is alone—except for Horatio, who is more a comforter than a strategist. No one else in Denmark seems to be worried about the change in the occupancy of the throne, or even about Fortinbras and his Norwegian horde on their doorstep.

Now, if you will permit us a theatrical "aside," we would suggest that *Hamlet* may contain a great Shakespearean pun, lurking in plain sight and sound. It involves a custom we mentioned earlier: In the Scandinavian cultures of the Viking age, disputes were settled, crimes adjudicated, and kings selected by a communal gathering of freemen known as the *thing*; such a political structure likely would have been in place in Hamlet's Denmark of 1050 (the era of the Amleth story).[51] Moreover, the *thing*-based system lasted well into the fourteenth century, although we cannot be certain that Shakespeare knew of it. Yet consider that Hamlet's use of *The Mousetrap* to elicit a guilty reaction from the king, in the presence of all the courtiers assembled for the play, has strong echoes of the *thing*. Perhaps, then, the Bard was punning, as he was wont, when he wrote Hamlet's famous lines, "The play's the thing / Wherein I'll catch the conscience of the king."

ONCE MORE INTO THE MIND OF THE BARD

Ironically, Shakespeare himself seems to be the anti-Hamlet: all about emotion—love, jealousy, fear, revenge, lust, power, hubris, betrayal, anger, treachery, and mirth. But was he also a man of reason? After all, he ignored the great revolution in reason occurring all around him: the development of science. Or did he?

His was the era of the English Renaissance, which saw the rise of *humanism* and *secularism.* Galileo and Shakespeare were born the same year; Francis Bacon, often called the father of the scientific method, was the Bard's contemporary. Copernicus's heliocentric conception of the universe had stirred up a huge controversy, adding fuel to the fires of the Reformation.[52] And it was the age that saw the first light of the coming Age of Reason—also known as the Enlightenment.

We must realize, however, that many of the momentous early discoveries, such as Galileo's upsetting observation of the moons orbiting Jupiter, came late in Shakespeare's life, yet there are definite traces in the Bard's work that indicate he was paying attention to the discoveries wrought by observation and induction. We have seen, for example, that he knew of the anatomical revelations of Leonardo and Vesalius, working them into his plays with mentions of the *pia mater* and the *ventricles* of the brain. He was also aware of the new lands discovered across the Atlantic by the likes of Columbus and Henry Hudson. But perhaps most stunning (for our purposes) is that we can consider William Shakespeare a psychologist-before-psychology, having anticipated discoveries that would be made by the science of psychology some centuries hence.

Paradoxically, Shakespeare was not always kind to learned men.[53] In *Love's Labour's Lost*, for example, he shows little patience with pedants, whom he lampoons as merely trying to assign things to categories:

> KING: Study is like the heaven's glorious sun
> That will not be deep-search'd with saucy looks:
> Small have continual plodders ever won
> Save base authority from others' books
> These earthly godfathers of heaven's lights
> That give a name to every fixed star
> Have no more profit of their shining nights
> Than those that walk and wot not what they are.
>
> —*Love's Labour's Lost*, act 1, scene 1

Here his message is clear. If we become enmeshed in the meaningless minutiae of mere classification, we destroy the wonder and beauty in nature.

Astronomers seem not to have taken offense, however, for they have long embraced the Bard. In 1787, when John Herschel first discovered moons orbiting Uranus, he named two of them Oberon and Titania (from *A Midsummer*

Night's Dream) and another Ariel (originally from Alexander Pope's *The Rape of the Lock*, but also the name of the wizard's sprite in *The Tempest*). With that, Herschel started a tradition followed by astronomers who discovered an additional slew of Uranian moons.[54] (The current list of the Shakespearean moons orbiting Uranus numbers twenty-five, including ten named after characters in *The Tempest*. In an endnote, we give their names, along with the plays from which the names were borrowed.[55])

As if to anticipate the favor, Shakespeare's plays are full of references to the stars, moons, planets, comets, and of course in those times, to astrology. The Bard knew that our Moon caused the tides, yet he knew of only five planets: Mercury, Venus, Mars, Saturn, and Jupiter. (Earth was not considered a planet until Copernicus prevailed, with considerable help from Galileo.) Galileo's discovery of Jupiter's moons occurred late in Shakespeare's career but arguably influenced the otherwise-puzzling Jupiter-dream scene in *Cymbeline* in which Posthumus dreams that the Roman god Jupiter is circled by four dancing ghosts. Writer Dan Falk has interpreted this scene as Shakespeare's awareness of Galileo's seminal discovery of the four moons circling the planet Jupiter.[56]

Lurking in *Hamlet*, we find one more source of evidence suggesting that Mr. Shakespeare was anything but oblivious to the scientific discoveries of his time. For that, let us return to the supernova of 1572 to explore the link between Shakespeare and the famous Danish astronomer Tycho Brahe, after whom the supernova of 1572 is named.[57] The link will take us on a convoluted path beginning in London and ending with a new context for understanding both Mr. Shakespeare and his *Hamlet*.

THE LONDON CONNECTION TO ELSINORE

Our playwright lived in many different places during his residence in London, but the location perhaps best confirmed was in the Cripplegate area, about two blocks from the home of the famous English mathematician and astronomer, Thomas Digges. And it happens that Will Shakespeare was a friend of Digges's son Leonard, a fellow poet. In fact, it was Leonard Digges who wrote the following in a preface to *The First Folio*, published as a memorial volume some seven years after Shakespeare's death:

> This Book,
> When brass and marble fade, shall make thee look
> Fresh to all ages: when posterity

> Shall loath what's new, thinke all is prodigy[58]
> That is not Shakespeare's;
> . . . .
> Be sure, our Shakespeare, thou canst never die
> But, crown'd with laurel, live eternally.
> —*First Folio*, Preface by Leonard Digges

It also happens that Leonard's father, Thomas Digges, was a friend of Tycho Brahe, to which their exchange of letters testifies.[59] So, it could easily have been Thomas who told Shakespeare about the astronomical location of the supernova in relation to the pole star.[60]

Digges was a rich man and also a celebrity for his scientific work, which included introducing the Copernican theory of the Sun-centered universe to England. In addition, he rejected the medieval idea that the Sun, stars, and planets moved on enormous crystalline spheres in the heavens. And he was the first to propose that the stars populated an "infinite space"—an idea that still boggles the mind. In *The Science of Shakespeare*, Dan Falk reminds us that Hamlet, speaking to Rosencrantz and Guildenstern, says, "O God, I could be bounded in a nutshell and count myself a king of infinite space—were it not that I have bad dreams" (2.2.273). Because the phrase "infinite space" was not a common expression at the time, Falk asserts that it likely came from Digges.[61]

As evidence of a Bard–Brahe connection, then, we have Tycho's star, the supernova, described in act 1 of the play; we have his friendship with Digges, whose son Leonard was Will's close friend and admirer; and we have the concept of "infinite space" that Shakespeare likely heard (or read) from Digges. What else?

We know that Tycho Brahe initially became famous for his careful observations of the supernova, which so impressed Denmark's King Frederick II that he gave Brahe an island in the strait separating Denmark from Sweden. There the king built for Brahe a lavish abode and an observatory known in Uraniborg as "Brahe's castle"—just across the water from another castle, newly refurbished by the king. Shakespeare called that castle "Elsinore." Danish astronomer Tycho Brahe would have been able to see Elsinore from his estate on the island of Hven in the Strait of Øresund. The nova (exploding star) mentioned briefly in act 1 of *Hamlet* is often called "Tycho's star" (see figure 9.2).

One might think Brahe an odd duck. Friends and acquaintances alike saw him as a brilliant eccentric with a reputation for a bad temper. When he was about twenty, he lost a substantial part of his nose in a sword duel (in the dark!)

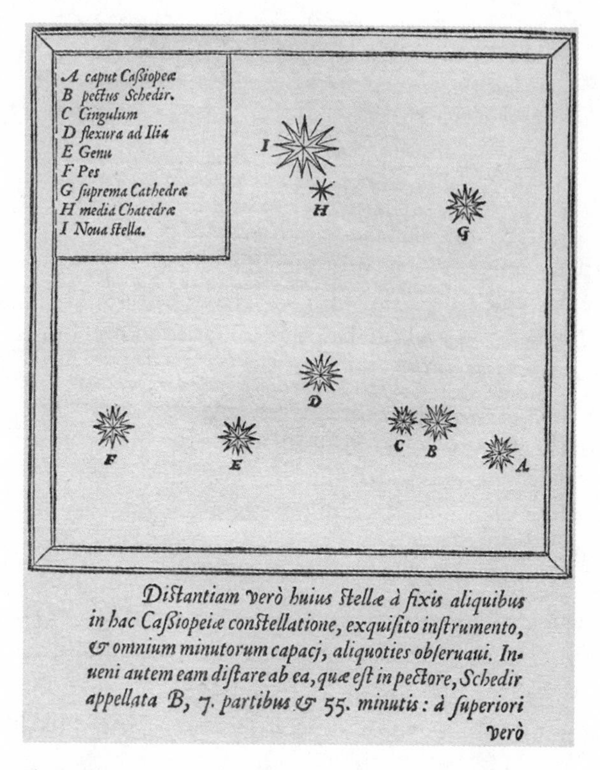

Figure 9.2 Tycho Brahe's drawing of the nova of 1572 (labeled "I" above) in the constellation Cassiopeia. The nova is also mentioned in *Hamlet*, act 1, scene 1. *Wikimedia Commons.*

with a fellow nobleman who was also his cousin; their dispute was over who was the better mathematician. Thereafter Tycho wore a prosthetic nose, initially one made of wax and later one that he fabricated from brass and repeatedly (perhaps obsessively) polished with oil.

Despite his partial proboscis, Brahe had a vain streak, exemplified by an engraving he commissioned of himself surrounded by an archway with the family crests of sixteen famous relatives. He sent copies of this print to friends in the scientific community of the time, notably one to Thomas Digges, who hung it in his house in Cripplegate. There Shakespeare is likely to have seen it while visiting his friend Leonard Digges. You can see a copy of it, too, in figure 9.3, where you will note that one of the crests carries the name Rosenkrans, and another, Guildensteren.

Figure 9.3 Etching of Tycho Brahe, with Rosenkrans and Guildenstern crests. *Wikimedia Commons.*

WHAT CAN WE LEARN ABOUT PSYCHOLOGY FROM *HAMLET*?

Like most of his great literary predecessors, Shakespeare wrote tales based on the observation that emotion can dominate reason without our realizing it; his prince of Denmark is the archetype for that most-human problem.[62] Moreover, he realized that what may seem to be *reason* is often a mishmash of what psychologists call *cognitive biases* and faulty *heuristics*—the flawed mental shortcuts served up by our unconscious minds.[63] Aristotle called them "tragic flaws." Will Shakespeare didn't use any of those terms, but he wrote many of those mental faults into his characters. Here are a few more examples:

We usually see what we expect or want to see (and the same goes for our other senses). That is the essence of *confirmation bias*, which we encountered earlier in a comedic scene in *Twelfth Night*, as Malvolio reads a counterfeit message that he presumes is from the woman with whom he is infatuated. Aside from some of Shakespeare's bawdiest text, the message contains these famous words, which Malvolio believes she meant to apply to him: "Some are born great, some achieve greatness, and some have greatness thrust upon 'em" (2.5.143).

In a much darker play, Shakespeare again gives an extended illustration of confirmation bias as the vengeful Iago plants suggestions in Othello's mind that Desdemona is having an adulterous affair with Lieutenant Cassio. To confirm these suspicions, Iago hatches a plan to obtain a distinctive handkerchief (which he calls a "napkin")—one that Othello had gifted to Desdemona—and hide it in Cassio's belongings.[64] Says Iago:

> I will in Cassio's lodging lose this napkin,
> And let him find it. Trifles light as air
> Are to the jealous confirmations strong
> As proofs of holy writ.
>
> —*Othello*, 3.3.367

So, when Othello sees Cassio with the handkerchief, he is certain that Desdemona and Cassio have been having a sexual relationship—or as Shakespeare calls it more graphically, "making the beast with two backs" (1.1.129). Othello accordingly sees what he has been led to expect. Not only is *Othello* a play all about the hazards of *confirmation bias*, but Shakespeare even uses the term *confirmations* when he has Iago say, "Trifles light as air / Are to the jealous confirmations strong" (3.3.367).

We find confirmation bias yet again in *Hamlet*—in the prince himself—when the angry King Claudius rises and demands an end to the performance of a play that parallels the murder of Hamlet's father. Hamlet selectively attends to information that already fits with his preconceptions, and so Hamlet leaps to the conclusion that the king is guilty (although the king's response only adds to suspicion, rather than providing absolute proof of his guilt). As he says to Horatio after seeing Claudius's reaction to the play, "I'll take the ghost's word for a thousand pound" (3.2.312).

King Claudius's reaction also suggests to the audience that he is guilty of killing his brother and grabbing the crown, even though the evidence, at first, is far from conclusive. And so, Shakespeare has cleverly "primed" the audience to accept Claudius's response to *The Mousetrap* as clear and public evidence of his guilt.[65]

In his "to be or not to be" soliloquy, Hamlet acknowledges that emotions can influence reason, for he speaks of heart-ache, dread, and a weary life, going on to say, "And thus the native hue of Resolution / Is sicklied o'er, with the pale cast of Thought" (3.1.64). Yet he seems to believe that he can hold emotion at bay while coming to a rational decision about how to handle his "sea of troubles" (3.1.64). He does not seem to realize (in Jonathan Haidt's terms) that he is also the rational "rider" of an intuitive "elephant" and that the difficulty he is having in making sense of events is also making him melancholy, or depressed. Consciously, he is feigning madness with his "antic disposition" (1.5.186), but the melancholy "elephant" of his unconscious is biasing the conscious thinking of the "rider." Haidt calls this the *rationalist delusion*.[66]

Hamlet is a bundle of cognitive biases that we all share to some degree. Consider this passage, in which he ceremoniously begs Laertes to pardon him for rashly killing Polonius, Laertes's father:

> HAMLET: And you must needs have heard, how I am punish'd
> With sore distraction. What I have done,
> That might your nature, honour and exception
> Roughly awake, I here proclaim was madness.
> Was't Hamlet wrong'd Laertes? Never Hamlet:
> If Hamlet from himself be ta'en away,
> And when he's not himself does wrong Laertes,
> Then Hamlet does it not, Hamlet denies it.
> Who does it, then? His madness: if't be so,
> Hamlet is of the faction that is wrong'd;
> His madness is poor Hamlet's enemy.
>
> —*Hamlet*, 5.2.240

Psychologist Mark Cleaveland suggests that Hamlet is displaying a *self-serving bias*, which is a way of denying responsibility by blaming one's mistakes on something beyond one's control. Says Cleaveland:

> Hamlet's understanding of his own emotions is essentially . . . "I don't know what came over me," or the "it wasn't me, it was the alcohol speaking." It's what will lead philosophers to place feelings and emotions outside of reason. Animals might possess emotions, but we humans are capable of reasoning . . . except when those pesky animalistic emotions get in the way.[67]

Like all of us, Hamlet sees psychological flaws in others more clearly than he does in himself, as when he lectures his mother about the immorality of marrying Claudius. He fails to realize how much he is driven by emotion, while at the same time believing that others are foolish and irrational. Psychologists call this the *blind spot bias*.[68] We glimpse some of this in his brooding and melancholy character. It is not just his put-on "antic disposition." We see it also in his flashes of anger, as when Hamlet draws his sword and fatally stabs a figure (that turns out to be his girlfriend's father) hiding behind the drapes in his mother's room. Here Shakespeare reveals an angry, impulsive side of the prince, showing us a Hamlet who has a streak of hair-trigger unpredictability in his nature.

At some level, however, Hamlet recognizes his own bias for elevating reason and discounting the role of emotion in his thinking. As he says to Rosencrantz and Guildenstern, too much thinking can sometimes make things worse:

> HAMLET: Denmark's a prison.
> ROSENCRANTZ: Then is the world one.
> HAMLET: A goodly one, in which there are many confines, wards,
> and dungeons, Denmark being one o' th' worst.
> ROSENCRANTZ: We think not so, my lord.
> HAMLET: Why, then, 'tis none to you, for there is nothing either
> good or bad, but thinking makes it so. . . .
>
> —*Hamlet*, 2.2.262–70

EPILOGUE

Psychology, Shakespeare, and Beyond

We set out to learn what Will Shakespeare had to say about psychology, and we found his psychology everywhere. Among the first clues was his felicitous phrase *nature-nurture*, raising the issue of heredity vs. environment that now lies at the heart of modern psychology. But we found that he wrote entire plays that explored controversies still raging within psychology today. *Measure for Measure* and the Henrys tetralogy address the person vs. situation issue; *A Midsummer Night's Dream* deals with the interaction of consciousness and the unconscious, while *Love's Labour's Lost* and *Hamlet* look at the emotion vs. reason issue from different perspectives.

We noted that much of the Bard's psychology comes from the ancient humor theory that psychology has long since abandoned. Yet he was amazingly prescient in his description of a dozen mental disorders that were unknown to the medicine of his time, his anticipation of ego defense mechanisms, and an understanding of sophisticated concepts that we now call, for example, *negative reinforcement*, *confirmation bias*, and *cognitive dissonance*. Moreover, we discovered that his views on psychological development mirrored those of Erik Erikson.

We cautiously attempted some analysis of Shakespeare himself through the lens of modern psychology. Using Howard Gardner's framework, we noted that the Bard was a man of high verbal, intrapersonal, and interpersonal intelligence. Socially, we have seen that he was beloved by his friends in the London theatrical scene (except in one celebrated case). We showed that, as he matured, our playwright's psychology of "redemption" became more and more important. And although he spent most of his working life in London, his heart seems to have remained in the countryside, to which he returned in retirement.

In terms of the Big Five traits, we found that he was open to experience, conscientious, introverted (in the classical Jungian sense), agreeable enough to be loved by his peers, and perhaps just a little off-center on the anxiety/neuroticism dimension.

What we have *not* done so far is attempt an analysis of Shakespeare on his own terms—that is, through the lens of his humor theory. We propose to rectify that omission now, in brief, because so little is known about him. To do so, we turn again mainly to comments made about him by his contemporaries and filter them through the concepts associated with the four humors: blood, black bile, yellow bile, and phlegm.

Nearly all who knew Mr. Shakespeare (and left written records of their opinions) spoke of him in positive terms. We used those positive comments to "diagnose" the Bard as *agreeable*. In humoral terms, agreeableness would translate as *sanguine*, or having an abundance of blood among his humors. We also noted that poets were assumed to have a touch of *melancholy*, so that would have suggested that he had more than the usual amount of black bile (melancholer) in his system. As for *yellow bile*, we guess that his amiable character hinted that he was low in choler and the anger associated with it. That leaves *phlegm*, a humor of which he must have been nearly devoid. No one with a life's work approaching that of the Bard could have been considered sluggish or phlegmatic.

Yet, as young Will Shakespeare soon discovered during his sojourn in London, the humors don't tell the full story of a personality or a person's life, nor does a mere list of traits. (See figure E.1.) We can only hope that further details about the man will eventually emerge, giving us a narrative of the Bard's life along the lines described by Prof. Dan McAdams.

What else can we say about Will Shakespeare in terms of his own psychology? Judging from his plays, he seems to have seen life in terms of stages involving crises to be resolved, much as Erikson claimed. Intellectually, he was a man of quick wit and enormously broad interests endowed with an astonishing level of creativity. Remarkably, his portrayals of Rosalind, Beatrice, Portia, Viola, and others betray a most uncommon understanding of women for his time.

In sum, we declare that Will Shakespeare was much as we had hoped: an affable fellow with wide-ranging interests, a quick sense of humor, and a razor-sharp intellect—a gentleman with whom we would have enjoyed sharing some hours and, perhaps, a few pints.

Figure E.1 The Cobbe Portrait of a young-looking William Shakespeare. Compare the hair in this image with the balding pate in later portraits. This portrait is believed (by some, not all) to be the only painting of the Bard made during his lifetime. *Wikimedia Commons.*

Finally, we would remind you why we originally set out to do this book. Initially, our reason was that while the Bard had been subjected to recurrent psychological analysis, no one had attempted a comprehensive look at Shakespeare's own psychology. That, in itself, was intriguing. But one other motive was even more compelling: to suggest how psychology is near the center of a web that connects nearly every aspect of human endeavor. It is a profound idea, yet one that, in our experience, is rarely taught or, perhaps, even more rarely learned. We have written *Psychology According to Shakespeare*, then, to make the notion of connections among ideas and disciplines more salient.

Psychology brought us to Shakespeare, but we don't want to leave the impression that Shakespeare is all about psychology. In fact, his work is about nearly everything. No matter what your interests or specialization, it is likely that you can find some connection with Shakespeare. Shapiro's book *Tyrant*, for example, links Shakespeare and politics, just as Barash did with Shakespeare and evolutionary psychology. And it was Carl Sagan who told us that the minerals in our teeth were formed of elements produced by ancient explosions of supernovae—and, lo! We find that Shakespeare mentions the first naked-eye nova seen by earthlings in nearly four hundred years.[1] That's two degrees of William Shakespeare![2]

If you are a teacher or professor or parent, you have the opportunity to help your charges see Wm. Shakespeare's work from multiple perspectives. We give, in an endnote, some examples from several disciplines.[3]

It was C. P. Snow who first wrote about the "two cultures" of the sciences and the humanities, fearful that the twain were drifting ever farther apart. But it's no longer just a gulf between the sciences and the humanities. As people communicate more and more with others in their own capsule of knowledge, they interact less frequently with those of other specializations, even within their own disciplines. So, experts on Shakespeare talk to each other, but they may have no time for those who specialize in Beowulf or Homer or math or biology—or psychology. Our ulterior motive, then, is to reestablish the idea of *connections* among diverse disciplines—to show that different fields and different perspectives can inform each other. To that end, we think we have shown that psychology can be useful in understanding literature—but the reverse is also true.

If we succeed, our book will be a minor point of light sharing the firmament with the brighter light of Jared Diamond, whose *Guns, Germs, and Steel* taught millions about the connections linking global inequities in health, wealth, technology, and geography. We would also be honored to be a mere twinkle compared to the star of Lin-Manuel Miranda, who linked early US history and hip-hop to create his *Hamilton* masterpiece. Or consider Don McLean, who connected his music ("Vincent," or "Starry Night") with the paintings of Vincent van Gogh to show us things about the painter's mind that cannot be found either in textbooks or the *DSM*-5 psychiatric manual. Add physicist Murray Gell-Mann, whose brilliant "eightfold way" model of elementary particles that form the stars, as well as ourselves, was inspired by the eightfold way pattern found in Buddhism—which led to the discovery of "quarks," which he named after a term he found in James Joyce's *Finnegan's Wake*.[4]

Walter Isaacson, author of a recent biography of the quintessential polymath, Leonardo da Vinci, wrote: "The ability to make connections across disciplines—arts, sciences, humanities, and technologies—is a key to innovation, imagination, and genius."[5] We would add, for those who seek to brush up their Shakespeare, it wouldn't hurt to brush up their art, music, culture, biology, literature, language, linguistics, astronomy, mythology, politics, poetry, history, philosophy, and religion . . . and psychology, as well.

ACKNOWLEDGMENTS

While this book describes the discoveries we made that connect Mr. Shakespeare with modern psychology, much of the material has been around for a long time—coming from dozens of scholars and teachers whose work is scattered in books, journals, lectures, and websites across the internet. We were surprised to find that it had never been brought together in one volume. To these myriad sources we are deeply grateful and have endeavored to give them the credit they are due.

We also drew heavily on a small group of well-known literary experts whose work we would commend to readers who want to learn more about the Bard. These include Stephen Greenblatt, author of the seminal *Will in the World*; James Shapiro, especially his brilliant volume *A Year in the Life of William Shakespeare—1599*; Jonathan Bate, who so thoughtfully wrote *Soul of the Age*; and the late Harold Bloom for his formidable tome, *Shakespeare: The Invention of the Human*.

But the more we read with the idea in mind that Shakespeare was a psychologist-before-psychology, the more we realized that Shakespeare belongs to everyone and to all fields of study. Among the psychological scientists who have influenced us, we would single out Daniel Kahneman for his book *Thinking Fast and Slow* and Jonathan Haidt for *The Righteous Mind*. Both have masterfully explicated the new science of the *dual-process mind*, which we argue Will Shakespeare anticipated. We have also borrowed ideas (with attribution, of course) from evolutionary psychologist David Barash, who wrote *Homo Mysterious* and penned, with his daughter Nanelle (a literary scholar), *Madame Bovary's Ovaries: A Darwinian Look at Literature*. Among journalists, we extend our gratitude to Dan Falk for *The Science of Shakespeare* and to Bill Bryson for the eminently readable little biography of the Bard entitled *Shakespeare: The World as Stage*. Finally, we are grateful to the late Isaac Asimov for a trove of material explaining the historical context of Shakespeare's plays in *Asimov's Guide to Shakespeare*.

And then there are those, nearer to us, who have helped us shape the manuscript into what we hope is a compelling form. Our go-to Shakespearean expert, Susan Stitham, of Southern Oregon University, read our every word and flagged our blunders, omissions, and points of confusion. Thank you, Susan! And, of course, we extend our thanks to our enthusiastic agent, John Willig, who also made many excellent suggestions that got the attention of editor Jonathan Kurtz at Prometheus Books, where he and his remarkable and patient editorial assistant, Brianna Soubannarath, guided the manuscript into production.

Phil Zimbardo also extends his praises to two special individuals: his wife, Christina Maslach, and his writing partner, Bob Johnson. Says Phil, "Christina has supported me in many ways for more than fifty years during our wonderful marriage. She also qualifies as a hero who forced me to terminate the Stanford Prison Experiment earlier than planned by challenging my authority openly. And Bob has been a wonderful collaborator over many years during which we wrote many editions of a best-selling psychology textbook. However, Bob has continually amazed me with the depth and breadth of his intimate knowledge of all of Shakespeare's canon."

For his part, Bob Johnson is ever so thankful for the guidance, patience, and support of his wife and life partner, Michelle, who had the savvy and courage to tell us when and where our writing needed further work. She is the best DE (developmental editor) ever! Boundless gratitude and thanks, too, to Rebecca Graham, his daughter, for an extensive, sharp-eyed, and wise proofreading of the manuscript. Bob would also like to thank their dear friend and Shakespeare fan, Mary Bagshaw, for reading and commenting on each chapter (sometimes several times) as a stand-in for our intended audience of intelligent and educated readers who are interested in psychology and Shakespeare but who are not necessarily experts in either field. And certainly not least, reciprocal thanks go to Bob's long-time colleague, friend, and writing partner, Philip Zimbardo, for the opportunity to collaborate with him again on this fascinating project.

Finally, as psychologists, rather than experts in English literature, we acknowledge and confess that each of us has one foot outside our customary boundaries. We want all readers to know that we do not intend this book to replace the work of those trained in literature. Rather, we hope that our efforts will help frame another perspective—a psychological window on the Bard's works. With this book, we aspire to attract a new audience to the old Bard and to inspire other scholars, like ourselves, to explore interdisciplinary connections between the sciences and the humanities—perhaps as an academic analog of the dual-process mind.

NOTES

PROLOGUE

1. The Necker Cube is an ambiguous figure named after geologist Louis Necker (1786–1861), who first recognized the illusion while studying rectangular crystals and subsequently published it in the *London and Edinburgh Philosophical Magazine and Journal of Science* in 1832. See: "Necker Cube," Wikipedia. Last modified January 23, 2023, https://en.wikipedia.org/wiki/Necker_cube.

INTRODUCTION

1. See a demonstration of original pronunciation in a YouTube video featuring a director and an actor from The Globe. "Shakespeare: Original Pronunciation," https://www.youtube.com/watch?v=gPlpphT7n9s&t=6s.

2. This reference indicates that the passage in *As You Like It* comes from act 2, scene 7, lines 25–30. All references to line numbers in Shakespeare's plays refer to the New Folger Library editions, unless otherwise noted. Line numbers may be slightly different in different editions.

3. Shakespeare's trademark is the pun: a play on words, usually ones with similar sounds that produce a double meaning, as in "hernshaw" and "handsaw." The literal meaning in this punning speech is this: When facing north, away from the wind, the sun was at one's back, making it easier to distinguish one bird from another. Metaphorically, of course, Hamlet was saying that he was in full possession of his faculties.

4. This quote comes from Ben Jonson's eulogy for Shakespeare, appearing in the Preface to the *First Folio*. See: Shakespeare Online, "Preface to The First Folio."

5. Well, there were thirty-eight when we commenced this project, and now there are two more! No, Shakespeare has not reappeared, Elvis-like. Rather, two more of his works have been discovered—or, more accurately, two works that were previously known are now attributed, in part, to collaborations between Shakespeare and others. They are *Edward III* and *Double Falsehood*. So, now the canon has expanded to two score.

6. In Gaelic, *bard* is the term for a minstrel, storyteller, or poet. "the Bard" (capitalized and with the definite article) has come to be the honorific title for Shakespeare as the preeminent poet of the English language.

7. Spoken by Hamlet's friend Marcellus—not by Hamlet (1.4.100)!

8. Barash and Barash, *Madam Bovary's Ovaries.*

9. According to Harold Bloom, Shakespeare had likely read a translation of *Don Quixote.* See: Harold Bloom, "Knight in the Mirror."

10. *Hamlet* (3.2.24-25), quoted in Barash and Barash, *Madam Bovary's Ovaries.*

11. In modern terms, psychology is the science of individual behavior and mental activity. The founding of scientific psychology is usually credited to Wilhelm Wundt, who established the first psychology laboratory in 1879. In Shakespeare's time, the scientific method, involving predictions and careful observations made under controlled conditions, had only recently been conceived (by Bacon and others), but it had not yet been applied to human behavior and mental processes.

12. Harold Bloom, *Shakespeare: Invention of Human.*

13. Oedipus was a Greek king who inadvertently killed his father and unwittingly married his mother, as told in *Oedipus Rex*, a Greek tragedy by Sophocles. Freud taught that all young male children must deal with an *Oedipus complex*: an unconscious antipathy for the parent of the same sex and an erotic attraction for the parent of the opposite sex. Children (said Freud) must work through this by identifying with the same-sex parent and transferring their love to a mate of the opposite sex. The Oedipus complex was Freud's solution to the developmental problem of gender identity and sexual orientation. For the connection to Shakespeare, see: Freud, *Interpretation of Dreams* and Jones, *Hamlet and Oedipus.*

14. Contestants used blunted weapons. Nevertheless, casualties were as high as 10 percent. See: Elizabethan Era, "Elizabethan Tournaments"; Cartwright, "Sports, Games and Entertainment in the Elizabethan Era."

15. Bowling greens were everywhere in the city. "There's the rub" is Hamlet's famous allusion to the sport: "rub" refers to a flaw in the green's surface that interferes with the trajectory of the ball.

16. There is a notable exception for which the Bard got in trouble when he inadvertently offended an important family's ancestor. We'll explain that further when we focus on Falstaff in chapter 3.

17. Shakespeare is known for playing fast and loose with historical facts, especially in the interest of staying in the good graces of Queen Elizabeth. Most historians now say that Richard III was a rather good king, who had the misfortune to be killed on Bosworth Field by Henry VII, founder of the Tudor dynasty, of which Elizabeth I was the current ruler in Shakespeare's time. Richard III was the last of a long line of Plantagenet rulers from the houses of Lancaster and York, going back over three centuries to Henry II (father of Richard I, the Lion Heart) in 1154. Shakespeare also is responsible for penning Richard's plea offering to swap "my kingdom for a horse!"

18. You will find more examples of Shakespearean neologisms here: Betts, "40 Common Words"; Mabillard, "Words Shakespeare Invented."

19. The original text reads, "For the apparel oft proclaims the man," which is oft rendered in modern English as "Clothes make the man."

20. In the original, Hamlet says "heart of heart."

21. This play-within-a-play is also known as *The Mousetrap*.

22. Other ego defense mechanisms were at work here, as well. Shakespeare was implying that Queen Gertrude was *identifying* with the queen in the play. Moreover, she was using *projection*, another defense mechanism in which a person expresses his or her own desire while attributing it to someone else. None of this, however, requires that Shakespeare was assuming that these defense mechanisms required a Freudian-style unconscious mind.

23. For a more complete explanation of the *self-serving bias*, see *self-serving bias* in Wikipedia. The opposite tendency—for people to insist that others who make mistakes should be blamed for personal deficiencies or character flaws—is called the *fundamental attribution error.* See *fundamental attribtion error* in Wikipedia. Both biases are more common in Western individualist cultures than in the collectivist cultures of Asia. An interesting exception has been found among people in northern China. See: Biello, "Rice Farming."

24. Barkan, "Alzheimer's Caregivers Can Learn"; Furness, "Could King Lear Have Suffered Lewy Body Dementia?"

25. George Bernard Shaw coined the term *Bardolatry*, meaning an excessive and overweening reverence for Shakespeare.

26. The Seven Deadly Sins were first iterated by Pope Gregory I in AD 590; the Seven Deadly Sins include wrath, avarice, sloth, pride, lust, envy, and gluttony. Shakespeare never used the term "Seven Deadly Sins," yet each one appears in his plays. For example, in *Othello*, Shakespeare calls *envy* "the green-eyed monster"; similarly, Falstaff is the paragon of *sloth*.

27. A notable exception is the Deadly Sin of *lust*—thanks to the obsession of Sigmund Freud.

28. "Anon" = "soon."

29. We promise not to get into the authorship controversy. We see no reason to suspect that Shakespeare was not really the bard from Stratford. But if that is your interest, we suggest you get a copy of James Shapiro's *Contested Will: Who Wrote Shakespeare?* For an opposing view, see: Elizabeth Winkler, *Shakespeare Was a Woman and Other Heresies*.

30. Most of the solid facts we have about the Bard himself come from legal documents: real estate transactions, taxes, legal cases, and wills, along with certificates of births, baptisms, marriages, and burials. See: Folger Shakespeare Library, "Family, Legal, and Property Records."

31. Here are a few good sources commenting on Shakespeare's grammar-school education: Bate: *Soul of the Age*; Greenblatt, *Will in the World*; and Garber, *Shakespeare After All*.

32. We recommend a readable and informative secondary source on Shakespeare's life: Bill Bryson, *The World as Stage*.

33. Books were becoming widely available in Shakespeare's time; printers in London produced more than seven thousand titles during Elizabeth's reign. Presumably, Shakespeare had no access to university libraries, but he may well have borrowed books from friends and fellow playwrights; Ben Jonson was known to have an extensive library. His well-to-do patron, the Earl of Southampton, undoubtedly owned many books. Shakespeare was also acquainted with Richard Field, owner of a major London publishing house, with whom young Will grew up in Stratford; it is likely that Field loaned him copies of books. Moreover, we must remember that, at the grammar school he almost certainly attended, Shakespeare would have read classic texts, such as Ovid's *Metamorphoses*, Virgil's *Aeneid*, and Caesar's war *Commentaries* in the original Latin. For more information on Shakespeare's access to books, see: Bate, *Soul of the Age*, 131–146; Folger Shakespeare Library, "Books and Reading in Shakespeare's England"; Greenblatt, *Will in the World*, 193; Shapiro, *Contested Will*, 274–75; Wood, *Shakespeare*, 49–55.

34. Maids of honor were attendants to the queen, junior to the lady-in-waiting.

35. Shapiro, *1599: A Year in the Life*.

36. Cervantes's *Don Quixote* may have been the first modern novel. See Harold Bloom's argument, "Knight in the Mirror."

37. Francis Bacon was a founder of the scientific method of empirical observation and reasoning by induction.

38. Among many other scientific achievements, Galileo discovered the moons of Jupiter, which upset the Church-supported view that the Earth was the center of the universe. An interesting irony: although Shakespeare may (or may not) have taken note of the revolutionary discoveries made by Galileo and other contemporary astronomers, nearly all the moons of Uranus are named after Shakespearean characters, including Titania, Oberon, Ophelia, Miranda, Desdemona, Juliet, Portia, Rosalind, Mab, Caliban, Sycorax, Prospero, and Puck. (The idea that Shakespeare ignored or was unaware of the birth of science during his time is disputed, as we will see in chapter 9.)

39. Descartes made rational thought, instead of religious doctrine, the basis for scientific exploration of the natural world. He is renowned in psychology for his distinction between mind and body.

40. Montuori and Purser, "Deconstructing the Lone Genius Myth."

41. This quote is most famously attributed to Newton, but some variation of it had been used by other sources previously. See, for example, https://en.wikiquote.org/wiki/Isaac_Newton.

42. In *The Republic*, Plato presents his famed "The Allegory of the Cave," in which he describes imaginary prisoners, chained all their lives facing the cave wall. All they can see of the outside world are the shadows of people passing through the light behind them. The idea is that our perceptions are an imperfect representation of reality. You can find a copy of "The Allegory of the Cave" online at Project Gutenberg.

43. The "music of the spheres" was a concept going back to Pythagoras and continuing into the Middle Ages. This belief was dashed by the teamwork of Copernicus and Galileo.

44. Some would dispute this, seeing veiled references to discoveries by Galileo and others in Shakespeare's works. See: Falk, *The Science of Shakespeare*, and Elliot, "Shakespeare's Worlds of Science."

45. Some scholars note that "Bermoothes" may refer instead to a tawdry district in Shakespeare's London where the "dew" might refer to strong drink. It is also possible that the reference is a pun on both the Bermuda islands and the lowbrow locus of London. But see: Delahoyde, "*The Tempest*." Concerning Shakespeare's inspiration for these lines coming from the discovery of the New World, see Isaac Asimov, *Asimov's Guide to Shakespeare.*

46. We must pass on a little-known fact here. Gutenberg's original family name was Gensfleisch ("Gooseflesh" in English)—a name he detested. He later took his mother's family name.

47. For a more complete discussion of the religious conflict in Elizabethan England, see Greenblatt, *Will in the World.*

48. Shapiro, *Contested Will*; Bate, *Soul of the Age*; Garber, *Shakespeare After All*; Stephen Greenblatt, "Shakespeare Doubter"; and Friedlander, "Five Myths."

49. Jones, *Hamlet and Oedipus.*

50. Freud, *The Interpretation of Dreams.*

51. Harbage, *Conceptions of Shakespeare*, 1966; Wagenknecht, *The Personality of Shakespeare.*

52. Paster, "*Much Ado About Nothing*: A Modern Perspective."

53. This is Norman Holland's view, as explicated in: Holland, *Shakespeare's Personality*, 7–9.

54. Kahneman, *Thinking Fast and Slow*; Haidt, *The Righteous Mind.*

55. Crane, *Shakespeare's Brain.*

56. Gardner, *Creating Minds.*

57. Gardner, *Creating Minds.*

58. A word of caution: Gardner's work involves case studies of just seven individuals, not controlled experiments involving scores of subjects. As such, we must be careful in generalizing the results to other creative individuals or to people-in-general.

59. We have no original manuscripts with dates on which they were written. So, scholars infer the chronology of the plays by registration with the Master of the Revels, records of performances, and by references to events (such as Jack Cade's rebellion) and references to other contemporary plays. Accordingly, different sources list the plays in somewhat different chronological orders. Readers, therefore, should take dates for Shakespeare's plays as approximations. See: Garber, *Shakespeare After All.*

60. Lewis, "Mixture of Styles."

61. For an opposing perspective, see Pressley, "Mrs. Shakespeare: Anne Hathaway."

CHAPTER ONE

1. The date is the year in which *The Tempest* was first presented, at the court of King James I of England.

2. Jamestown was named in honor of King James I, who ascended the English throne upon Elizabeth's death in 1603.

3. See: Dungey, "Shakespeare and Hobbes"; Bowling, "Theme of Natural Order"; Mowat, "*Tempest*: A Modern Perspective"; Takaki, "*Tempest* in the Wilderness."

4. The "state of nature" debate went on for a century or more, so not all of these individuals were contemporaries of Shakespeare. "Nature" had a double meaning, involving not simply heredity but also the question of how "man" lived before civilization: in a "state of nature." See: Stables, "Unnatural Nature."

5. Hobbes, *Leviathan*.

6. The historian Herodotus recorded this in the second volume of his *Histories*. Herodotus was known to stretch the truth a bit, so we should add a bit of salt to his account of Psamtik I's experiment.

7. The ancient kingdom of Phrygia lay on the Mediterranean coast of Anatolia in what is now Turkey. Aside from being Psamtik I's kingdom, it was home to the legendary King Midas and to the famed Gordian Knot. It is not known whether the Phrygians made good *bekos*.

8. See a theatrical and historical analysis of these lines from *The Tempest* by actor Philip Voss via an internet search for "Philip Voss: Prospero from an Actor's Eyes."

9. The heredity-environment issue lies at the heart of psychology, where partisans still debate the relative influence of hereditary nature vs. nurture (also known as *experience* or *learning*) arising from our environment, including culture and upbringing.

10. Galton, *English Men of Science*.

11. Galton, by the way, was Charles Darwin's cousin.

12. Sometimes "nature" in this passage is misinterpreted as the mystical effect of a "wilderness" experience—not at all what the context of the passage suggests.

13. The legendary context of the Trojan War is also important in understanding Ulysses's speech. The Greeks are losing their war with Troy largely because they lack the services of their greatest warrior Achilles, who is sulking in his tent and refusing to fight. It's all about wounded pride: Agamemnon, leader of the Greek armies, has pulled rank on Achilles by taking his slave girl Briseis. Further, Achilles senses that the other Greek soldiers are ignoring him and making jokes behind his back. This is the context, then, in which Ulysses is trying to cajole Achilles to get back into the fight. In this speech, then, Ulysses is telling him that memory is short, and it is human nature for people to forget great deeds of the past and focus on new accomplishments ("new-born gawds"). This is the "one touch of nature" that people everywhere have in common. (Here Shakespeare is positing a universal trait common to all humanity.) Ulysses argues, to maintain his reputation as a great warrior, Achilles must continue to burnish it by getting back in the fight.

14. And so we can say that the Bard anticipated the *recency effect*, well-known by psychologists who study memory. The recency effect simply states that memory is better for recent events than for earlier ones. Lest you think this is merely "common sense," another discovery shows that it is not: while the most recent items on a list of

words or numbers are remembered better than those at the beginning of the list, the effect is not strictly chronological, for items in the middle are remembered most poorly of all. (Shakespeare left these nuances to be discovered by modern psychologists.)

15. We often refer to Shakespeare's times as "Elizabethan," after Queen Elizabeth I (1558–1603). Yet he outlived her by about seventeen years and wrote at least thirteen more plays under King James I, famed for the version of the Bible he commissioned. That period is called the Jacobean era because Jacob is the Latin rendition of James.

The Jacobean era was also noteworthy because of the infamous Gunpowder Plot of 1605, in which a group of Catholic rebels tried to blow up the Parliament building, with James inside. The king had the instigator, one Guy Fawkes, unpleasantly hanged, drawn, and quartered. For reasons understood mainly by them, the Brits even today enjoy celebrating events surrounding the Gunpowder Plot on Guy Fawkes Day, November 5. Shakespeare may have been moved by the Gunpowder Plot to write *Macbeth.* See: Mabillard, "Shakespeare and Gunpowder Plot."

16. Berg, "Shakespeare as a Geneticist."

17. God's presumed approval and blessing of the monarch and all of his (or her) actions was referred to as the Divine Right of Kings. We find the divine right as a central issue in *Richard II*, where Richard's cousin Henry Bolingbroke usurps the throne. This sets the stage for the Wars of the Roses between the houses of York and Lancaster. Shakespeare wrote an eight-play serial featuring the principal players in this conflict: *Richard II*, the two *Henry IV* plays, *Henry V*, the *Henry VI* trilogy, and *Richard III*. Brace yourself for more of this in chapter 3.

18. Hook, "Shakespeare, Genetics, Malformations."

19. It wasn't until about fifty years after *The Tempest* that the Dutchman Antonie van Leeuwenhoek invented a microscope that could see single cells. Leeuwenhoek's device, however, was not nearly powerful enough to resolve DNA, the long, ladder-like molecule that contains our genes.

20. Darwin proposed a theory, called Pangenesis, stating that all cells in the body shed minute *gemmules* that contained hereditary characteristics. These gemmules then circulated in the blood and collected in the gonads, whence they transmitted hereditary information to the offspring. See: Liu, "New Perspective on Darwin's Pangenesis," and Dobzhansky, Robinson, and Griffiths, "Heredity."

21. Cruttwell, "Physiology and Psychology." The National Library of Medicine also has an excellent discussion of humorism in Shakespeare's day: "And There's the Humor of It." And, in fact, there is much more of it in chapter 3, where we look at Shakespeare's understanding of personality.

22. In 1953, Watson and Crick announced their research showing that a molecule of DNA is shaped like a "double helix," much like a twisted ladder. Further, they proposed that genes are segments of DNA, analogous to beads in a necklace. (Or, using the ladder analogy, the genes are like a series of rungs on the twisted ladder.) Each gene produces a template for a protein, which, in the aggregate, form the structures of all the cells in an organism. See also: Pray, "Discovery of DNA."

23. The patronage of the Crown may have caused Shakespeare some cognitive dissonance. Elizabeth's father, Henry VIII, had feuded with the Roman Church and replaced it with his own Anglican variation. Stephen Greenblatt suggests that Shakespeare may have had Catholic sympathies (as his family certainly did), which was risky in Elizabethan England, not only to one's occupation but to liberty and life. See also: Greenblatt, *Will in the World.*

24. Greenblatt. *Will in the World.* For some recent discoveries about Will seeking the coat of arms that eluded his father, see also: Schuessler, "Shakespeare: Actor. Playwright."

25. Readers interested in the authorship controversy might enjoy James Shapiro's book *Contested Will* and Elizabeth Winkler's *Shakespeare Was a Woman.*

26. We cannot give Shakespeare full credit, however, for he seems to have been inspired by a Roman play called *Menaechmi*, written by Plautus. Shakespeare modified and embellished the play by adding another set of twins and much more comedic confusion; *The Comedy of Errors* was likely written around 1594.

27. For *Twelfth Night*, Shakespeare probably borrowed and adapted the plot from an Italian play known as *Gl'Ingannati (The Deceived)*. See: "*Gl'Ingannati*," British Library.

28. Kelly, "Division, Harmony, and Medical Mistakes."

29. To learn more about this case and others, see the Minnesota Center for Twin and Family Research website at https://mctfr.psych.umn.edu/index.html. There are also many popular articles about twin research on the internet. See, for example: Than, "Brief History of Twin Studies," and Grimes, "Jack Yufe."

30. Woo, "Jack Yufe Dies."

31. Certainly it was just a coincidence that they were both named Jim by their adoptive parents. Likewise, it is hard to imagine a hereditary reason that each was married twice, first to women named Linda and second to women named Betty, or that one had a son named James Alan and the other a son named James Allan. It seems coincidental, too, that both had owned dogs named Toy.

32. Readers curious about public dissections and autopsies in the early-modern period will find more information at these sources: Gamblin, "Reading between Bloodied Lines"; Ghosh, "Human Cadaveric Dissection"; Hurren, *Dissecting the Criminal Corpse*; Imbracsio, "Corpses Revealed"; Keating, "The Performative Corpse."

33. Nelson, "Elizabethan Incomes," and Nelson, "Elizabethan Money."

34. Leonardo da Vinci was also interested in anatomical detail, and he also produced amazingly detailed drawings based on dissection of cadavers. Unfortunately, Leonardo never got around to publishing his drawings. Happily, they were preserved and became widely available only much later, in the nineteenth century. Thus, Shakespeare could not have been influenced by Leonardo's anatomical work, even though they were roughly contemporaries. Many of these drawings are now easily accessible online at: "A Rare Glimpse of Leonardo da Vinci's Anatomical Drawings," on the Brainpickings .org website, and at "Leonardo" on *Wikipedia*. The best available anatomical studies in

Shakespeare's day were excellent drawings by the Flemish anatomist, Andreas Vesalius, also known as the "father of anatomy." Many of his renderings are available online at Wikicommons and The British Library.

35. *Pia mater* is a Latin translation of an Arabic term meaning "gentle mother" and is pronounced either PEE-Ya MAH-ter or PIE-ya MAY-ter. The *ventricles* consist of four interconnected cavities inside the brain, produced during embryonic development when the upper end of the neural tube expands and folds in on itself to become the brain. The ventricles are continuous with a hollow portion of the spinal cord and filled with cerebrospinal fluid. In his studies of the brain, Leonardo injected molten wax into the hollow spaces or *ventricles* of cadaver brains in order to study their shape. For much more information on da Vinci's studies of the brain, see: Gross, *Brain, Vision, and Memory.*

36. Oddly, Dr. John Hall was a prolific diarist, yet he never mentions his father-in-law.

37. But then, in Shakespeare's *Romeo and Juliet,* the bride was only thirteen when she and Romeo were married.

38. We recommend another book that connects many of Shakespeare's plays to the discoveries of modern neuroscience: Matthews and McQuain, *The Bard on the Brain.*

39. *Biopsychology* (or biological psychology) is part of a larger interdisciplinary enterprise called *neuroscience* that looks at the brain, mind, and behavior from multiple perspectives, including biology, behavior, chemistry, and medicine.

40. In poetic language worthy of the Bard, the pioneering neuroscientist Sir Charles Sherrington described activity in the brain awakening to the world each morning:

> That where hardly a light had twinkled or moved, becomes now a sparkling field of rhythmic flashing points with trains of traveling sparks hurrying hither and thither. The brain is waking and with it the mind is returning. It is as if the Milky Way entered upon some cosmic dance. Swiftly the head mass becomes an enchanted loom where millions of flashing shuttles weave a dissolving pattern, always a meaningful pattern though never an abiding one; a shifting harmony of subpatterns.

Sherrington's image/metaphor here is of a sophisticated nineteenth-century mechanized loom, programmed with punch cards, much like a computer in the 1970s. Sherrington's "enchanted loom" piece was originally published in: *Sherrington, Man on His Nature.*

41. Early brain stimulation studies showed many parts of the brain that were "silent," seemingly having no function. This probably gave rise to the myth that we use only 10 percent of our brains, a notion that has long since been shown to be false. Indeed, we use the whole brain, but not typically all at one time. Simultaneous activation of very large regions of the brain can produce a grand mal epileptic seizure.

42. Much of our understanding of the brain we can attribute to brain scanning technology, which includes fMRI (functional magnetic resonance imaging) and PET scans, both of which are used to map active regions of the brain when subjects are engaged in various tasks.

43. Or to mix a metaphor (as Shakespeare frequently did), we might think of the brain as a sort of neural Swiss Army knife—a collection of highly specialized neural tools, each tasked for distinct sensory, motor, cognitive, emotional, or motivational functions.

44. *Synapses* are the junctions between neurons. Learning and memory apparently make changes in synapses so that communication between nerve cells flows either more or less efficiently. Nearly all the psychoactive drugs affect the flow of *neurotransmitters* at synapses in the brain. These chemicals that neurons use to communicate across the synaptic gaps come in dozens of varieties, some of which double as hormones in the bloodstream. Hence their similarity to the humors of old.

45. We still are largely ignorant of the biology of mental illness, even though we have found drugs that ameliorate the symptoms of many mental disorders, such as depression, bipolar disorder, and even schizophrenia. Ironically, neuroscience understands more about how the drugs work (usually by changing the chemistry of synapses between neurons) than it understands about the underlying disorders the drugs are treating.

46. Gall may have had some biases with regard to skull and brain size, for portraits invariably render him with a large head.

47. Driscoll and Leach, "No Longer Gage." See also: Fleischman, *Phineas Gage.*

48. This led Broca to take sides in the controversy over *localization of function*. His most famous case was nicknamed "Tan," after the only word the patient could speak. When the man died, Broca performed an autopsy that revealed extensive damage to a posterior region of the left frontal lobe (now known as Broca's area) that controls the motor movements involved in speech. Subsequent research determined that damage to this region produces *expressive aphasia*, while damage to a nearby region (Wernicke's area) produces *receptive aphasia*, a condition that interferes with the understanding of speech.

49. Pop psychology has latched onto this idea of hemispheric differences and made much more of it than the evidence warrants. In fact, the idea that people are "right-brained" or "left-brained" is nonsense. Everyone uses both sides of the brain all the time.

50. We have learned much about the brain from battlefield casualties and from people whose brains were ravaged by accidents and diseases, as we saw also in the case of Phineas Gage.

51. Vision is somewhat more complicated: The lenses of the eyes turn images upside down and backward on the retinas. Then the complex wiring in the brain delivers the image on the left side of the visual field *of each eye* to the brain's right hemisphere and the image on the right half of the visual field *of each eye* to the left hemisphere. A diagram will help you visualize this, and many are available on the internet by searching for *vision and optic chiasma.*

52. Renaissance folk should have paid more attention to the rats, however, but for a different reason. It was those critters that brought wave after wave of the Black Death in England. The plague closed the London theaters and claimed over a third of the

population. Again it was ignorance, based on medieval beliefs: medicine had not yet developed the germ theory, so physicians had no way of knowing that the ubiquitous rats carried the fleas that transmitted the plague bacillus.

53. Penfield, "The Interpretive Cortex."

54. Just the boldfaced words are Shakespeare's; the sentences were otherwise in modern English. The researchers feared that Shakespeare's four-hundred-year-old English might introduce another variable that could overwhelm or confound the effects of the functional shift.

55. *Functional magnetic resonance imaging*, or *fMRI*, is a type of brain-scanning technology that detects tiny magnetic changes in hydrogen atoms, associated with blood flow in the brain (or other soft tissues). With the help of a computer, these magnetic emissions produce detailed three-dimensional images of the brain. Of special interest to neuroscientists are changes in blood flow patterns in particular brain regions seen when subjects are engaged in various mental tasks: in this case, reading sentences with Shakespeare's functionally shifted words "light up" only certain parts of the brain. These changes show which brain regions are most active while a person is engaged in the task.

56. Keidel et al., "How Shakespeare Tempests the Brain."

57. We cannot but wonder whether Joseph Haydn, who occasionally included Shakespeare's words in his compositions, might have found inspiration for his "Surprise Symphony" (No. 94) from Shakespeare's functional shift.

58. Freud also borrowed these ideas from Darwin and made them part of the "instinctual" desires of the unconscious.

59. In our sense, "folk psychologists" are those who use an understanding of human behavior absorbed from their culture and personal observations rather than from a scholarly perspective or clinical training. They are likely to refer to their views as a "common sense" understanding of "human nature." See: Carroll, "An Evolutionary Approach to Shakespeare's *King Lear*."

60. These problematic behaviors are all featured in Pope Gregory's famous list of Seven Deadly Sins. A mnemonic (memory) device for remembering the Deadly Sins list is **WASPLEG,** standing for **W**rath, **A**varice, **S**loth, **P**ride, **L**ust (or Lechery), **E**nvy, and **G**luttony. We can surmise that Pope Gregory wasn't entirely a kill-joy, for he is also known for his beautiful Gregorian chants, which we might consider as a sort of seventh-century precursor of hip hop. Musicologists may dispute this suggestion.

61. We will address these gender differences in sexuality more completely in chapter 8: "Emotion, Motivation, and Elizabethan Love."

62. Homosexuality is yet another biological mystery, not yet fully explained by evolutionary psychology—although several explanations have been offered. The problem is that while gays and lesbians may engage in sex, their orientation does not produce offspring, as required by a simple Darwinian evolutionary theory. Our suspicion is that the solution to this mystery involves benefit to survival and reproduction of the group. For further discussion of this issue, see David P. Barash's excellent book, *Homo Mysterious: Evolutionary Puzzles of Human Nature*. (We find it interesting that creationists and other anti-evolutionists never focus on evolutionary theory's most obvious

unsolved evolutionary mysteries surrounding sexuality. Perhaps they do not read the scientific literature.)

63. See also: Barash and Barash, *Madame Bovary's Ovaries*.

64. See: Carroll, "An Evolutionary Approach to Shakespeare's *King Lear*."

65. Gall was *partly* right about this: the *inside* of the skull does reflect the shape of the brain, which imprints faint impressions of the cortical lobes. He wasn't right, however, about mental abilities leaving bumps or depressions that could be read on the outside of the skull, nor was he right about brain regions that correspond to character traits.

66. One Italian observer likened the cerebral cortex, instead, to a plate of macaroni. See: Messbarger, "Anna Morandi's Wax Self-Portrait."

67. Ramachandran, "Adventures in Behavioral Neurology."

68. Reuters, "Radar Scan of Shakespeare's Grave."

CHAPTER TWO

1. We give the date as 1599, the year when the play was first performed. The location, the Forest of Arden, is a fictitious place. Yet, knowing Shakespeare, one might guess that the name more likely involves a pun. His mother's family name was Arden, and the Ardens were well-to-do landowners who lived near Shakespeare's hometown of Stratford. But the play is actually set in France, where there was a well-known Arden Wood.

2. English law of the time forbade women on stage, so the actors were all men (or boys, who often played the female parts). Much humor, therefore, involved the audience knowing that boys were playing women disguised as men. Shakespeare uses this device not only in *As You Like It* but in *The Merry Wives of Windsor*, *Cymbeline*, *The Merchant of Venice*, and *Twelfth Night*. See: Stigler, "Gender Swaps."

3. Remember: Duke Senior, who professes to be lonely, is encamped in the Forest of Arden with an entourage of servants, supportive nobles, and gentlemen from the court.

4. You will note that, although Jaques says "men and women," his speech gives us nothing else about developmental stages in women—especially remarkable in view of his patron being Queen Elizabeth I. We will return to this thread at the end of the chapter.

5. "Pard" is a shortened form of *leopard*.

6. A "good capon" refers to a castrated cockerel. This line, therefore, may have two meanings. The mention of castration suggests that this man's days as "the lover" are over. The second meaning, involving the "fair round belly," suggests the man of this age makes his belly fat by overindulgence in rich food.

7. The term "pantaloon" refers to comfortable lounging clothes, consisting of slippers and pants with wide legs—or to persons who wore them. The *Pantaloon* or *Pantalone*, who played the money-grubbing old fool, was an important stock character

in the *commedia dell'arte* or Italian theater, which was popular across Europe in Shakespeare's day.

8. Harcum, "The Ages of Man."

9. Shakespeare likely would have first been introduced to the concept of life ages or stages in a Latin poem by Marcellus Palingenius Stellatus in the book *Zodiacus Vitae*, which was part of the curriculum in the new English grammar schools. See: Bate, *Soul of the Age*, 93–95. The idea of *seven* ages also harkens back to the Middle Ages, when lists often came in sevens, as in the Seven Deadly Sins, the Seven Heavenly Virtues, and the Seven Sacraments. Nor was the metaphor of the world as a stage new. You may remember that Shakespeare also used it in Macbeth's melancholy soliloquy, upon hearing of his wife's suicide:

> Tomorrow, and tomorrow, and tomorrow,
> Creeps in this petty pace from day to day,
> To the last syllable of recorded time;
> And all our yesterdays have lighted fools
> The way to dusty death. Out, out, brief candle!
> Life's but a walking shadow, a poor player
> That struts and frets his hour upon the stage
> And then is heard no more: it is a tale
> Told by an idiot, full of sound and fury,
> Signifying nothing.
>
> —*Macbeth*, 5.5.20

10. See, for example: Thorpe, "Alas, Poor Hamnet."

11. As evidence, Blake also cites Macduff's children who are gratuitously slaughtered in *Macbeth*, the princes murdered in the tower in *Richard III*, young Arthur who jumps off the castle wall in *King John*, Lucius in *Titus Andronicus*, the Boy in *Henry V*, and Brutus's page in *Julius Caesar*. See: Blake, "Children and Suffering."

12. Fleance's age is indeterminate, but Banquo calls him a boy.

13. Dramatically, children were not considered especially interesting for audiences consisting mostly of adults. Hence, Shakespeare wrote no substantial roles for them. It is worth noting, however, that the England of Shakespeare's day had a much larger proportion of children than in modern times, despite the high childhood mortality rate.

14. These princes were Richard of Shrewsbury and the never-crowned King Edward V. The Duke of Gloucester sent them to the Tower of London ostensibly to prepare for Edward's coronation, but they never emerged; Gloucester was soon crowned King Richard III.

15. British Shakespeare Scholar, Jonathan Bate, has written a clever and important book about the Bard's development: *Soul of the Age: A Biography of the Mind of William Shakespeare*. His approach involves looking at the events and culture that characterized Shakespeare's life at each of the seven ages. In particular, he describes the rigorous curricula of the English grammar schools that young William would have attended. See: Bate, *Soul of the Age*.

16. The other Lovers we identified:

Juliet and **Romeo**, the tragic "star-crossed" pair who would die for each other.

Kate and **Petruchio** (*The Taming of the Shrew*): She is married to him against her will, which turns out to be quite strong.

Beatrice and **Benedick** (*Much Ado About Nothing*), whose classic banter obscures their love—until it comes time to kill Claudio.

Troilus and **Cressida** (in the play of the same name) become two sides of a triangle stretching across enemy lines during the Trojan war.

Posthumus, Imogen, and **Iachimo** (a would-be lover or rapist) are among the protagonists in *Cymbeline*, a production in which *machismo* in the Age of the Lover nearly brings ruin on everyone.

Prince Hamlet: Shakespeare suggests that Hamlet was betrothed to Ophelia when he left for college. But he does not act like a lover when he returns home to find that Dad had been murdered by Uncle Claudius.

Prince Hal belongs in this category before his developmental "awakening" as the Soldier, the slayer of Hotspur. Shakespeare gives us only the barest hints about Hal as a Lover, but the prince's romantic antics in wooing Princess Katherine certainly fit into the same "ridiculous" category as singing a woeful ballad to an eyebrow.

Ferdinand, **Berowne**, **Longaville**, and **Dumaine**, of *Love's Labour's Lost*, certainly qualify as lovers-in-denial, as the men vow to retire from the world, and especially from women, to study for three years. Happily, the vow lasts only a few hours.

There are a few more female characters who, like **Ophelia**, don't often get their due as lovers, including **Desdemona**, the unjustly accused (and unjustly murdered) wife in *Othello*; the thwarted **Helen** in *All's Well*; **Portia**, the faux barrister who, in *The Merchant of Venice*, cleverly prevents Shylock from getting his "pound of flesh" and who manages to navigate the silly stipulations of her father's will so she can marry the man she loves; and **Princess Katherine** of France, who briefly steals the show in the wooing scene with King Henry V.

17. Greenblatt, *Will in the World*; Bate, *Soul of the Age*; Bryson, *Shakespeare: The World as Stage*; Dickson, "Shakespeare's Life."

18. The plague played a huge role in determining mortality in Shakespeare's time. Medicine knew nothing of germs, such as the bacillus responsible for the Black Death that killed a third of England's population, including Will's three sisters. Persons surviving into adulthood, however, had a 50 percent chance of living until sixty. See: Whitbourne, "Shakespeare's Seven Ages."

19. While we learn nothing about Lear's visual acuity or dental status, the term *second childishness* certainly fits Shakespeare's depiction of him as a tormented and demented man. According to a paper by Shakiya Snipes, Lear's case reveals the way the elderly were treated with disrespect, looked down upon as past their prime, of "diminished power," and childlike—rather than repositors of wisdom. You can find

Snipes' paper online through the link in our bibliography: Snipes, "Treatment of the Elderly."

20. Not everyone agrees with Erikson, but there is consensus that he was an influential pioneering figure in developmental psychology. Erikson wrote extensively on stages of *psychosocial development* across the lifespan, and although he was a psychoanalyst, he broke from Freud by emphasizing the role of consciousness and of social interactions, rather than unconscious conflicts. Other developmentalists, like Jean Piaget, have studied cognitive development, the psychology of discovering how children's minds mature. Still others look at emotional development, including "emotional intelligence," or moral and social development. Most recently, evolutionary psychology has been contributing to our understanding of how nature and nurture interact in their contributions to all periods of psychological development.

21. For a more complete introductory discussion of cultural factors shaping the emergence of adolescence as a distinct stage of life, see Jaworska and MacQueen, "Adolescence" and *Psychology Today* Staff, "Invention of Adolescence."

22. And, yes, Shakespeare's King Henry V was a lover, albeit an awkward one. See *Henry V*, 5.2.101. (As we noted earlier, the Bard only obliquely mentioned Prince Hal's amorous exploits.)

23. Psychologists now generally prefer to speak of "stages" of development, rather than "ages." The meaning is essentially the same, but the term *stages* sidesteps the problem of tying development to rigid chronological age groupings. Incidentally, there is a continuing debate within psychology about whether the transition between stages is smooth or abrupt. The latter seems to make for more interesting theater.

24. Here Gloucester is accusing Winchester and his clerics of being hypocrites, of not praying sincerely for the recovery of the deceased Henry V; he is suggesting that the bishop really wanted a boy-king that he could manipulate for his own purposes. A few lines later, Gloucester adds, "Name not religion, for thou lov'st the flesh, / And ne'er throughout the year to church thou go'st / Except it be to pray against thy foes" (*Henry VI, Part 1*, 1.1.41).

25. Henry VI married Margaret of Anjou in 1455, when he was about twenty-three years old, a move designed to solidify England's claim to the French throne. Resumption of the Hundred Years War, however, led to England's defeat; that outcome was not solely Henry's fault, although he did not provide strong leadership. Meanwhile, conflict among the English nobles led to the civil wars known as the Wars of the Roses, leaving the English in a weakened position for reigniting the conflict with France.

26. Barton, *The Hollow Crown*, 33.

27. On the following page we suggest an Eriksonian "developmental crisis" central to each of the crisis plays (table 2.4). We expect that readers with different interpretations of the plays may disagree; this is the stuff of future discussions. It is also important to note that a character may be currently struggling with a crisis at that stage of development or dealing with a failure to deal effectively with a developmental crisis at an earlier stage. Moreover, some plays have multiple characters in the throes of different developmental crises.

Table 2.4 Eriksonian Crises in Shakespeare's Plays

Play	Character(s)	Eriksonian Crisis
Antony and Cleopatra	Antony and Cleopatra	intimacy vs. isolation
Coriolanus	Coriolanus	identity vs. role confusion
Hamlet	Prince Hamlet	identity vs. role confusion
Henry IV, Pt. 1	Prince Hal	identity vs. role confusion
Henry IV, Pt. 2	Prince Hal	identity vs. role confusion
Henry V	King Henry V	intimacy vs. isolation
Julius Caesar	Caesar and Brutus	generativity vs. stagnation
King Lear	Lear	integrity vs. despair
Macbeth	Macbeth and Lady Macbeth	generativity vs. stagnation
Measure for Measure	Angelo	intimacy vs. isolation
Merchant of Venice	Shylock	identity vs. role confusion
Othello	General Othello	intimacy vs. isolation
Richard II	Richard and Bolingbroke	generativity vs. stagnation
Richard III	Richard	intimacy vs. isolation
Romeo and Juliet	Romeo and Juliet	intimacy vs. isolation
The Tempest	Prospero	generativity vs. stagnation
The Winter's Tale	Leontes	intimacy vs. isolation
Timon of Athens	Timon	intimacy vs. isolation
Troilus and Cressida	Troilus and Cressida	identity vs. role confusion

28. The "invention" of adolescence testifies to the fact that no stage theory can be universal. See: Kett, "Discovery and Invention."

29. Arain et al., "Maturation of Adolescent Brain"; Johnson, Blum, and Giedd, "Adolescent Maturity."

30. Agnes Strickland was an English writer specializing in poetry and historical subjects. Her 1827 book, *The Seven Ages of Woman*, is still available both in print and electronic formats. See: Strickland, "Seven Ages of Woman."

31. We are *not* suggesting here that the male should be the standard against which females should be assessed or judged. For much more on this topic, we recommend *The Mismeasure of Woman*, a book by our friend Carol Tavris.

32. By doing so, Shakespeare comes much closer to a modern view of development than we had initially thought—in that people become more diverse and heterogeneous in outlook, behavior, and personality as they age.

33. Boyce's book is a good reference to own if you are a Shakespeare buff. (The quotes are from p. 269.) You will find the full reference in our bibliography: Boyce, *Shakespeare A to Z*.

34. You may remember from chapter 2 that Fastolfe was a real person who was falsely accused of cowardice during the military campaigns in France, when England and France were engaged in the Hundred Years War.

CHAPTER THREE

1. The dates refer to the historical reigns of Henrys IV and V—from the time that Bolingroke deposed his cousin, King Richard II, and became King Henry IV, to King Henry V's glorious victory at Agincourt—as told in the plays *Richard II*, *Henry IV, Parts 1* and *2*, and *Henry V.* For these tales, Shakespeare relied heavily on Holinshed's *Chronicles.* See also: Rosen, "Arts and Culture."

2. Shakespeare's primary source for his history plays was Raphael Holinshed's *Chronicles*, published in 1587. The Bard also borrowed material from this three-volume source for *King Lear*, *Macbeth*, and *Cymbeline*. See: Holinshed, *Chronicles.*

3. The "nutshell" metaphor probably originated with the Roman Pliny, but Shakespeare borrowed and adapted it when he had Hamlet say to Rosencrantz, "O God, I could be bounded in a nutshell and count myself a king of infinite space, were it not that I have bad dreams" (*Hamlet*, 2.2.273).

4. *Sack*, Falstaff's favorite fluid, is a sherry, or fortified sweet white wine, to which brandy or other spirits have been added. Often, sack was imported from Spain.

5. *Kine* is an archaic term for cattle. Here Falstaff is making a biblical reference to Pharaoh's dream in which seven lean kine emerged from the river and ate seven fat kine. Joseph interpreted the dream as foretelling seven years of good harvests followed by seven years of famine.

6. Another play, *Sir John Oldcastle*, includes a character, Sir John of Wrotham, who has many characteristics that Shakespeare infused in Falstaff (who was originally dubbed Oldcastle). This play, however, came out a year after Shakespeare's *Henry IV, Part 1* and is seen by scholars as a probable attempt by other playwrights to cash in on the success of Shakespeare's Falstaff. A second-edition quarto of *Sir John Oldcastle* lists Shakespeare as a coauthor, a presumptive fact disputed by scholars. See: Asimov, *Guide to Shakespeare*, 327.

7. The real Oldcastle's life was an interesting story in itself, with a sad ending. Oldcastle was a Lollard, an early Protestant group who were disciples of John Wycliffe. This was dangerous because, at the time, England was thoroughly Catholic. (Henry VIII's conflicts with the pope came later—after which England became a Protestant country.) So, even though Oldcastle was a friend of Henry V, he was found out, declared a heretic, and sentenced to death. Henry graciously gave him a forty-day stay of execution, with the chance to change his mind and save his life. Instead, the unrepentant Oldcastle escaped and hid in Wales for four years, where he staged an inept rebellion aimed at kidnapping the king. Eventually, he was recaptured. An old play, *The Famous Victories*, told of Oldcastle's end as it really was. He suffered the death deemed appropriate for heretics of the time: slow roasting suspended over an open fire. There are many modern sources for the Oldcastle story, but our favorite is *Asimov's Guide to Shakespeare*, 327–28.

8. Yes, *that* Isaac Asimov, noted more for his science fiction. Asimov, however, wrote or edited some five hundred books spanning nine of the ten categories of the Dewey Decimal Classification. *Asimov's Guide to Shakespeare* gives the historical and

cultural context behind the plays. For another perspective, see Williams, "Fastolf or Falstaff."

9. Indeed, Wikipedia reports that "Falstaffe" was an Elizabethan-era pun for erectile dysfunction. See: Wikipedia, "John Falstaff."

10. Shakespeare briefly introduced a cowardly character named Sir John Fastolfe in the play *Henry VI, Part 1* (dealing with the much-later Wars of the Roses). Many experts believe that the Fastolfe of *Henry VI*, written in 1591, inspired the Falstaff of the *Henry IV* plays, written some five or six years later. The problem with that theory, however, is that the first two folio editions of *Henry VI, Part 1* name our man as Falstaffe. So, the matter remains unsettled. See also Harold Bloom's version of Falstaff's origin: Bloom, *The Invention of the Human*, 274–75.

11. Epstein, *The Friendly Shakespeare*.

12. Bloom, *Shakespeare: The Invention of the Human*, 282.

13. In Elizabethan times, the words *disposition* and *temperament* were synonymous. In his works, Shakespeare used *disposition* forty-six times. *Temperament* appears only in its truncated version, "temper," which he uses thirty-four times. In the same era, the word *personality* referred to the essence of being a person, rather than an animal. And, yes, phlegm is "mucous," including the substance familiarly known as "snot," a word that probably relates to "snout" and comes to us from Middle Dutch or German via Middle English. See these terms in the online *Oxford Living Dictionaries*.

14. Blood and phlegm are real, of course, but we now know that there is only one kind of bile (which is a yellow-greenish fluid produced in the liver and stored in the gallbladder).

15. Stephen Greenblatt notes that humoric stereotypes can be an effective theatrical device:

> Jonson's plays are sometimes called, "The Comedies of Humors," the humor being your characterological disposition, type of personality you have. And it's part of the pleasure of watching Jonson, to watch the characters come back, over and over again, to their characteristic humor, whether it's anger or jealousy, folly, of one kind or another. . . . If you think in the terms of Seinfeld, people are what they are; they always turn out to be exactly what you expect them [to be]—but behind that is probably a set of philosophical ideas about what constitutes fully achieved personhood. In this view, someone who has reached the point of stable personhood is someone who probably will not change.

See: "Stephen Greenblatt: Shakespeare's Life Stories," Folger Shakespeare Library.

16. Minton, "Every Man in His Humor."

17. Ekström, "Shakespeare and the Four Humours."

18. See: Draper, *The Humors and Shakespeare's Characters*. See also: PAEI, "Hippocrates & Galen."

19. Was it a coincidence that Shakespeare's contemporary, Cervantes, also brought to life the complex personality of Don Quixote? Or was there something new in the air in those Renaissance days?

20. In *Shakespeare: The Invention of the Human*, Bloom clarifies this by saying, "Shakespeare's originality was in the *representation* of cognition, personality, and character," xviii.

21. The play was known as *The Murder of Gonzago*, but Hamlet tells his uncle Claudius that it is called *The Mousetrap*. Hamlet goes on to say, "The play's the thing / Wherein I'll catch the conscience of the king."

22. For more on the psychology of Shylock, see: National Library of Medicine, "The Case of Shylock."

23. Examples: the clown Lancelet reads his own palm in *The Merchant of Venice* (2.2.155), and in *The Two Noble Kinsmen*, the Gaoler's Daughter says, "Give me your hand . . . I can tell your fortune" (3.5.92).

24. This was an orderly universe envisioned by Ptolemy in the second century CE. For a brief, readable overview of Ptolemy's cosmology, see: Miller, "Four Humors."

25. The word *planet* comes from a Greek word for "wanderer." The seven planets all followed the path of the Zodiac, which was defined as the path in the sky blazed by the Sun. The Zodiac itself consists of the constellations lying in this path. The astrological connections with medicine, temperament and behavior, the four humors, and the seven days of the week formed a complex system. Each planet was associated with an element or a metal and with a day of the week. (Monday was "moon day," for example.) Six of the seven planets correlated with the four humors. Mercury was the exception; it represented *balance*, but it was linked to the slippery metal quicksilver—a fitting symbol for the god of thieves. See: Tate, "The Ancient Planets."

26. The name of the pickpocket Autolychus (pronounced aw-TOL-icus) means "lone wolf." In Greek mythology, Autolycus was a trickster, a con man, and a thief. He plays much the same role—and also the clown—in *The Winter's Tale*.

27. Since astronomers of the day had not yet accepted the idea of a sun-centered solar system, this illusion was puzzling. In a sun-centered universe, however, the retrograde path of the planets suddenly made sense because the apparent backward motion occurred when the Earth, in its orbit, caught up to and passed a planet that seemed to be moving more slowly. The heliocentric universe was proposed by Copernicus (1543) and supported by evidence from Galileo's observations (1610), near the end of Shakespeare's career.

28. Shakespeare seems to have made this up. Similarly, he has Calpurnia warn her husband, Julius Caesar, "When beggars die there are no comets seen; The heavens themselves blaze forth the death of princes" (*Julius Caesar, Part 1*, 2.2.31).

29. Barash and Barash, "Biology as a Lens."

30. In 1593, Thomas Kyd was arrested and tortured on suspicion of treason and heresy relating to fliers questioning the divinity of Christ that appeared anonymously around London. Told that heretical documents had been found in his room, he blamed his former roommate, Christopher Marlowe, who was then also arrested. (Kyd added

the claim that Marlowe believed Jesus was a homosexual, which didn't go down well with the authorities either.) Marlowe was allowed out on bail pending the disposition of his case, but he was killed in a bar fight before the judgment was handed down. Kyd was eventually released, although he never regained his acclaim as a playwright.

31. Cocaine, for example, amplifies messages carried in the brain's norepinephrine pathways, producing a stimulant effect. Likewise, LSD mimics serotonin, strongly affecting mood, as well as the brain's sensory processing circuits. And alcohol similarly activates dopamine, stimulating the brain's reward circuitry. Many prescription drugs, too, have neurotransmitter-like effects on communication between nerve cells, including the popular antidepressant Prozac, the antipsychotic Thorazine, and the antianxiety drug Valium. Such drugs, then, become artificial "humors" in the body.

32. To see more clearly how the Five-Factor theory works, you can take one of many Big Five personality "tests" online by simply searching for "Big Five personality test" or "Five-Factor model." And if you want to remember the names of the Big Five traits, think of the acronym OCEAN, standing for the first letter of each trait dimension. If you seek another water-based acronym, you can use CANOE instead.

33. These same five factors pop out of a variety of personality assessment "tests" employing a complex mathematical technique called *factor analysis.* Trust us on this.

34. For a review of the cross-cultural evidence, see: Carver and Scheier, *Perspectives on Personality*; Zimbardo, Johnson, and McCann, *Psychology: Core Concepts.*

35. McAdams and Pals, "A New Big Five."

36. "Greenblatt: Life Stories," Folger Shakespeare Library.

37. McAdams, "American Identity"; McAdams, "Exploring Psychological Themes."

38. Woike, "State of the Story."

39. *Generative* is a term that McAdams borrows from psychoanalyst Erik Erikson's developmental theory of personality. Generativity is one path through the midlife crisis; it is found in healthy, mature adults who have successfully dealt with the midlife crisis of generativity vs. stagnation.

40. See: McAdams, *The Redemptive Self.*

41. Freud's suggestion that Hamlet had an outsized Oedipus complex appears in *The Interpretation of Dreams.* Other examples of psychoanalytic literary criticism can be found in abundance in Holland, Homan, and Paris, *Shakespeare's Personality.*

42. Freud was profoundly influenced by Darwin—who was his contemporary. We can see the parallels with Darwin's theory in Freud's twin drives of sex and aggression. See, for example: Marcaggi and Guénolé, "Freudarwin: Evolutionary Thinking."

43. Many psychologists no longer accept Freud's notion that the ego defenses operate as Freud taught: beneath the surface of consciousness in an ego that was at war with the impulses of the id. Cognitive psychologists these days are more likely to use the term *self-serving bias,* which has evolved as a self-protective mechanism in the wiring of our brains. Incidentally, those who, like Hamlet, suffer from melancholy (or as we call it now, *depression*) are less likely to display a self-serving bias.

44. David Barash and Nanelle Barash offer a lively look at literature from the standpoint of evolutionary psychology in their little book, *Madame Bovary's Ovaries: A Darwinian Look at Literature.*

45. Perhaps we belabor the obvious, but we would point out the kinship between the *heredity* vs. *environment* issue and the debate about history being the story of "great men" vs. challenging *situations*.

46. The familiar Martin Droeshout portrait greets those who open the *First Folio*. The modern writer Bill Bryson is not a big fan of the Droeshout portrait, which he describes as follows:

> Nearly everything about it is flawed. One eye is bigger than the other. The mouth is curiously mispositioned. The hair is longer on one side of the subject's head than the other, and the head itself is out of proportion to the body and seems to float off the shoulders like a balloon.

This quote comes from Bill Bryson's witty little volume, *Shakespeare: The World as Stage*.

47. This comes from a second-hand report by John Aubrey, a seventeenth-century figure who collected anecdotes about famous British men. He published his collection in a two-volume work, *Brief Lives*. Here is his full anecdotal description of Shakespeare, based on interviews with people who had firsthand knowledge of the man.

William Shakespear (1564–1616)

> Mr. William Shakespear was borne at Stratford upon Avon in the county of Warwick. His father was a butcher, and I have been told heretofore by some of the neighbours, that when he was a boy he exercised his father's trade, but when he kill'd a calfe he would doe it in a high style, and make a speech. There was at that time another butcher's son in this towne that was held not at all inferior to him for a naturall witt, his acquaintance and coetanean [of the same age], but dyed young.
>
> This William being inclined naturally to poetry and acting, came to London, I guesse, about 18; and was an actor at one of the play-houses, and did act exceedingly well (now B. Johnson [here Jonson's name is misspelled] was never a good actor, but an excellent instructor).
>
> He began early to make essayes at dramatique poetry, which at that time was very lowe; and his playes tooke well.
>
> He was a handsome, well shap't man: very good company, and of a very readie and pleasant smooth witt.
>
> The humour of the constable, in Midsomernight's Dreame, he happened to take at Grendon in Bucks—I thinke it was Midsomer night that he happened to lye there—which is the roade from London to Stratford, and there was living that constable about 1642, when I first came to Oxon: Mr. Josias Howe is of that parish, and knew him.
>
> Ben Johnson [*sic*] and he did gather humours of men dayly where ever they came. One time as he was at the tavern at Stratford super Avon, one Combes, an old rich usurer, was to be buryed, he makes there this extemporary epitaph,

> Ten in the hundred the Devill allowes,
> But Combes will have twelve, he sweares and vowes:
> If any one askes who lies in this tombe,
> 'Hoh!' quoth the Devill, ''Tis my John o Combe.'
>
> He was wont to goe to his native countrey once a yeare. I thinke I have been told that he left 2 or 300 li. per annum there and thereabout to a sister. Vide his epitaph in Dugdale's Warwickshire.
>
> I have heard Sir William Davenant and Mr. Thomas Shadwell (who is counted the best comoedian we have now) say that he had a most prodigious witt, and did admire his naturall parts beyond all other dramaticall writers. He was wont to say (B. Johnson's [*sic*] Underwoods) that he 'never blotted out a line in his life'; sayd Ben: Johnson [*sic*], 'I wish he had blotted-out a thousand.'
>
> His comoedies will remaine witt as long as the English tongue is understood, for that he handles *mores hominum* [human customs]. Now our present writers reflect so much upon particular persons and coxcombeities [referring to a conceited, foolish person], that twenty yeares hence they will not be understood.
>
> Though, as Ben: Johnson [*sic*] sayes of him, that he had but little Latine and lesse Greek, he understood Latine pretty well, for he had been in his younger yeares a schoolmaster in the countrey.—from Mr. Beeston

The full text of Aubrey's *Brief Lives* is available from the Project Gutenberg website.

48. These comments come from a volume of anecdotes and observations by playwright and actor Ben Jonson; it was titled (oddly enough) *Timber* or *Discoveries.* We give you the full text of his comments on Shakespeare:

> I remember the players have often mentioned it as an honor to Shakespeare, that in his writing, whatsoever he penned, he never blotted out a line. My answer hath been, "Would he had blotted a thousand," which they thought a malevolent speech. I had not told posterity this but for their ignorance, who chose that circumstance to commend their friend by wherein he most faulted; and to justify mine own candor, for I loved the man, and do honor his memory on this side idolatry as much as any. He was, indeed, honest, and of an open and free nature; had an excellent fancy, brave notions, and gentle expressions, wherein he flowed with that facility that sometime it was necessary he should be stopped. Sufflaminandus erat, [He should have been restrained] as Augustus said of Haterius. His wit was in his own power; would the rule of it had been so too.
>
> Many times he fell into those things, could not escape laughter, as when he said in the person of Caesar, one speaking to him: "Caesar, thou dost me wrong." He replied: "Caesar did never wrong but with just cause"; and such like, which were ridiculous. But he redeemed his vices with his virtues. There was ever more in him to be praised than to be pardoned.

49. If you have read this far, you would surely enjoy *The Book of Will*, a modern play, written by Lauren Gundersen, recalling how the *First Folio* came to be. The play tells how Shakespeare's friends Henry Condell, John Heminges, Ben Jonson, and others compiled and published the *Folio*, without which half of Shakespeare's plays might have been lost forever. While this volume is commonly known today as the *First Folio*, its actual title is *Mr. William Shakespeare's Comedies, Histories, & Tragedies.* It contains thirty-six of his (perhaps forty) plays.

50. John Aubrey seems to contradict himself on a slip of paper found tucked into his copy of *Brief Lives*. There, in Aubrey's hand, inscribed above Shakespeare's name, is a note: "not a company-keeper." While Shakespeare lived frugally and alone in London lodgings, he was no recluse, according to his friends. Consequently, Aubrey's note—"not a company-keeper"—has puzzled modern Shakespearean fans and scholars. Was Shakespeare a friendly and outgoing type—as Ben Jonson and others suggest—or was he a loner? Recently, Christopher Matusiak has disputed this supposed standoffishness. He makes the persuasive argument that Aubrey's "not a company keeper" notation refers to another man—one William Beeston—and that the fact that it was not included in *Brief Lives* strongly suggests that Aubrey did not think it applied to Shakespeare. See: Matusiak, "Not a Company Keeper?" Stephen Greenblatt opines that if the Bard had *not* been a bit of a loner, "it would be hard otherwise to imagine how Shakespeare could have done what he did—learn his parts and perform them onstage, help to manage the complex business affairs of the playing company, buy and sell country real estate and agricultural commodities, compose exquisitely crafted sonnets and long poems, and for almost two decades write on average two stupendous plays a year" (Greenblatt, *Will in the World*, 70).

51. In Shakespeare's England, women were forbidden by law to be actors. Consequently, English travelers who attended performances in continental Europe were often shocked to see women on stage. Shakespeare took full theatrical advantage of the fact that the parts of women were often played by boys, so there was a humorous irony in the many scenes he wrote in which a female character, played by a boy, disguises herself as a young man—as Viola does in *Twelfth Night* and Rosalind does in *All's Well that Ends Well.*

52. Bryson, *Shakespeare: The World as Stage*, 71.

53. *A Groats-worth of Witte*, written in 1592, not long after Shakespeare moved to London, was a short, rambling pamphlet in which Greene groused about other theatrical figures, as well. A "groat," by the way, was a small coin worth about four pence. The full text of Greene's screed can be found at the oxford-shakespeare.com website.

54. Ben Jonson, "Shakespeare."

55. Yes, there were two Jo(h)nsons. Ben Jonson (no "h") was the one we have dealt with so far; he was Shakespeare's bombastic and friendly rival, both as a playwright and an actor. Dr. Samuel Johnson lived about a century later and was an early Shakespearean scholar and lexicographer who compiled the first extensive dictionary of the English language. Those who intend to pursue an interest in Shakespeare are advised to keep these two straight in their minds, for they are frequently cited and easily confused.

56. Jonson, "Shakespeare," Prologue.

57. Keep in mind, however, that "comedy" had a more expansive meaning than just "funny." In Shakespeare's day, a comedy was a light-hearted play with a happy ending. It usually had funny elements, although brief "comic relief" was also a hallmark of Shakespearean tragedies. Examples include Shakespeare's use of the Fool in *Lear*, the Porter's "Knock! Knock!" scene in *Macbeth*, and portions of the gravedigger scene in *Hamlet*.

58. His vocabulary is reputed to have been enormous—admittedly, in part, because he is credited with inventing some seventeen hundred words and terms that we still commonly use, although this count, some say, is exaggerated. See: Misra, "1,700 English Words." And for a sample of his linguistic inventiveness, see: Mabillard, "Words Shakespeare Invented."

59. Chambers, *William Shakespeare*.

60. See Stephen Greenblatt's excellent little book, *Tyrant*, on Shakespeare's politics.

61. In the England of Shakespeare's time, *recusants* were secret Catholics who refused to attend Anglican Protestant services, as mandated for everyone by the Crown.

62. Rowse, "The Personality of Shakespeare."

63. Will's father, John Shakespeare, was still living when Will moved back home to Stratford.

64. It is a common misunderstanding to confuse *introversion* with *shyness* or an aversion to social situations. Rather, introversion and extroversion are defined by psychologists in terms of a person's preferred focus on the internal or external world. So, introverts may, or may not, be shy; extroverts may be outgoing—or not.

65. We are indebted to one Andrew Esquibel, an amateur blogger, for pointing out several of these traits and that Shakespeare's ability to read and write extensively requires isolation and introversion. Mr. Esquibel is a bartender at Terminal Gravity Brewing in Enterprise, Oregon.

66. Others would include Rosalind in *As You Like It*; Portia in *The Merchant of Venice*; Kate in *The Taming of the Shrew*; Isabella in *Measure for Measure*; Cordelia in *King Lear*; Beatrice in *Much Ado About Nothing*; Desdemona in *Othello*; Viola in *Twelfth Night*; Margaret of Anjou in *Henry VI, Pts. 1, 2*, and *3*, as well as *Richard III*; Hermia in *Midsummer Night's Dream*; and Imogen, Miranda, Tamora, Cressida, Titania, Volumina, Helena . . .

67. Rowse, "The Personality of Shakespeare."

68. In one atypical year, he knocked out four major hits, including *Henry V*, *Julius Caesar*, *As You Like It*, and *Hamlet*. James Shapiro tells the story of this remarkable accomplishment in his book *A Year in the Life of William Shakespeare: 1599*. (Some critics point out that the dating of the plays is imprecise, which may cast doubt on Shapiro's claim that all four of these plays were written in 1599).

69. Greenblatt, *Will in the World*.

70. Bryson, *Shakespeare: The World as Stage*.

71. For evidence about the relationship between birth order and achievement, see: Paulhus, Trapnell, and Chen, "Birth Order Effects"; Sulloway, *Born to Rebel*; and Zajonc and Marcus, "Birth Order."

72. Greenblatt, *Will in the World*, 311; Joynes, "Shakespeare's Friends."

73. The Italian plays include *Romeo and Juliet*, *The Merchant of Venice*, *The Two Gentlemen of Verona*, *Love's Labour's Lost*, *The Taming of the Shrew*, *Much Ado About Nothing*, *As You Like It*, *Twelfth Night*, *All's Well That Ends Well*, *Othello*, *Measure for Measure*, and *The Tempest.* A case for Shakespeare having sojourned in Italy during some of his "lost years" is presented by Roe, "Shakespeare Guide to Italy."

74. His two younger brothers moved to London, as well. There Edmund, the youngest, also pursued a career in theater as an actor, but he died at age twenty-seven. The middle brother, Gilbert, became a haberdasher.

75. McAdams, *The Redemptive Self.*

CHAPTER FOUR

1. We date this play according to the year in which it was probably written, although the script suggests that the the play was set in the High Middle Ages. See: Lever, "Date of *Measure.*"

2. Note that, while the play is set in Austria, all the characters have Italian names, but we all know that this confusion is really a smokescreen for the *real* setting, which is London. (There are places in northern Italy and southern Austria that have been ruled by both countries, so Austrian characters with Italian names is not entirely unrealistic.)

3. It is also noteworthy that bawdy houses were common in London, just outside the city walls and in the theater district south of the Thames, where The Globe was located. To complicate matters of morality, there were two sets of courts, civil and ecclesiastical, that were authorized to punish sexual misconduct. Shakespeare's audiences were accustomed to living in a world where such laws could be applied unevenly and arbitrarily. See, for example: Friedberg, "Policing Sex"; Mullan "*Measure* and Punishment."

4. While fornication would have been technically illegal, it would not ordinarily have been a problem for couples intending to be married in Shakespeare's England: when couples were formally "betrothed," they were considered married in the eyes of the Church, and it was common for them to take up residence together. So, Shakespeare's audiences would have sympathized with Claudio, as most of us would today, when the worst punishment for sex outside of marriage is likely to be a frown and a wink. It is noteworthy that Shakespeare, in lawyerly fashion, removes Claudio's culpability by pointing out that Claudio had slept with Juliet "upon a true contract," and he considers her his "wife" (act 1, scene 2). See also: Dolan, "Shakespeare and Marriage," 620–34.

5. In Shakespeare's day, "cuckold" was a term for a husband whose wife had been sexually unfaithful to him. The word comes from the Old French term for "cuckoo" referring to a bird that lays its eggs in other birds' nests for them to raise. Drawings of cuckolds, however, didn't picture them as birds but as buffoonish men with horns. The association of horns with cuckolds is not clear, although the Devil seems to have assumed a role in cuckoldry, as we see in *Much Ado About Nothing*, when Beatrice says, "There will the Devil meet me, like an old cuckold, with horns on his head." The cuckold's horns also seem to be the origin of the notion that a man in need of sex is

"horny" (*Much Ado*, 2.1.43). See: Swift, "History of Cuckold's Horns"; British Library, "Broadside Ballad on Cuckolds."

6. In the bed trick, the intended sexual partner trades places with another woman in the darkness of the bed chamber. The bed trick, by the way, was not a Shakespeare original; he borrowed it from Boccaccio or possibly from an even earlier source: the Bible, Genesis 29.

7. Sadly, the dowry was lost in a shipwreck, along with her brother. And then she lost Angelo, who backed out of the marriage contract, showing that he was even more of a cad than the audience had realized.

8. Or is it Mariana who is punished by having to marry Angelo? And punished for what? Shakespeare doesn't seem to look at it from her perspective, but modern audiences probably would.

9. See, for example, this piece on the history of the play: "*Measure for Measure*," Folger Shakespeare Library. It is an interesting irony that Vienna is also the birthplace of Freudian psychoanalysis—and Freud often quoted Shakespeare (although not *Measure for Measure,* as far as we know). Some productions of the play, however, pun on this coincidence. The sexuality aspect has (unfortunately, in our opinion) led many literary critics to apply Freudian analysis too liberally to this play.

10. Whitehall Palace was the main residence of the English kings and queens during Shakespeare's time, much as Buckingham Palace is today. Whitehall was located in the Westminster area of London.

11. Shakespeare apparently began writing the play almost immediately after the death of Queen Elizabeth I in 1603. He probably did not invent the "disguised monarch" device, for there were several other plays employing the same idea between 1604 and 1606. This literary trick goes back at least to Odysseus. See: Kreider, "Mechanics of Disguise," 167–80.

12. Technically, the Stanford Prison Experiment was not a true experiment but a "pilot study" or a demonstration because it had no comparison group. More information about the research can be found in the following sources: American Psychological Association, "Power of Social Situations"; Wilson, "When Evil Deeds."

13. The results of both "experiments" foreshadowed the more recent torture of inmates at the prison in Abu Ghraib run by the US Army and at the Guantanamo naval base, as well as the even more recently publicized abuses of police power against minorities.

14. See: Zimbardo, "Revising the Stanford Prison." As we noted, Dr. Zimbardo was not aware at the time of the parallels between his proposed experiment and the one Shakespeare described in *Measure for Measure.* It was therefore ironic that Shakespeare gives the duke a rationale nearly identical to Zimbardo's when Vincentio says, "Hence shall we see, / If power change purpose" (1.3.137). (And, as far as we can determine, it is purely coincidental that there is a "bard" in Zimbardo.)

15. More details and images of the Stanford Prison Experiment are available here: https://www.prisonexp.org/. Zimbardo's extensive explanation of the study is contained in his 2007 book, *The Lucifer Effect: Understanding How Good People Turn Evil,* published by Random House.

16. Zimbardo assumed the role of "prison superintendent," so he was able to observe the experiment up close, even though (unlike Vincentio) he was not disguised.

17. Christina Maslach has since become a prominent psychologist in her own right. (In Zimbardo's defense, we note that all the participants were thoroughly debriefed and followed up to detect possible lasting ill effects. No such effects were found.) Oh, and there is one more tangential detail: Zimbardo and Maslach were married a few months later and have been happily together ever since.

18. Nobody was actually punished in these experiments because the supposed recipient was a confederate of the experimenter, trained to respond as though he had been given an electric shock. Such deceptive tactics are common in experiments conducted by social psychologists.

19. The supposed recipient of the shocks was a confederate of the experimenter and was playing the role of just another naive participant in the experiment. As for the controversial nature of the experiment, the following is an accessible explanation: Romm, "Rethinking One of Psychology's Most Infamous."

20. In another Twilight Zone–like coincidence, Philip Zimbardo and Stanley Milgram were classmates at the same high school in the Bronx.

21. The situation was complicated in the English mind by the fact that James's mother was Mary Queen of Scots, also known as Mary Stuart, who had a claim to the English throne as the direct descendant of Henry VII (Henry Tudor, her grandfather). However, when England's Queen Mary I ("Bloody Mary") died, her sister Elizabeth seized the throne and later had Mary Queen of Scots beheaded for treason. Ironically, when Elizabeth died, leaving no heir, it was King James VI of Scotland, the son of Mary Queen of Scots, who claimed the crown and founded the House of Stuart.

22. Scholars believe that "Lucifer" was a mistranslation of the Hebrew term for "morning star," mentioned in the biblical book of Isaiah. The war between Lucifer/Satan and God is religious lore with no substantial support in Hebrew or Christian scripture. It was Milton's poem *Paradise Lost* that fleshed out the story of Lucifer from scant Biblical references by describing the aftermath of the great war between the fallen angel and the deity. To carry the metaphor a bit further, one could say that the entire Genesis account of the creation of Adam and Eve can be seen, through a psychological lens, as an early version of the Stanford Prison Experiment located in the Garden of Eden. For those interested in more details on the devil, many relevant sources are available in the Wikipedia article "Lucifer" at https://en.wikipedia.org/wiki/Lucifer. We also recommend the following satanic pieces: Simon, "What's So 'American'"; Delbanco, *The Death of Satan*.

23. The Stanford Prison Experiment has become controversial for several reasons. One, as we have seen, is that it was not technically an *experiment* in that it had no control group. And, as Zimbardo has readily agreed, it was really more a demonstration or "pilot study." Another part of the controversy surrounds the ethics of putting volunteers through such a humiliating process. Zimbardo's response is that, at the time—before the Stanford Prison Experiment was performed—no one guessed the outcome would have been so extreme.

24. Darley and Latané, "Bystander Intervention in Emergencies."

25. Asch designed his experiments to study the psychology of *conformity*. To do so, he would ask a group of people to look at pairs of cards in which one card had a single line; the other had three lines labeled A, B, and C. The task was to say which of the three lines was the same length as the single line on the other card. The process was then repeated through a run of eighteen card pairs. There was only one real subject in the experiment; the rest were confederates of the experimenter, trained to give wrong answers on certain trials. The real subject always responded last. The object of the experiment, then, was to study the conditions under which people would yield to the social pressure they felt when all or part of the group gave an obviously wrong answer.

26. "Glassy essence" refers to the image seen in a mirror. The "angry ape" reference is remarkable because Shakespeare seems to be suggesting a kinship between apes and humans, some three hundred years before Darwin. See Steven Pinker's comments on this in an interview found here: Fassler, "Shakespeare, One of the First."

27. Pinker, *Better Angels of Human Nature.*

28. Angelo's choice is brutally simple: should he attempt to force sex on Isabella to satisfy his own desires—or not? Isabella's choice is more complex, but no less extreme: she must choose between her chastity and her vows, on the one hand, and her brother's life and her feelings of horror, disgust, and guilt on the other.

29. In 1896, Shakespeare critic F. S. Boas first applied the term "problem plays" to *All's Well That Ends Well*, *Measure for Measure*, and *Troilus and Cressida*, works that fell somewhere between comedies and tragedies. (You will recall that a *comedy*, in Shakespeare's time, was a play that ended happily, usually with nearly everyone falling in love and getting married.) Another aspect the "problem plays" have in common is that the plot revolves around a central moral issue—a characteristic they share with several of the Bard's other works that critics have added to the list of Shakespeare's problem plays. These include *The Winter's Tale*, *Timon of Athens*, and *The Merchant of Venice*. Your authors would add *Twelfth Night* for its approving abuse of Malvolio and *Taming of the Shrew* for its misogyny (both of which, perhaps, could only be seen as "problems" in modern times).

30. In England, the arch-moralist Puritans were opponents of the theater in Elizabethan and Jacobean times. See: "Globe Theater and the Puritans."

31. Pinker, *The Blank Slate*, 224.

32. Moral Foundations Theory originally posited five dimensions, but Jonathan Haidt, one of its leading advocates, has proposed a sixth foundation that he calls "Liberty." See: Graham et al., "Moral Foundations Theory."

33. Here is a more complete description of each of the dimensions in Moral Foundations Theory:

- **Care/harm**—Caring involves protection and avoidance of harm for self and others of one's family and in-group, as well as promoting reciprocal altruism.
- **Fairness/cheating**—Valuing fairness promotes cooperation, marital fidelity, and justice.

- **Loyalty/betrayal**—Loyalty encourages cohesiveness, patriotism, and self-sacrifice for the good of the group.
- **Authority/subversion**—Respecting authority fosters hierarchies of authority and obedience, as well as promotes peaceful relationships, deference to social superiors, and respect, but it can lead to disdain for those lower in the hierarchy.
- **Sanctity or purity/degradation or disgust**—Valuing purity encourages cleanliness, avoidance of persons or objects associated with disease; evolutionary psychology suggests that purity became associated with following the commandments of the gods.
- **Liberty/oppression**—Valuing liberty encourages individuality and freedom but also may encourage rebellion against authority perceived as distant, limiting, arbitrary, and oppressive.

For a more complete description of Moral Foundations Theory, see Graham, Haidt, Kaivola, et al., cited above.

34. Just a reminder: "Jacobean" refers to King James I of England. It is a Latinized term that presumably makes discourse about those days sound more sophisticated than does "Jamesean." The word did not come into use until some two centuries after the demise of its Jacobean namesake.

35. There is also Hermione's "trial" for adultery in *The Winter's Tale*, suggesting that justice can be corrupted by irrational jealousy and other emotions. Shakespeare gives us even more extreme examples of justice *cum* revenge in *Titus Andronicus*.

36. James's mother, Mary Queen of Scots, was a Catholic, but he was brought up in a Scottish Protestant church, known as The Kirk.

37. Watson, "False Immortality"; Skulsky, "Pain, Law, and Conscience."

38. Yes, it is the same James I, although Shakespeare could not have pulled these words from the King James Bible because it was not published until some seven years after the play. Most likely, he used the Geneva Bible, which renders the passage in nearly identical wording. The Geneva Bible was the first printed version accessible to the literate public, which amounted to perhaps 30 percent of men and 10 percent of women. James detested the Geneva Bible because it contained study aids that he felt might lead to questioning his authority. So, he commissioned his own version, first published in 1611.

39. On nearly all accounts, *Titus Andronicus* is an outlier. It contains multiple acts of betrayal, subterfuge, brutality, mayhem, murder, wanton sex, oppression, and treason. Oh, and there is a scene in which Tamora feeds her enemy a meat pie into which she has baked his butchered sons—a scene that fits squarely on the moral dimension of purity vs. disgust. Such violent plays often hark back to the Roman dramatist, Seneca, and were common on the London stage early in Shakespeare's career. Shakespeare seems to have wanted to demonstrate that he could top them all; he never again wrote a play the likes of *Titus Andronicus*.

40. At the time of James I, the English were still living in the twilight of a feudal society, when individual freedom and liberty were simply not an option for most people.

Even the castaways in *The Tempest* lived in a rigidly hierarchical micro-society. In *Measure for Measure*, Shakespeare uses the term "liberty," but in the negative sense of rudeness, "license," or perhaps lasciviousness—a sense of the term foreign to moderns. We first hear the term when Lucio asks Claudio why he is going to prison, and Claudio responds:

> From too much liberty, my Lucio, liberty:
> As surfeit is the father of much fast,
> So every scope by the immoderate use
> Turns to restraint. Our natures do pursue,
> Like rats that raven down their proper bane,
> A thirsty evil, and when we drink, we die.
> —*Measure for Measure*, 1.2.122

We hear of liberty again when the duke tells Friar Thomas that "liberty plucks justice by the nose" (1.3.20) because the immoral behavior of his subjects is flaunting the laws. Jacobean audiences, therefore, might have understood "liberty" as an abuse of freedom by taking it too far in violation of norms and laws. In that sense, Jacobeans would have understood "liberty" more as a moral foundations theorist would think of "authority."

41. Cognitive dissonance involves holding two conflicting ideas in mind, and the need to resolve the dissonance can be powerfully motivating. It can also take many forms, one of which involves merely two conflicting pieces of information, as when your checkbook does not jibe with your bank statement or when you read a report that a favorite politician is involved in a scandal. In the five plays we are discussing here, however, the dissonance is between one's own morals or values and one's own behavior.

42. Notice, too, that he is not tempted as much by her femininity but by her virtue, and so by Isabella as "forbidden fruit," all the more tempting because it is proscribed. This refers, of course, to the temptation of Adam and Eve. You may be surprised to know that there is solid research support for the "forbidden fruit hypothesis." See: PsychCentral, "Forbidden Fruit in Relationships," and DeWall, "Forbidden Fruit."

43. In early Scotland, a *thane* was a clan chief. Cawdor is the name of a town in northern Scotland that is the location of Cawdor Castle, which still stands.

44. Roman law provided for two *consuls*, who occupied the highest elective offices in the Roman republic. The consuls and tribunes held veto power over the Senate.

45. No, Shakespeare did not use the term "high crimes and misdemeanors," nor was it original with Thomas Jefferson. But, for your information, the term dates to a 1386 decree by the English parliament, some two hundred years before the Bard.

46. Historians also debate the same issue, where some emphasize that history is driven by circumstances (such as migrations, climate change, or the invention of gunpowder), while others favor the rise of "great men" (and, more recently, great women).

47. See: Boyce, *Shakespeare A to Z*, 518–20.

48. Goldberg, "Markers for the Big-Five." You can take the inventory by clicking on "Big Five Personality Test" at the Open Psychometrics website.

49. This is a controversial dimension that some say should be split in two. It is important to realize that the Big Five inventories were not designed to measure IQ but more of an intellectual disposition or interest pattern. See: Oleynick et al., "Openness/Intellect," 55–130.

50. Kahneman, *Thinking Fast and Slow.*

51. From *Twelfth Night*, 2.5.143. This line is actually a spoof. Malvolio is reading these words from a counterfeit message, intended as a practical joke, that he presumes is from the woman with whom he is infatuated. Malvolio, of course, believes the message is legitimate and that greatness is being thrust upon him: the psychological term for this selective credulity—believing what you *want* to believe—is *confirmation bias.*

CHAPTER FIVE

1. The War of Cyprus, between Venice and the Ottoman Turks, spanned the years 1570–1573.

2. An *ensign* is the lowest-ranking commissioned officer—below lieutenant, the rank to which Cassio had been promoted.

3. A legal marriage must be consummated. See: Cleland, "Fast Married."

4. Shakespeare was always on the lookout for a good metaphor, and this one (alluding to sexual intercourse) he probably borrowed from Rabelais's *Gargantua and Pantagruel*, a work in French that had been translated and published in English in 1593, coincidentally at the time when Shakespeare was beginning his career as a playwright in London. The Bard seems to be the first English writer to use the expression. See: *Wikipedia*, "Beast with two backs."

5. Othello is a composite that Shakespeare derived from several sources, but primarily from the Italian writer Cinthio's story, "Un Capitano Moro" ("A Moorish Captain"), published in 1565 when Shakespeare was one year old. See: Dollimore, "Shakespeare's Sources."

6. Asimov, *Asimov's Guide to Shakespeare*, 609–34.

7. Before it appeared on stage at The Globe, *Othello* debuted at Whitehall Palace in London, where it was likely that King James I was in the audience.

8. The poem was written in 1585 when he was King James VI of Scotland. It was reissued when he became James I of England.

9. Appelbaum, "War and Peace."

10. Karim-Cooper, "Strangers in the City." See also: Lantern Theater, "Venetian Mercantile Empire."

11. *Moor* was a Christian term originally applied to the Berbers of northern Africa, who also swept across the Strait of Gibraltar to conquer the Iberian Peninsula, Malta, and Sicily, leaving their indelible mark on language, customs, and culture. While they were of African extraction, Moors were not necessarily Black. Shakespeare seems to have cast the Moor primarily as an exotic, but because of a comment by Roderigo, a Venetian gentleman, shortly after the play opens, we can be more confident that Shakespeare also conceived of Othello as a Black African:

RODERIGO: What a full fortune does the thick-lips owe
If he can carry't thus! [meaning, if his marriage to Desdemona succeeds]
—*Othello*, 1.1.72

It is noteworthy that Shakespeare highlights the differences in appearance between Desdemona, the daughter of a White patrician family, and Othello, the Moor. In the following passage, we hear Iago warn Desdemona's father of miscegenation, as Iago calls to him at night from the street:

IAGO: Zounds, sir, y'are robbed! For shame. Put on your gown!
Your heart is burst, you have lost half your soul
Even now, now, very now, an old black ram
Is tupping your white ewe. Arise, arise;
Awake the snorting citizens with the bell,
Or else the devil will make a grandsire of you.
—*Othello*, 1.1.94

12. We psychologists would call it an *approach-avoidance conflict*.

13. For a readable and reliable description of the elements in a Shakespearean tragedy, see: Rafiq, "Shakespearean Tragedy?"

14. We are all susceptible to *confirmation bias*: selectively attending to information that fits with (confirms) our preconceptions and prejudices while ignoring information that does not. As we have seen, it is part of our cognition that operates mainly beneath the level of consciousness. Kahneman called it "thinking fast" or System 1 thinking. See: Kahneman, *Thinking Fast and Slow*.

15. Brutus is the protagonist and a flawed hero in *Julius Caesar*, and it is Brutus who goes to war with Marc Antony, hoping to save the Roman Republic—although his efforts reveal his own lust for power. See, for example: Mabillard, "*Julius Caesar*."

16. There is no agreed-on definition of a "hero." Discussions of dramatic heroes often hark back to Aristotle, who defined the *tragic hero* as a man of noble birth who, because of a "tragic flaw," experiences a severe reversal of fortune that may leave him dead, destitute, or an abject failure in the eyes of others. Our point, however, is that Shakespeare wrote of another kind of hero, too—such as Henry V, who had no tragic flaw or downfall.

17. As we have seen, historians generally agree that *Richard III* was Shakespeare's hatchet job on the last Yorkist king, whose death marked the founding of the Tudor dynasty. By the standards of kings in the fifteenth century, Richard was not nearly so nasty as Shakespeare portrayed him. But Shakespeare also knew on which side the butter was spread on his bread, and so he made the narrative more favorable to his Queen Elizabeth's grandfather, who purportedly slew Richard.

18. This reference is puzzling, although editors often interpret "liberal" to mean "free and without restraint," while "the north" refers to the cold and blustery north wind. Remember, too, that Shakespeare was writing in what we now call "the Little Ice Age," when temperatures across Europe averaged several degrees lower than they do today, making "the north" especially cold. See: Markley, "Shakespeare in the Little Ice Age."

19. Franco and Zimbardo, "Banality of Heroism."

20. Norwood, "Rosa Parks."

21. Turnbull, "Underrated Heroines."

22. Turnbull.

23. Shylock is a Jewish moneylender, so the play has generated controversy over what seems to be apparent antisemitism. In an impassioned speech, Shylock lets us know that he is the victim of prejudice, saying, "If you prick us, do we not bleed? If you tickle us, do we not laugh? If you poison us, do we not die? And if you wrong us, shall we not revenge?" (3.1.52) Portia does nothing to compensate Shylock, and so we leave it to your judgment whether she deserves hero status.

24. The three are only called "Witches" once in the play, while being referred to as "Weird" or "Weird Sisters" six times. In Old English, the term *wyrd* meant "fate," so Shakespeare may also have had the three fates of Greek mythology in mind.

25. Here is our list of minor characters who play pivotal heroic roles in Shakespeare's plays:

Men	**Women**
Kent—*King Lear*	Emilia—*Othello*
Edgar—*King Lear*	Cordelia—*King Lear*
The Fool—*King Lear*	Paulina—*The Winter's Tale*
Camillo—*The Winter's Tale*	Miranda—*The Tempest*
Escalus—*Measure for Measure*	Portia—*The Merchant of Venice*
	Imogen—*Cymbeline*
	Lavinia—*Titus Andronicus*
	The Weird Sisters—*Macbeth*
	Cassandra—*Troilus and Cressida*

26. Zimbardo, *The Lucifer Effect*.

27. Nor does this notion imply that there are not genuinely bad people—psychopaths, sociopaths, and others who promulgate evil. Perhaps the hardest lesson to learn from the Stanford Prison Experiment is that all of us who think we are "good people" are susceptible to the *power of the situation*.

28. Milgram, *Obedience*.

29. Darley and Latané, "Bystander Intervention."

30. Franco and Zimbardo, "The Banality of Heroism."

31. For more information, see: Heroic Imagination Project, https://www.heroicimagination.org/.

32. There are twelve, if one counts *Cymbeline* a tragedy, as the *First Folio* does.

33. Shakespeare does give us some other antiheroes in plays that are not classified as tragedies. They include Richard III and Antiochus in *Pericles*.

34. If you have read Daniel Kahneman's book *Thinking Fast and Slow*, you may also recognize Macbeth's tragedy turns on several mistaken *heuristics*, including *the sunk-cost heuristic*, which involves continuing figuratively to dig deeper the hole into which one has fallen.

CHAPTER SIX

1. A *changeling* is a child believed to have been exchanged by fairies for one of their own.

2. "Mechanicals" are skilled laborers. In *Midsummer Night's Dream*, the mechanicals are a group of six who put on a play-within-a-play, hoping to have their act chosen as part of the entertainment at the wedding party for Theseus and Hippolyta.

3. The four sub-plots: (a) events surrounding the wedding of the Athenian nobles, Theseus and Hippolyta; (b) the consequences of Egeus's demand that Hermia marry Demetrius and her refusal, complicated by her love for Lysander and her friend Helena's love for Demetrius; (c) the humorous and inept rendition of the Pyramus and Thisbe story (play-within-a-play) by the "rustic mechanicals"; and (d) the marital squabbles of the fairy king and queen, Oberon and Titania, plus the clever tricks played by their servant, Puck.

4. Cline, "Midsummer."

5. Warren, "Biology in Shakespeare"; Fang, "Renaissance Sense of Sleep."

6. Hirai, "Renaissance Medicine"; Crutwell, "Psychology in Shakespeare's Age."

7. The quote here is a description of Magnus's position, in: Camden, "Sleep and Dreams."

8. Oddly, these lines do not appear in the Folger Library edition of *Love's Labour's Lost*, yet they are found in many internet sources. Here we give the text and line number used by the Internet Archive. See Internet Archive, full text of "Love's Labour's Lost."

9. Note that this passage is from *Quarto 1*, and we have modernized the spelling. The *Folio* has slightly different wording, leaving out the segment about the "vital spirits." For the full text of the Quarto 1 edition, see: Internet Editions, Romeo and Juliet (Quarto 1, 1597).

10. That is, "sleep" or one of its variants, such as *sleeps*, *sleeping*, or *slept*, appears in all the plays.

11. Chandler, "Shakespeare and Sleep," 255–60.

12. These include *Henry IV, Pts 1 and 2*, *The Tempest*, *Hamlet*, *Romeo and Juliet*, *Macbeth*, *A Midsummer Night's Dream*, *Measure for Measure*, *Julius Caesar*, *The Winter's Tale*, *Othello*, *King Lear*, and *Richard III.*

13. More recently, dreams have been conceived as the source of creativity, as when the famed chemist Michael Faraday claimed he saw, in a dream, the ring-shaped structure of the benzene molecule, instantly solving a long-standing problem in organic chemistry.

14. In Genesis 41:4, Jacob interprets the Pharaoh's dream about seven fat kine (cattle) and seven lean ones as foretelling seven years of bountiful harvests followed by seven years of famine.

15. Mandel, "Dream and Imagination," 61–68.

16. This was the first use of the EEG, or electroencephalograph, to record brain activity during sleep. For the full story of Aserinsky's discovery, see: Brown, "Stubborn Scientist."

17. Aserinsky and Kleitman, "Regularly Occurring Periods of Eye Motility," 273–74.

18. This is sometimes called the "old hag syndrome." Here is the source, in which Mercutio mentions Queen Mab and describes her influence on sleep:

MERCUTIO: O, then, I see Queen Mab hath been with you.
She is the fairies' midwife, and she comes
In shape no bigger than an agate-stone
On the fore-finger of an alderman,
Drawn with a team of little atomies
Athwart men's noses as they lies asleep;
Her wagon-spokes made of long spinners' legs,
The cover of the wings of grasshoppers,
The traces of the smallest spider's web,
The collars of the moonshine's wat'ry beams,
Her whip of cricket's bone; the lash of film;
Her waggoner a small grey-coated gnat,
Not half so big as a round little worm
Pricked from the lazy finger of a maid:
Her chariot is an empty hazelnut
Made by the joiner squirrel or old grub,
Time out o' mind the fairies' coachmakers.
And in this state she gallops night by night
Through lovers' brains, and then they dream of love;
O'er courtiers' knees, that dream on court'sies straight,
O'er lawyers' fingers, who straight dream on fees,
O'er ladies' lips, who straight on kisses dream,
Which oft the angry Mab with blisters plagues,
Because their breaths with sweetmeats tainted are:
Sometime she gallops o'er a courtier's nose,
And then dreams he of smelling out a suit;
And sometime comes she with a tithe-pig's tail
Tickling a parson's nose as a' lies asleep,
Then dreams, he of another benefice:
Sometime she driveth o'er a soldier's neck,
And then dreams he of cutting foreign throats,
Of breaches, ambuscadoes, Spanish blades,
Of healths five-fathom deep; and then anon

Drums in his ear, at which he starts and wakes,
And being thus frighted swears a prayer or two
And sleeps again. This is that very Mab
That plaits the manes of horses in the night,
And bakes the elflocks in foul sluttish hairs,
Which once untangled, much misfortune bodes:
This is the hag, when maids lie on their backs,
That presses them and learns them first to bear,
Making them women of good carriage:
This is she—

—*Romeo and Juliet*, 1.4.58

19. Ficca and Salzarulo, "What in Sleep Is for Memory," 225–30.

20. Eckert and Malhotra, "Adult Obstructive Sleep Apnea," 144–53.

21. We all have brief sleep apnea episodes, at least occasionally; five or fewer apneas per hour is considered normal.

22. Sorry!

23. See: Ergoflex, "Genius Sleep."

24. Fang, "The Renaissance Sense of Sleep."

25. Freud, *The Interpretation of Dreams.*

26. Haidt, *The Righteous Mind.*

27. Kahneman's view and Jonathan Haidt's, as well, are called *dual-process theories*. For a good summary of these concepts, see: Evans and Stanovich, "Dual-Process Theories."

28. System 1 and System 2 (intuition and rational thought) are not discrete entities in the brain; there is no "brain center" for either one. Some of the structures they use in the brain are, indeed, distinct, but there is also much overlap.

29. Here is a more detailed comparison of the two parts of the "dual-process" mind proposed by Kahneman and Haidt:

Kahneman's System 1 (which is also Haidt's "elephant" of the mind) is fundamentally our long-term memory structured around an ancient response system designed for survival and procreation. Importantly, it is vast—but not rational; it is also mostly unconscious. Its "thinking" consists of following links among memories embedded in a network of associations where the strongest memories—the memories most likely to pop up in consciousness—are linked to strong emotions. Accordingly, you are more likely to remember your first kiss than your first car ride, and because most people are programmed to be "risk averse," you are more likely to respond to a threat than to an opportunity.

Thus, we can think of unconscious intuition as a web of facts, experiences, genetic tendencies, and emotions that sends out queries and ripples throughout the brain, returning "gut" reactions to our consciousness. No modern computer can match its speed. But it is not rational or necessarily devious or evil. It makes us love and support our children and grandchildren. It drives us to excel, as best we can, because it makes us more attractive mates and leads to worldly resources that help us and our descendants to thrive.

These are the features of System 1, the *unconscious* mind:

- Long-term memory (LTM): stores memories "tagged" with associated emotions and motives
- Intuitive: unlike consciousness in System 2, System 1 cannot "think" rationally.
- Associative: contents of System 1 are linked by similarities of patterns involving all the senses, emotions, and motives.
- Main job of System 1: to alert consciousness to events that affect survival and reproduction and to do so very quickly

In contrast, the conscious mind (Kahneman's System 2 and Haidt's "rider") is where reason can occur. But the conscious mind constantly draws on emotion-tinged memories, so its ability to reason often succumbs to the emotional biases of our unconscious memories. System 2 is often called short-term memory: information, if not used, is lost after about thirty seconds.

These are the features of System 2, the *conscious* mind:

- Small capacity: holds about seven items (or "chunks") of information
- Capable of "thinking" manipulating contents and forming new associations that can be sent to LTM
- Rational but also *rationalizing*: capable of rational thought, but often follows emotional or motivational urges proposed by System 1 and then rationalizes its decisions.

In their research, Kahneman and Haidt have each compiled a great deal of evidence for this dual-process view of the mind, and we suggest that this new paradigm is one of psychology's greatest achievements in the last half-century.

30. See Dement's delightful book, *Some Must Watch While Some Must Sleep*, the title of which is taken from *Hamlet*, 3.2.297.

31. In Freud's view, the mind was tripartite, with the *id* being the unconscious source of desires and repressed memories, while the *ego* acted as the conscious and rational controller of the unconscious, drawing on the moral resources of the mind that he lumped together as the *superego*.

32. Hammond, "Shakespeare's Poisons"; Shakespeare Lives, "Shakespeare Lives in Science."

33. Recent testing of ancient hair samples shows that mandrake and several other hallucinogens have been used at least since the Bronze Age. See: Jacobs, "Tripping in the Bronze Age."

34. Elizabethans knew that water was associated with dysentery, typhoid, and cholera, so they generally avoided drinking it. Milk was available as an alternative, but so were beer and ale, where the alcohol was a preservative. Low-alcohol ales were often given to children. See: Folger, "Fooles and Fricassees," and Cartwright, "Food and Drink."

35. Pollard, "A Thing Like Death."

36. Edwards, "Paracelsus."

37. Homeopathy, by that name, was unknown as a form of medical treatment in Shakespeare's day. It was conceived at the end of the eighteenth century by a German

physician who taught that the more dilute the medicine, the more powerful it would be. Indeed, many homeopathic treatments have been diluted (usually with water or alcohol) so much that there is no detectable trace of the original substance in it. Neither of the two main ideas on which the theory of homeopathy is based—"like cures like" and dilution increases the potency of the treatment—has solid science to support it.

38. In this scene, Richard (who wants the public to embrace him as their king) is advised by the Duke of Buckingham to make a show of piety and pretend to be reluctant to accept the crown:

> And look you get a prayer-book in your hand,
> And stand betwixt two churchmen, good my lord;
> For on that ground I'll build a holy descant:
> And be not easily won to our request:
> Play the maid's part, still answer nay, and take it.

We cannot help but note the parallel between Richard's holy charade and that of a certain recent American politician.

39. Machovec, "Shakespeare on Hypnosis," 73–88.

40. In a personal communication, Csikszentmihalyi explained that his interesting name is of Transylvanian origin that translates as "the church of St. Michael on the hill."

41. From the viewpoint of dual-process theory (involving Kahneman's Systems 1 and 2 or Haidt's "intuition" and "reason"), we judge that *flow* may be a process in which both the unconscious associative memory of System 1 and the limited capacity of working memory of System 2 are completely occupied with an interesting task.

42. al'Thor, "Does Portia Subconsciously Influence?"

CHAPTER SEVEN

1. 1485 marks the battle of Bosworth in which Richard was slain by Henry Tudor, who founded the Tudor dynasty as King Henry VII. In the play, Henry Tudor is referred to as the Earl of Richmond.

2. Some 527 years after his death, Richard III's bones reappeared in 2012, unearthed beneath a parking lot in Leicester. Identification was possible by means of DNA matching with living descendants. The bones revealed at least ten wounds, including eight to the head. One blow—probably the lethal stroke—went deep into the skull. This fits with eyewitness accounts saying that Richard was dispatched by a halberd—a nasty instrument composed of an axe on a pike. His bones also gave evidence of spinal scoliosis, which Richard is known to have had, making the identification even more certain. See: King et al., "Identification of Remains"; English Monarchs, "Remains of Richard III"; and Smith, "Richard III Killed."

3. The relationships among the prominent nobles in English history are complicated, especially during the English civil wars of the late 1400s. Known as the Wars of the Roses, these bloody skirmishes were essentially a long and deadly family feud,

pitting the Yorkists (Hotspur's and Richard's side) against the Lancastrians (including Henrys IV, V, and VI). These two factions were diverging branches of the House of Plantagenet, originally from the Anjou region in France. Their quarrel involved a dispute over succession to the throne after Edward III died in 1377. Some three hundred years earlier, the Plantagenets, led by William I (also known as William the Conqueror), became rulers of England after winning the Battle of Hastings in 1066.

4. At the time, the Tower of London was one of the royal castles, rather than simply a prison and an execution venue. As a castle, it would have a prison/dungeon, where, according to Shakespeare, the Duke of Clarence was drowned in a butt of Malmsey wine. (*Malmsey* refers to a genus of wine grapes known as *Malvasia*, grown in the Mediterranean region; a *butt* or barrel held about 105 gallons.)

5. This pair is often referred to as "the princes in the Tower"—the Tower of London, in which they were held hostage and from which they never reemerged.

6. Apologists point out that Richard was not exceptionally brutal in comparison to the warlord-nobles in those days. Even more to his credit, Richard instituted some reforms that benefitted the common people, such as the legal presumption of innocence until proven guilty. We can attribute some of the Bard's historical inaccuracies to poetic license, as well as to erroneous and biased sources, such as Holinshed's *Chronicles*. Even so, Shakespeare does seem to have cast him in a bad light, no doubt to portray the Tudors more charitably. The following examples illustrate some of the historical errors and anachronisms in *Richard III*.

According to some sources, the wine-drowning incident contained a kernel (or, perhaps, a gallon) of truth. But it was on the orders of another brother, King Edward IV, because Clarence switched sides during the Wars of the Roses. Moreover, at the time of Clarence's death and for some years afterward, Richard was living in northern England, far from the Tower and the lethal wine-butt.

Neither was Richard guilty of murdering his wife Anne Neville's first husband, Edward of Westminster (the son of Henry VI), who actually died in the battle of Tewkesbury. Nor is there evidence that Anne herself was murdered on Richard's orders. Rather, she is thought to have died of tuberculosis.

The Bard seems also to have falsely implicated Richard in the death of a cousin, King Henry VI, from the rival House of Lancaster. None of Shakespeare's sources support that claim. Instead, historians now believe Edward IV was again the guilty party.

Regarding the putative palace coup following Edward IV's death, our Will seems to have taken considerable poetic license. Richard was staying in the north of England at the time of Edward's demise, only returning from the North to rule as Lord Protector, according to the dying wishes of Edward. Following a Plantagenet family tradition, the king-to-be, Edward V (the elder of the two "princes in the Tower"), was to stay in the royal apartments in the Tower of London while awaiting his coronation. No one knows what happened to the two princes, and it is possible that Richard *was* guilty of their disappearance.

At the Battle of Bosworth, Richard III did not fight Henry Tudor in single combat. Rather, Richard found himself unhorsed, on foot, and surrounded by enemy soldiers after his mount stumbled in a boggy area. At the end, he cried "Treason!" rather than "My kingdom for a horse!" because he saw some of his allies switching sides when they realized he was in trouble. Surrounded by attackers, he died of numerous wounds. See: Ingram, "Death of Richard III."

As we noted earlier, an excavation fortuitously uncovered Richard's remains in 2012, which made it clear that he suffered multiple wounds in his final battle at Bosworth. While his skeleton also showed that he did have scoliosis (curvature) of the spine, the degree and direction of the curvature was not so serious that he would have been called a "hunchback." Contemporary accounts merely mention that one shoulder was higher than the other.

7. Elizabeth also had Shakespeare's company, The Lord Chamberlain's Men, perform at court, which was an honor and, not incidentally, lucrative for the company. She was never, technically, Shakespeare's patron. After her death, King James did become the company's patron, and they became known as The King's Men.

8. The term *lunacy* derived from *lunar* or references to the moon, which was believed (by some) to be responsible for mental illness. Even today, people still believe that phases of the moon influence madness, despite the absence of anything but anecdotal evidence.

9. According to Schmidt, *Shakespeare Lexicon*.

10. This pronunciation was the source of our word *bedlam*, referring to a locus of noise and confusion. The original hospital was part of the Priory of St. Mary of Bethlehem, so it was known as St. Mary Bethlehem Hospital. The locals most commonly abbreviated the name to Bethlem or Bedlam.

11. *Amain* is an archaic word for "forcefully."

12. We found this quotation from Burton's book (Burton, *Anatomy of Melancholy*) at Project Gutenberg.

13. Shakespeare's theater was called, simply, "The Theater."

14. Here's a collection of short pieces illustrating conceptions of mental illness in Shakespeare's time, published online: Shakespeare Birthplace Trust, "Method in the Madness."

15. For the history and psychology of witchcraft beliefs, see: Zimbardo, *Lucifer Effect*.

16. Bartholomeus Anglicus was a Franciscan monk, famous for his encyclopedic tome, *De proprietatibus rerum* (*On the Properties of Things*), written some 350 years before Shakespeare. Here is our own quasi-modern English translation of his description of melancholy, as given by Dimitrijevic, "Being Mad in Early Modern England":

> Melancholy is a humor, boisterous and thick, and is bred of troubled [impurities] of blood. . . . Of this humor having mastery in the body these be the signs and tokens. First, the color of the skin changes into black or blue. Sour savor, sharp—and earthy is felt in the mouth. By the quality of the humor the patient is faint—and fearful in heart without cause, and oft sorry. . . . Some dread enmity of some man: Some love and desire death.

17. See: The British Library, *The Discovery of Witchcraft*.

18. Tasca et al., "Women and Hysteria," and Waardt, "Witchcraft, Spiritualism, and Medicine."

19. Neeley, "Documents in Madness."

20. Hanson, "Tudor England Mental Illness," and Zlatar, "When the Body Is Ill."

21. "Mental retardation" is one of those terms that has been buffeted by the tempests of fashion and political correctness. Some object to the term, preferring *general learning disability* or *impaired intellectual functioning* because they contend that "mental retardation" is a pejorative term. Our view is that *mental retardation* is an improvement over the terms used in the early days of psychology, when people with intellectual deficits were called "dull," "feebleminded," or "stupid." As IQ testing became popular, the official terms became, in order of increasing intellectual ability, idiot (IQ 0–25), imbecile (IQ between 26 and 50), and moron (IQ 51–70).

22. Not one to ignore a knocking opportunity, Shakespeare wrote a play about a Scottish king and a band of witches when James VI of Scotland ascended the English throne as King James I. James was so concerned with witchcraft and demonic possession that he wrote a book, *Daemonologie*, describing how Satan deviously used these tools to capture our souls. Oddly enough, James was skeptical of lycanthropy, suggesting that the supposed victims were delusional because they had an overabundance of the melancholy-inducing humor, black bile.

James's views on witchcraft were probably influenced by a harrowing experience at sea, leading to a series of high-profile witch trials in Scotland. It seems that a storm had destroyed James's fleet as he ferried his bride across the North Sea from Denmark, a disaster that James believed had been conjured by some seventy accused witches from North Berwick. (Most of the accused confessed, under torture.) Shakespeare borrowed some of James's own words from *Daemonologie* and inserted them into the mouths of the "weird sisters" when he wrote *Macbeth*.

Surprisingly, King James's book has not gone out of print but is available from many booksellers. See: King James I, *Daemonologie*. See also: Dimitrijevic, "Being Mad."

23. Jayatunge, "Shakespearean Work."

24. Lagay, "Legacy of Humoral Medicine."

25. Owens, "Shakespeare and Medicine."

26. Owens, "Shakespeare and Medicine."

27. "Shakespeare and Insane Asylums" is an interview by Rebecca Sheir with Professor Benjamin Reiss, author of *Theaters of Madness: Insane Asylums and Nineteenth-Century American Culture*. See Sheir, "Shakespeare and Insane Asylums," in the bibliography.

28. Reiss, *Theaters of Madness*, and Folger, "Madness of Hamlet and King Lear."

29. This was a heyday for Darwin's theory of natural and sexual selection, and Freud's theory of the sexual basis of mental illness seemed a natural extension of Darwin's ideas.

30. Freud, *Interpretation of Dreams*.

31. For a look at how a modern-day Freudian views Shakespeare's work, see: Jayatunge, "Shakespearean Work."

32. Cognitive-behavioral therapy is a research-validated form of psychotherapy that combines elements from traditional talk therapies with techniques borrowed from behavioral psychology, particularly operant learning and classical (Pavlovian) conditioning. The result is a therapeutic approach that helps patients both understand their condition and practice new, more adaptive behaviors. There are many online reviews of cognitive-behavioral therapy, such as Hoffman et al., "Efficacy of Cognitive Behavioral Therapy."

33. National Institute of Mental Health, "Genetics and Mental Disorders."

34. *DSM-5* is short for the *Diagnostic and Statistical Manual of Mental Disorders, 5th ed.*, produced and published in 2013 by the American Psychiatric Association. The *DSM-5* describes nearly six hundred mental problems in twenty-two categories (listed below).

Neurodevelopmental Disorders (e.g., Intellectual Disability, Language Disorder, Autism Spectrum, ADHD, Tourette's)
Schizophrenia Spectrum and Other Psychotic Disorders
Bipolar and Related Disorders (formerly "manic-depressive" illness)
Depressive Disorders
Obsessive-Compulsive and Related Disorders
Anxiety Disorders (e.g., Phobias, Panic Disorder)
Trauma- and Stressor-Related Disorders (e.g., PTSD)
Dissociative Disorders (e.g., Dissociative Amnesia and "multiple personality")
Somatic Symptom Disorders (e.g., Psychogenic Weakness and "hypochondria")
Feeding and Eating Disorders (e.g., Anorexia and Bulimia)
Elimination Disorders (e.g., Bedwetting)
Sleep-Wake Disorders (e.g., Insomnia, Narcolepsy, Sleep Apnea)
Sexual Dysfunctions (e.g., Erectile Disorder, Female Orgasmic Disorder)
Gender Dysphoria (e.g., "transgender")
Disruptive, Impulse-Control, and Conduct Disorders (e.g., Antisocial Personality, Oppositional Defiant Disorder, Pyromania)
Substance-Related and Addictive Disorders
Neurocognitive Disorders (e.g., Delirium, Alzheimer's Disease)
Personality Disorders (e.g., Paranoid Personality, Borderline Personality, Histrionic Personality)
Paraphilic Disorders (e.g., Pedophilia, Fetishism)
Other Mental Disorders (unspecified)
Medication-Induced Disorders
Other Conditions That May Be a Focus of Clinical Attention (e.g., Bereavement, Parent-Child Relational Problem)

The *DSM-5* is used primarily in the United States. Another manual, the *ICD-11*, is used internationally by nations belonging to the World Health Organization.

35. The first four categories—hallucinations, delusions, disorganized thinking, and abnormal motor behavior—involve so-called "positive" symptoms of schizophrenia

in that they are mainly exaggerations or additions to normal behavior. Here is a brief description of each:

- *Hallucinations* are false sensory experiences, such as hearing voices or seeing objects that do not exist.
- *Delusions* involve persistent false beliefs, particularly about oneself, such as delusions of persecution by enemies or delusions of grandeur, as in the belief that one is God or some famous individual.
- *Disorganized thinking* may involve a "flight of ideas," moving rapidly and disjointedly from one topic to another, or it could involve repetitive or entirely nonsensical speech, known as "word salad."
- *Disorganized motor behavior*, at one extreme, could manifest as inappropriate silliness or agitation or its polar opposite of unresponsiveness or stupor.

In contrast, *negative symptoms*, including social withdrawal and impoverished emotional responses, are those that seem to "subtract" from normal functioning. They occur in a large proportion of schizophrenic patients and are signs of a poor long-term prognosis.

36. Psychotic symptoms also figure prominently in *The Winter's Tale*, where Leontes grows mad with jealousy, and in *King Lear*, where the old king also is suffering from senile dementia.

37. In our view, this creeping mental health hegemony has a dark side. It may be one reason that so many conditions associated with poverty have come to be seen as rooted in a lack of individual responsibility and gumption, which gives us a convenient excuse to ignore problems lying within our culture and economic system.

38. Shakespeare gave us many lethal characters, including Macbeth, Othello, Coriolanus, Titus, and even Hamlet. All had mental issues—to the extent that we might consider them, at some time in their plays, to be "lunatics" or "insane"—at least in the Elizabethan sense of the terms. Today, however, *insanity* has a different meaning; it is a legal term, not a psychiatric or psychological diagnosis. In American law, it means either not knowing the difference between right and wrong or not being unable to resist the impulse to commit an unlawful act. The popular confusion between mental illness and insanity compounds the problems of those with mental illnesses, particularly because the general public may consider them to be dangerous. While there are some psychopaths and other disturbed folks who are dangerous, people suffering from depression, phobias, schizophrenia, PTSD, and other mental afflictions are far more likely to harm themselves than others. Shakespeare's presentation of a melancholy and vengeful Hamlet, for example, describes a rare complication: while depression often *does* lead to suicide, it rarely leads to homicide.

39. If we were to do a clinical workup on a Falstaff today, we could check his blood alcohol level—but even then, we would not diagnose him with a substance-abuse disorder unless we took his history and found a pattern of overindulgence.

40. Warren, "Biology in Shakespeare."

41. Myers, "Psychopath and So-Called Sociopath," explains how psychopathy is diagnosed via personality traits, and antisocial personality is diagnosed via behaviors.

42. Richard has mixed emotions—as do most of Shakespeare's villains. He expresses anger at others who are more successful at love, especially at his brother Edward, who holds the position he covets. This anger, however, is mixed with melancholy. In these frames of mind, he sees an opportunity to seize the throne.

43. Shakespeare borrowed much of the plot for *Titus Andronicus* from Ovid's story of the rape of Philomel, told in the *Metamorphoses.* And he even weaved a conspicuous mention of Ovid's story into the *Titus Andronicus* script. See: Royal Shakespeare, "Dating *Titus.*"

44. The cannibalism is inadvertent on the part of the diners, but it was intentional on the part of Queen Tamora. The statistics on mayhem come from SparkNotes, *Titus Andronicus.*

45. Theatrical tradition warns against saying "Macbeth" inside the theater, except during a performance, lest one tempts bad luck. Instead, actors traditionally refer to *Macbeth* as "the Scottish play." However, if the actor slips and utters the forbidden name, there is a remedy. The wayward individual must dash outside, spin around three times, spit, ask for forgiveness, and beg reentrance to the theater. See: Farley, "Qualitative Analysis."

46. The historical Macbeth's surname was "mac Findlaech." See: Jaramillo, "Macbeth: The Real King of Scotland." According to the Wikipedia entry for "Macbeth, King of Scotland," the anglicized version of the name would be "MacFinlay."

47. A *thane* in Scotland was a feudal lord who held land given him by the king for his military service. His position in the noble hierarchy depended on both the extent of his lands and his favor with the king. Generally, this could mean that a thane ranked somewhere between the equivalent of a knight and a baron. Cawdor Castle, by the way, is a real place near the coast in northern Scotland. It exists today as a tourist attraction, except when the royal family is in residence.

48. We deliberately do not list psychopathy or antisocial personality because Macbeth's feelings of guilt and remorse seem to disqualify him, on a technicality. Psychopaths typically do not feel guilty about their misdeeds.

49. In *Troilus and Cressida*, Shakespeare gives us another interesting woman, Cassandra, who is thought to be "mad"—yet she was not. Following Homer's account in *The Iliad*, Cassandra raves maniacally about how the long war with the Greeks will end in the fall of Troy, but everyone considers her mad. Apollo had given her the gift of prophecy but added the curse that no one would believe her warnings.

50. Auditory hallucinations ("hearing voices") are the most common form of hallucinations among psychotic patients, as compared with hallucinations involving the other senses: vision, touch, smell, and taste. Macbeth's hallucination is visual.

51. The ghost of Hamlet's father represents something different from Banquo's ghost. Hamlet is not hallucinating because the script tells us that several other witnesses see the ghost, not just the prince. The ghost, then, is a real representation of his father's spirit. In contrast, Macbeth *does* have a hallucination. We know this because no one else sees Banquo, but rather an empty stool. Both plays were plausible to the audiences of the time because the belief in ghosts was widespread, as was the distinction between ghosts and hallucinations.

52. Macbeth calls her "chuck" (3.2.51), but the "c" is lower case. The historical Lady Macbeth was a woman named Gruoch, of uncertain lineage. See: Jaramillo, "Macbeth: The Real King of Scotland," and History of Royal Women, "Gruoch."

53. When it occurs regularly, sleepwalking may be diagnosed as one of the *parasomnias* that occur in non-REM sleep—that is, outside of the sleep stage in which the eyes move rapidly under the eyelids and when dreams most commonly occur. Full disclosure: in his youth, one of us (Johnson) walked in his sleep on occasion, usually during stress. For more information regarding the association of sleepwalking and other disorders, see "Parasomnias" in: American Psychiatric Association, *Diagnostic Manual*, 399–404, and also Wood, "Sleepwalking."

54. Unlike her husband's vision of the dagger, Lady Macbeth's visions of blood on her hands do not qualify, technically, as psychotic hallucinations but rather nightmare images. Yet they are descriptive of OCD.

55. See: Jayatunge, "Shakespearean Work."

56. The *DSM-5* further describes obsessional jealousy (see: *Diagnostic Manual*, 263–64). Clinicians often call it the *Othello syndrome.*

57. *Hamlet* (2.2): Shakespeare seems to have borrowed elements of *Hamlet*, including the idea of feigning mental illness, from Thomas Kyd's *The Spanish Tragedy* and from an old Danish tale, *Almeth.* The literary device of feigning mental illness goes back at least as far as Sophocles. See: Alou, "Spanish Tragedy and Hamlet," 52–57; Best, "Amleth/Hamlet"; and Sabio, "Mask of Madness." It is also worth noting that many Shakespearean plays rely on people pretending to be who they are not—for examples, Rosalind is Ganymede in *As You Like It*, Viola poses as Cesario in *Twelfth Night*, Imogen disguises herself as a boy in *Cymbeline*, and arrogant men are fooled by "the bed trick" (borrowed from Boccaccio), which appears in both *Measure for Measure* and *All's Well.*

58. Nor is Hamlet the only one in the play who deals with depression. The most obvious other is his ex-ladyfriend Ophelia, whom Hamlet taunts and abuses. Tragically, she ultimately takes her own life by drowning (act 4, scene 7).

59. Alou, "Spanish Tragedy and Hamlet."

60. Wheeler, "Deaths in the Family."

61. Tosh, "Shakespeare and Madness."

62. If a King Lear ever existed, it was in Britain's remote past, and no credible trace of him remains. Shakespeare found the legend of King Lear in Geoffrey of Monmouth's volume, *The History of the Kings of Britain*, written in the twelfth century. The book was taken seriously in Elizabethan-Jacobean times, but historians now pooh-pooh it.

63. The diagnosis of phrenitis appears here: Andreasen, "Artist as Scientist."

64. In his "All the world's a stage" speech, Jacques, too, describes the final stage of life: "Last scene of all, / That ends this strange eventful history, / Is second childishness and mere oblivion, / Sans teeth, sans eyes, sans taste, sans everything" (*As You Like It*, 2.7.146).

And was it mere coincidence that, at the same time, Cervantes was writing about Don Quixote, another delusional and demented old man? For further speculation on this, see the following sources: Kipen, "Secret Connection," and Todorov, "Shakespeare, Cervantes."

65. In psychiatric parlance, "senile" refers to the *senium* or one's advanced years—from the Latin word for old and feeble.

66. For a medical perspective on Lear, see the following sources: Ottilingam, "Psychiatry of King Lear," and Tosh, "Shakespeare and Madness."

67. In Shakespeare's day, "fond" meant credulous or foolish.

68. Freud made his reputation treating hysteria, which he also called "psychosomatic illness." His 1895 book (with Josef Breuer), *Studies on Hysteria*, marked the launch of psychoanalysis. In it he laid out his theory that the symptoms were caused by repressed sexual issues. See: Bogousslavsky and Dieguez, "Freud and Hysteria."

69. Dimitrijevic, "Being Mad."

70. The term *conversion disorder* dates back to Freudian times, in the late nineteenth and early twentieth centuries, when Dr. Freud declared it to be caused by "conversion" of unconscious conflicts into physical symptoms. See Ali et al., "Conversion Disorder"; Morris, "Shakespeare's Minds Diseased"; Tasca et al., "Women and Hysteria."

71. We suspect that Shakespeare's audiences would have included Kate, from *Taming of the Shrew*, in this list, but we think most audiences today extend more sympathy to her character.

72. Folger Shakespeare Library, "How Shakespeare Describes."

73. See: Will I Am Shakespeare, "Dr. William Shakespeare, M.D."

74. See: Tosh, "Shakespeare and Madness"; Neely, "Documents in Madness."

75. The Boar's Head Inn existed in the London of Shakespeare's time near London Bridge—but not during the reign of Henry IV.

76. The marriage was arranged with Kate's father, as was common in many cultures of the time.

77. Audiences would have known that mutton was believed to increase irascibility (by boosting choler, or yellow bile) in a couple already angry with each other—adding to the humor of the play. Incidentally, wine was also thought to promote choler, so it was often given to phlegmatic persons.

78. There is some controversy over whether Caesar's "falling sickness" was due to epilepsy or to mini-strokes (TIAs). The following sources give both sides of that issue. Either way, sources agree that he developed the condition toward the end of his life. See: Hughes, "Dictator Perpetuus," and Kilgrove, "Julius Caesar's Health Debate."

79. See: Shakespeare Birthplace Trust, "Treating Mental Illness."

80. Owens, "Shakespeare and Medicine."

81. See: Shakespeare Birthplace Trust, "Treating Mental Illness."

82. Cognitive-behavioral therapy (CBT) is a form of psychotherapy that deals with the mental and environmental conditions that cause or maintain undesirable behaviors, rather than with childhood trauma, unconscious motives, and other psychodynamic processes that were once believed to underly mental problems. In *As You Like It*, Rosalind (disguised as Ganymede) employs role-play, a common technique in CBT.

83. Other studies had demonstrated the Lady Macbeth Effect, but this research was the first to study it in a population with OCD. See: Reuven, Liberman, and Dar, "Effect of Physical Cleaning."

84. A controversial meta-analysis by Zuckerman et al. claimed to show that the Lady Macbeth Effect is not real. (Meta-analyses compare a group of studies done with essentially the same methodology.) The study reported by Reuven and her colleagues was *not* included in that meta-analysis by Zuckerman's group. Moreover, other studies published subsequently have supported the original findings about the Lady Macbeth Effect. See: Siev, Zuckerman, and Siev, "Immorality and Cleansing."

85. The following are popular summaries of the literature on the Lady Macbeth Effect: Herber, "Damned Spot"; Schieffelin, "Lady Macbeth Effect"; Risen, "Lady Macbeth Effect."

CHAPTER EIGHT

1. *Love's Labour's Lost* was one of Shakespeare's earliest plays—probably written before 1593. It features the king of Navarre, a country that, by Shakespeare's time, had ceased to exist independently. The play is set in medieval times when Navarre was the realm of Ferdinand of Aragon. (This was the same Ferdinand who later married Isabella of Castile; together they sent Christopher Columbus on his fateful voyage that included the discovery of the "New World"—in 1492, as you will recall—which is the date we arbitrarily assign to the story.)

In the Bard's day, Navarre existed as a buffer between the two superpowers of France and Spain—both of which had a claim to its territory but neither of which relished going to war over the tiny kingdom. When Shakespeare was a boy, Navarre had a king known as Henry III—which was confusing because France also had a King Henry III. Further, the tangle of royal intermarriages in Europe, placed Henry III of Navarre next in line for the crown of France. So, in 1589, when France's Henry III was assassinated by a crazed Catholic monk, Navarre's Henry III became Henry IV of France.

The English rejoiced when the papist Henry III of France was murdered. Initially, Henry of Navarre was Protestant. Consequently, he was not embraced by the French Catholic majority. To appease them, Henry reluctantly converted to Catholicism. This did not sit well with England's thoroughly anti-Catholic Queen Elizabeth. Thus, we can understand that *Love's Labour's Lost* would have been of topical interest to Shakespeare's audiences, who would have interpreted the play as a thumb of the nose to France's King Henry IV. Shakespeare apparently decided against adding yet another King Henry in this play, reverting instead to Ferdinand, who *did* rule Navarre in earlier times. See: Asimov, *Asimov's Guide to Shakespeare*, 421–24.

2. See, for example, Bloom, *The Invention of the Human.*

3. We see what Asimov means by "Latinity" in the following exchange between the pageboy Mote and the witty bumpkin Costard, who have been listening to the pedant Holofernes trying to impress a friend with his scholarly acumen:

> MOTE: [aside to Costard] They have been at a great feast of languages and stolen the scraps.

> COSTARD: [aside to Mote] O, they have lived long on the almsbasket of words. I marvel thy master hath not eaten thee for a word; for thou art not so long by the head as *honorificabilitudinitatibus*. Thou art easier swallowed than a flapdragon.
>
> —*Love's Labour's Lost*, 5.1.38–44

Perhaps a bit of explanation of *those* words is in order, as well: A "flapdragon" was a raisin or plum in a dish of burning brandy—not an easy object to swallow. As for "honorificabilitudinitatibus," it is a Latin word meaning "honorableness," but the point is that it is a long word—the longest in Shakespeare's works, and he used it here merely for its length and satirical value. In this form (the ablative plural), this monstrosity of a word is reputed to be the longest word in the Latin language. Explaining the ablative case is, however, far beyond the scope of this book, and we think it wise that most languages have abandoned it.

4. When *Love's Labour's Lost* was written, the English had recently (in 1588) and miraculously (and with a lot of help from the weather) repelled the Spanish Armada—and, with it, a Catholic takeover.

5. Here's our list of seventeen plays that deal with romantic love: *All's Well That Ends Well*, *As You Like It*, *Cymbeline*, *Hamlet*, *Love's Labour's Lost*, *The Merchant of Venice*, *Measure for Measure*, *A Midsummer Night's Dream*, *Much Ado About Nothing*, *Othello*, *Pericles*, *Romeo and Juliet*, *The Taming of the Shrew*, *Troilus and Cressida*, *Two Gentlemen of Verona*, *Two Noble Kinsmen*, *The Winter's Tale*. There is room for quibbling on one or two others.

6. Note that the scholar Holofernes, the pedant who is all reason and no emotion, serves as a counterpoint to the lovesick men. Shakespeare seems to have inserted Holofernes here to emphasize the idea that extremes either of emotion or reason are both undesirable and even ludicrous.

7. Burton, *Anatomy of Melancholy*; Choi, "Melancholy Humor."

8. For more information on the relationship between love and madness, as understood by people in Shakespeare's time, see: Walthall, "The Lunatic, the Lover."

9. Lyon, "Wooing and Wedding."

10. These announcements were called "the reading of the banns." (The term comes from an old English word meaning "proclamation.") The purpose was to allow anyone to voice a legal objection, which might include a previous marriage that had not been annulled or a prohibited kin relationship, as between first cousins.

11. For more details on the issuance of marriage licenses in those days, see the following website: Family Search, "Marriage Allegations."

12. Here are our nominations for the strong, rebellious, and defiant women in Shakespeare's plays:

- **Imogen**—(*Cymbeline*): defies her father, the king, by refusing to marry Cloten and secretly marrying Posthumus instead.
- **Lavinia**—(*Titus Andronicus*): refuses the emperor's proposal of marriage and ends up with her tongue removed and her hands amputated.

- **Hermia**—(*A Midsummer Night's Dream*): defies her father's demands, under a threatened penalty of death, for an arranged marriage to Demetrius.
- **Portia**—(*The Merchant of Venice*): cleverly deals with the restrictive marriage mandates in her father's will.
- **Juliet**—(*Romeo and Juliet*): loves Romeo despite the family feud with his family, the Montagues.
- **Kate**—(*The Taming of the Shrew*): resists demands first from her father and then from her husband Petruchio to submit to their dominance.
- **Helena**—(*All's Well That Ends Well*): will not stand for Bertram backing out on the bargain to marry her.
- **Cressida**—(*Troilus and Cressida*): uncertain about Troilus's commitment and so switches her love and fidelity from the Trojan hero to the Greek warrior Diomedes.
- **The Princess of France and her retinue**—(*Love's Labour's Lost*): refuse to marry the men from Navarre until they prove, for a year and a day, that they can keep their vows.
- **Cordelia**—(*King Lear*): refuses to fawn over her father when he demands that she tell him of her love.
- **Isabella**—(*Measure for Measure*): refuses the sexual advances of Angelo.
- **Emilia**—(*Othello*): exposes the dastardly scheme of her husband, Iago, at the cost of her life.
- **Paulina**—(*The Winter's Tale*): stands up to the paranoid King Leontes in defense of his wife Hermione's virtue.
- **Cleopatra**—(*Antony and Cleopatra*): not just a strong woman but the head of the Egyptian state.
- **Lady Macbeth**—(*Macbeth*): although not a particularly admirable woman, displays her strength by initiating a lethal plot by which her husband seizes the crown.
- **Viola**—(*Twelfth Night*): disguises and comports herself as a man to take charge of her fate in a new land.
- **Beatrice**—(*Much Ado About Nothing*): more than holds her own while matching wits against Benedick.
- **Rosalind**—(*As You Like It*): when banished from the kingdom by her usurping uncle, manages to woo and marry the man she loves.
- **The Merry Wives**—(*The Merry Wives of Windsor*): amuse themselves by playing tricks on Falstaff, who thinks of himself as a dashing suitor.

13. See Asimov, "Romeo and Juliet," 476.
14. Rasmussen, "Marriage and Courtship."
15. Kreps, "Paradox of Women."
16. Inheritance in England, as well as many countries on the Continent, followed the tradition of primogeniture, by which land and titles were passed down to the oldest male child. Countries varied, however, and some allowed inheritance by females if no male offspring were available.

An old Frankish law, dating back to the sixth century, was even more restrictive. This "Salic Law" demanded that no male could wear the crown if his claim relied solely on descent through the female line. As we see in *Henry V*, this custom could cause chaos upon the death or removal of a king, especially when there are multiple claimants for the crown.

By the time of Queen Mary, some other countries had accepted female heads of state, which made Mary's claim (and, later, Elizabeth's) more credible. Moreover, both were powerful women in their own right, and so, able to crush any opposition to their becoming successive queens "regnant"—that is, "reigning" queens with the full titles and powers of a male ruler.

17. Shakespeare took his inspiration for *Romeo and Juliet* from an Italian story translated by Arthur Brooke and published in 1562 under the name *The Tragical History of Romeus and Juliet*. In that earlier version of the tale, both Romeus and Juliet are sixteen years old.

18. Best, "Age of Marriage"; Feminae, "Isabella of France."

19. The "bed trick" is a literary device involving the substitution of one sexual partner for another. In *All's Well*, Bertram is reluctantly betrothed to Helena but unknowingly consummates their marriage when Helena substitutes herself (in the dark) for Diana, with whom Bertram thinks he has arranged a tryst.

20. The word *whore* appears some fifty times in Shakespeare's works, and we can add to that total if we include puns on sound-alike words (in the "original pronunciation" of Shakespeare's time), as in this speech by Jaques (also mentioned in the introduction):

> 'Tis but an hour ago since it was nine,
> And after one hour more 'twill be eleven;
> And so, from hour to hour, we ripe and ripe,
> And then, from hour to hour, we rot and rot;
> And thereby hangs a tale.
>
> —*As You Like it*, 2.7.12

21. For more on how words take on offensive meanings as cultures change, see: McWhorter, "How the N-Word."

22. American Psychological Association, "Science Briefs: Evolutionary Theory"; Buss, "Evolutionary Psychology."

23. Many people hold the false belief that they can "multitask," meaning they can consciously attend to and respond to multiple sources of information at the same time—such as in studying and watching TV simultaneously. In Kahneman's terms, doing so would require the conscious System 2 to focus attention on multiple streams of stimulation. This, however, is an illusion. While it *is* possible for the preconscious System 1 and the conscious System 2 to deal with different things, it is impossible for the conscious System 2 to attend to two different things simultaneously. In reality, people who believe they are multitasking are switching their attention back and forth at a considerable cognitive cost. You will find more information on so-called multitasking at these links: Hamilton, "Multitasking"; American Psychological Association, "Multitasking: Switching Costs."

24. Kahneman, *Thinking Fast and Slow.* We will look at the preconscious System 1 and the conscious-thinking System 2 (the dual-process mind) more closely in chapter 9.

25. The Trojan war and Medea's infanticides are also typical of System 1's impulsive thinking and the reluctance (or laziness) of the brain's (conscious) System 2 to override it.

26. This, too, is typical of System 1 thinking.

27. Barash, *Homo Mysterious*; Barash and Barash, *Madame Bovary's Ovaries.*

28. According to the online *Oxford English Dictionary*, the word *emotion* did not exist until the mid-sixteenth century—when it referred to creating a public disturbance. The current meaning, involving feelings and temperament, emerged in the early 1800s.

29. See: Warren, "Biology in Shakespeare."

30. Black bile was the only one of the four humors that does not exist (although there is a substance called *melanin*). As for yellow bile, we now simply call it "bile," and it does indeed come from the gall bladder.

31. There never was full consensus on the specifics of the humoral theory. Notions about the humors, therefore, existed in several incarnations, with most disagreement centering on the role of the internal organs and their role in producing the humors. Notably, Galen and Aristotle had expressed somewhat different ideas about the functions of the liver, the heart, and the spleen; partisans of each debated their views well into the 1800s. The first real blow to humorism occurred not long after Shakespeare's death, when Harvey published his discovery of blood circulation, and medicine recognized the heart's function as a pump for the blood. Nevertheless, the humor theory lingered as an accepted explanation for disease until it was finally undone by the germ theory. See the following sources: Friedland, "Discovery"; Findlen, "Emergence of Medicine"; Cruttwell, "Physiology and Psychology"; National Library of Medicine, "There's the Humor."

32. Warren, "The Biology in Shakespeare."

33. Should you need them, perhaps for lyrics in a country-western song, here are a few more examples: When anxious or afraid one might sense that the heart is "racing" or "in one's mouth." A horror movie might make your "blood run cold," although if you are angry you might feel "hot under the collar" or your "blood boiling." A jilted lover may have a "broken heart." Your stomach may sometimes feel as if "in a knot" or full of "butterflies." A more severe case of anxiety can give you "the jitters." Worry or anticipation can feel like being "on pins and needles." A narrow escape may leave you "shaken up." Love can make someone "fall head over heels." Likewise, we say that a death can cause pain or feelings of emptiness. Disgust sometimes feels like nausea. Distraction makes us "spaced out." Exhaustion makes us feel "beat." We call exasperation "fed up." And annoyances get us "bent out of shape."

34. There is no physical response that indicates a lie with any certainty. Polygraph operators, however, rely on people believing that they can detect the physical signature of a fib—but they cannot.

35. Harrison and Loui, "Thrills, Chills."

36. Damasio, *Descartes' Error*; Damasio, *Looking for Spinoza.*

37. There were other lists of deadly sins, but by virtue of his popeship, Gregory had the clout to push his list to the fore.

38. Although Shakespeare never used the term *seven deadly sins*, he certainly would have known about them from the old medieval morality plays, still seen occasionally in Elizabethan and Tudor times.

39. There are exceptions to this admittedly simplified definition of emotions. For example, one can experience apprehension and fear about an internal pain of unknown origin. The main point is that emotions are feelings that (a) are generally directed at external objects and events, and (b) exclude the usual physical needs, such as hunger and thirst.

40. You will note that the two terms, *motive* and *emotion*, share the same root *moti-*.

41. See: Damasio, *Descartes' Error*. A summary of this case can be found in our textbook: Zimbardo, Johnson, and McCann, *Psychology: Core Concepts.*

42. Hamilton, "Why Brain Scientists."

43. Damasio, *Descartes' Error.*

CHAPTER NINE

1. We date the play in 1572 because of the mention in act 1, scene 1 of a "star that's westward from the pole." Recent research links that "star" to a *supernova*, an exploding star that was visible in England and Denmark in that year, when Shakespeare was a young boy. More about that "star" later in this chapter.

2. The journey was more than five hundred miles, requiring travel both overland and by boat.

3. Shakespeare is taking some poetic license here to make customs in Hamlet's Denmark sound familiar to Elizabethan audiences. Play-goers knew that *Hamlet* was supposed to be set in medieval Denmark, yet Shakespeare was describing the customs and laws as though they were English.

4. According to Asimov, the model for Hamlet's pique may have been King Edward the Confessor. When the Danish king Canute married Emma, the widow of the previous king (Æthelred the Unready), his wife's children—Edward among them—were excluded from succession. However, Edward managed to regain the throne and send Emma to a nunnery, where she remained for life. By the way, Wikipedia says that "unræd" meant "poorly advised" in Old English, and it is a pun on his name, Æthelred, which means "well advised." If Shakespeare, the punster, knew about that, he would have certainly been amused. See: Asimov, *Asimov's Guide to Shakespeare*, 2:95–96.

5. *Hamlet* is, perhaps, the most Christian play in the Shakespearean canon, not because it dwells on theology but because it turns on the assumption that Hamlet could be damned if he mistakenly kills an innocent Uncle Claudius, putting the prince in league with the Devil. This is especially interesting because Shakespeare usually

avoided theological matters. He gave clergymen prominent roles mainly when they were involved in politics, as in *Henry V* and *Henry VIII*.

6. Tom Stoppard's amusing, absurdist play called *Rosencrantz and Guildenstern Are Dead* looks at the story of *Hamlet* through the eyes of Hamlet's (supposed) friends, Rosencrantz and Guildenstern. It was also made into a delightful movie of the same title, starring Richard Dreyfuss.

7. The term *conscience* (used here in act 2, scene 2) had a double meaning in Shakespeare's time. One was the same as ours, referring to a sense of right and wrong. The other meant "awareness." It seems likely that Shakespeare was punning on both meanings.

8. Today's psychologists would say that this is an example of *confirmation bias*, in which Hamlet jumps to the conclusion he was expecting and presumably wanted to be true.

9. The High Middle Ages is a period that might be called the middle-Middle Ages or the mid-medieval era. The dates are not precise but generally range from about 1000 CE (overlapping the Viking age) and 1347 (when the plague known as the Black Death became widespread in Europe).

10. The book, *Gesta Danorum*, or *The History of the Danes* (available online from Project Gutenberg), was written about 1200 CE by a fellow with the Latin pen name Saxo Grammaticus. See also: British Library, "Saxo's Legend," and Neill, "Modern Perspective."

11. In the original source, *Gesta Danorum*, "Amleth" was Latinized as "Amlethus."

12. Asimov, *Guide to Shakespeare*, 2:79–80.

13. Falk, *Science of Shakespeare*.

14. Cassiopeia is the W-shaped constellation in the northern sky—the reclining Queen, who revolves around the pole star with her king Cepheus, a boxy, house-shaped constellation in the same path around Polaris.

15. Texas State University, "Star in Hamlet."

16. Falk, *Science of Shakespeare*; Falk, "What Shakespeare Knew"; Folger Shakespeare Library, "Shakespeare, Science, and Art"; Gambino, "Was Shakespeare Aware"; Harkup, "Remarkable Scientific Accuracy"; Stofan, "Shakespeare Meets Science."

17. English audiences in Shakespeare's day might have thought it unusual that a member of the Danish royal family had been sent off to a university. Traditionally, royals were educated at home by private tutors. Incidentally, the first prince in Britain with a university degree is our modern Prince *cum* King Charles, who attended Trinity College in Cambridge, where he studied history, archaeology, and anthropology and graduated with honors.

18. The real name of the castle described in the play is Kronborg. It has existed since the 1420s on the Danish island of Zealand, where it commands a strategic strait between Denmark and present-day Sweden. The name Elsinore, which Shakespeare used as the castle's name, comes from the adjacent town, Helsingør.

Originally Kronborg was a fortress guarding the entrance to the Baltic Sea, but some 150 years later King Fredrick II gave it a major makeover that changed it into a magnificent Renaissance castle. That project was completed when Shakespeare was about twenty-one years old and, tantalizingly, at the beginning of the second of his "lost years" periods. Kronborg castle is now a World Heritage site.

19. Wittenberg was also a hotbed of Copernican sympathizers who sought to overthrow the belief that the Earth was the center of the universe. This Copernican revolution added greatly to the religious upheaval of the Reformation.

20. Not long after Martin Luther nailed his grievances on the door of the Wittenberg church in 1577, England's King Henry VIII repudiated Luther's arguments in writing. This led to Henry being awarded the title of Defender of the Faith by Pope Clement II. A decade later and in need of an heir and a divorce (Henry blamed his wife for the inability to conceive a boy), he switched allegiances by defying the pope, outlawed Catholicism in England, and married Anne Boleyn. See: Rex, "English Campaign against Luther."

21. Audiences would have known, however, that England was once ruled by the Danes. King Canute of Denmark conquered England in 1016, toward the end of the Viking Age. Thus, he is listed as one of the early English monarchs.

22. The term (sometime spelled as "ting") has origins in Old Norse and Old English. See: Wikipedia, "Thing"; Ahlness, "Legacy of the Ting"; Turner, "Aspects of the Danelaw."

23. Henry VIII's many marriages stemmed, in part, from his desperate quest for a male heir, but his biological legacy included England's first two female heads of state. Initially the Catholic "Bloody" Mary ascended the throne, and after her death, the crown passed to her younger sister, the Protestant Elizabeth.

An early Germanic code, known as the Salic Law, prevented a man from inheriting the crown through a female line. It was this law that Henry V invoked to claim sovereignty over France. By Shakespeare's time, however, there had been several female monarchs in Europe, including Catherine of Aragon, the first wife of Henry VIII.

24. Levy, "Problematic Relation."

25. The play gives us many examples suggesting that Hamlet valued reason over emotion. One involves Hamlet's bitter censure of his mother, Gertrude, for surrendering her judgment to "compulsive ardor" (3.4.63). Another occurs in the graveyard scene, where Hamlet berates himself for succumbing to his emotions, saying, "But sure the bravery of his grief did put me / Into a tow'ring passion" (5.2.83). See: Levy, "Problematic Relation."

26. Haidt, *Righteous Mind*; Kahneman, *Thinking Fast and Slow.*

27. Parvini, "How Noble in Reason."

28. Parvini, "How Noble in Reason."

29. Parvini, *Shakespeare and Cognition.*

30. The notion that rationality should hold emotion in check harks back to a medieval school of thought known as Scholasticism—and most notably from Thomas Aquinas, who wrote, "All the passions of the soul should be regulated according to the rule of reason." See Aquinas, *Summa Theologica, Pt. 2*, Q 39.

Shakespeare's dithering, rational Hamlet was influenced by a movement that had been gaining momentum ever since the time of Dante, Ovid, Petrarch, Boccaccio, and Chaucer. Scholars now call it *Renaissance humanism.* At its core, it was a revival movement, seeking to rediscover and learn from the scholars, writers, and philosophers of antiquity—people like Aristotle, Plato, Herodotus, and Homer—who valued reason over impulse. Another goal of the Renaissance humanists was sharing this ancient knowledge with an increasingly educated citizenry. Underlying these objectives was a new respect for ordinary people's abilities to think and reason for themselves. Hamlet is the quintessential rational Renaissance humanist whose ideal world is one in which reason reigns. See: Levy, "Problematic Relation."

31. We confess to adding italics for emphasis.

32. Orbell, "Hamlet and Psychology."

33. The two better-known choices—to be or not to be—come later, as Hamlet tries to find a way out of his dilemma.

34. Table adapted from Orbell, "Hamlet and Psychology."

35. One effective tool in the treatment of depression by cognitive behavioral therapists is to teach patients how to *act* as though they are *not* depressed.

36. We introduced Jonathan Haidt's metaphor of the elephant and the rider in chapter 5. See also his book *The Righteous Mind.*

37. Psychologists would call this a "2 × 2 factorial design." In social psychology, such an experiment might involve comparing, say, the scores of liberals vs. conservatives (factor 1) on Jonathan Haidt's measures of Authority vs. Fairness (factor 2).

38. To be wonkish, the *experimental hypothesis* is that the Ghost is telling the truth. The prince's experiment, therefore, involves testing the experimental hypothesis against the *alternative hypothesis* that the Ghost is lying (*not* telling the truth).

39. Technically, it was not a well-controlled experiment because (among other issues) it relies simply on Hamlet's and Horatio's judgments of what they saw, without any explicit rubric for making their judgments. But remember, this was a prescientific age when Francis Bacon—often called the "father" of the *scientific method*—was just starting to publish his ideas.

40. A modern experimental psychologist would have one additional reason for constructing such a table: to avoid two types of erroneous conclusions from the experiment. One error, commonly known as a *false positive*, would occur in Hamlet's "experiment" if the Ghost were lying but he concluded that the Ghost was truthful. The other error is a *false negative*, which would occur if Hamlet were to believe the Ghost was *not* truthful when the Ghost was actually telling the truth. (Scientists also call the false positive a "Type 1 error" and a false negative a "Type 2 error.")

41. Folger Shakespeare Library, "Shakespeare, Science, and Art"; Usher, "Hamlet's Love Letter"; Usher, "Hamlet and Infinite Universe."

42. Although it is a pall that hangs over the whole play, the word *melancholy* is said only twice, first by Hamlet:

> The spirit that I have seen
> May be a devil; and the devil hath power
> T' assume a pleasing shape; yea, and perhaps
> Out of my weakness and my melancholy,
> As he is very potent with such spirits,
> Abuses me to damn me.
>
> —*Hamlet*, 2.2.575

The other time we hear "melancholy" is from Claudius, who says:

> There's something in his soul
> O'er which his melancholy sits on brood
>
> —*Hamlet*, 3.1.176

43. Ekström, "Hamlet, the Melancholic."

44. US National Library of Medicine, "History of Medicine."

45. Jorgensen, "Hamlet's Therapy."

46. Here we borrow a term from *Macbeth*, where the Thane of Cawdor pleads with the doctor to "pluck from the memory [of Lady Macbeth] a rooted sorrow" (5.3.49).

47. At first, neither Hamlet nor the audience can be sure that Claudius is guilty of murdering Hamlet's father, and Shakespeare does not spoil that tension until we hear Claudius in the chapel praying for forgiveness. Even then, Hamlet cannot be sure because he merely *sees* Claudius praying, but he doesn't hear his words. This lack of definitive evidence, then, leads Hamlet to have the players stage *The Murder of Gonzago.* On the other hand, the king apparently does not know that Hamlet has been listening to the Ghost's accusations. So, from the king's perspective, Hamlet could simply be mourning his father's death.

48. You will recall (from chapter 2) that Jaques's Seven Ages of Man were the infant, the schoolboy, the lover, the soldier, the justice, the lean and slippered pantaloon, and second childishness (*As You Like It*, 2.7.146).

49. See: National Library of Medicine, "History of Medicine."

50. It becomes a "tragic flaw" if we see the ending of the play as Hamlet's fault—a *personal* character flaw—rather than an inevitable consequence of an overpowering *situation*.

51. In the Germanic languages, "thing" came from a root word meaning "place." See: Wikipedia, "Thing (assembly)"; Biørnstad, "'Twas Dangerous to Insult."

52. Rowan-Robinson, "Shakespeare's Astronomy."

53. There were a few women who were scientists, including Danish astronomer Tycho Brahe's sister, Sophia Brahe, who was a horticulturist and a chemist, as well as an astronomer. Wikipedia also lists several other women of science from the sixteenth and seventeenth centuries. For more on Sophia Brahe, see: Wikipedia, "Sophia Brahe."

54. Franz and Paul, "Why the Moons of Uranus."

55. The Shakespearean moons of Uranus and the plays in which their eponymous characters appear:

- Titania, Oberon, Puck: *A Midsummer Night's Dream*
- Ariel, Miranda, Caliban, Sycorax, Prospero, Setebos, Stephano, Trinculo, Francisco, Ferdinand: *The Tempest*
- Cordelia: *King Lear*
- Ophelia: *Hamlet*
- Bianca: *The Taming of the Shrew*
- Cressida: *Troilus and Cressida*
- Desdemona: *Othello*
- Juliet, Mab: *Romeo and Juliet*
- Portia: *The Merchant of Venice*
- Rosalind: *As You Like It*
- Margaret: *Much Ado About Nothing*
- Perdita: *The Winter's Tale*
- Cupid: *Timon of Athens*

Nor is that all. In the asteroid belt, there is an object named 2895 Shakespeare, along with others known as 171 Ophelia (from *Hamlet*), 218 Bianca (*Taming of the Shrew*), 593 Titania (*A Midsummer Night's Dream*), 666 Desdemona (*Othello*), 763 Cupido (*Timon of Athens*), and 2758 Cordelia (*King Lear*). See: Wikipedia, "Moons of Uranus," and International Astronomical Union, "Skies Are Painted."

56. The first documented production of *Cymbeline* occurred in the spring of 1611, nearly a year after Galileo published *Siderius Nuncius*, in which he reported observing the four moons circling Jupiter with his telescope. See the following sources: Usher, "Jupiter and *Cymbeline*," and Falk, "Shakespeare, Galileo."

57. Brahe was one of the first to notice and report the 1572 supernova. Because of his early report—and because he was already famous—it is often referred to as "Tycho's star." Astronomers now regard it as a Type I supernova; its remnant is still visible in large telescopes.

Brahe was also one of the last astronomers who did not have the benefit of a telescope, an instrument first used by Galileo for astronomical observations in 1609, approximately ten years after Shakespeare wrote *Hamlet*.

58. An early meaning of "prodigy" is "monster."

59. Pumfrey, "Your Astronomers."

60. Coincidentally, a few years after Digges died, his wife Anne remarried a lawyer from Warwickshire, the county in which Stratford-upon-Avon is located. Her husband later became the executor of Shakespeare's will.

61. Falk, *Science of Shakespeare*.

62. For example, *hubris*, or pride, is an emotional theme running through much of classical literature, especially in Homer's *Iliad* and *Odyssey*.

63. Kahneman, *Thinking Fast and Slow*.

64. He does so by having his wife, Emilia, obtain it, which she does in her capacity as Desdemona's maid.

65. Shakespeare was quite clever in crafting *Hamlet*, for the prince and the audience are not always privy to the same evidence—yet both succumb to confirmation bias. Up to the chapel scene, where we hear Claudius asking forgiveness for the murder, the evidence is circumstantial, although it seems to point to Claudius as the murderer. Although Hamlet sees Claudius in the chapel, he does not hear the confession; only the audience does. Indeed, Hamlet *never* acquires conclusive evidence against his uncle.

Some critics have pointed out that Hamlet doesn't even know (nor do we!) how long Gertrude has been having sex with Claudius. It could have been an affair that started long before they were married. Did the prince imagine that Claudius could have been his real father? Did Shakespeare deliberately build that suspicion into the plot? Consider that, in the play that Hamlet called *The Mousetrap*, the murderer was not the uncle but the nephew—which was what Hamlet himself would have been had Claudius been his father. And then consider who kills Claudius in act 5. We owe this line of puzzlement to the following source: Flesch, "Would DNA Evidence."

66. Haidt, *Righteous Mind*; Auerbach, "One Line from Hamlet."

67. Cleaveland, "Emotions as Possession."

68. This is a form of *self-serving bias*, as is the *rationalist delusion*. See: *APA Dictionary of Psychology*, "Bias Blind Spot."

EPILOGUE

1. Asian astronomers recorded a supernova that they called a "guest star" in 1181 CE, and it was nearly four hundred years before humans saw another one—which was Tycho's star of 1572. Another naked-eye supernova, known as Kepler's Supernova, appeared just thirty-two years later, in 1604. Then there was another four-century hiatus, when Supernova 1987A was visible briefly.

2. See Stanley Milgram's famous study suggesting that a maximum of six links connect any one of us to anyone else in the United States. This has given rise to the popular notion of "six degrees of Kevin Bacon." See: *Psychology Today*, "Six Degrees of Separation: Two New Studies Test 'Six Degrees of Separation' Hypothesis."

3. Some connections between Shakespeare and sundry disciplines:

- **Botany:** Shakespeare's plays and poems mention more than two hundred different plants, especially flowers, and he often added details on where they grow. Suggested read: Gerit Quealy, *Botanical Shakespeare: An Illustrated Compendium of All the Flowers, Fruits, Herbs, Trees, Seeds, and Grasses Cited by the World's Greatest Playwright*.
- **Evolutionary biology:** See *Madam Bovary's Ovaries: A Darwinian Look at Literature*, written by evolutionary psychologist David Barash and his daughter, Nanelle Barash, a student of literature and veterinary medicine.

- **History:** For a deeper understanding of Shakespeare as a man of the Renaissance, have a look at James Shapiro's book, *The Year of Lear, 1599.* See also: Stephen Greenblatt's book *Will in the World: How Shakespeare Became Shakespeare*, as well as his "Shakespeare's Leap" on the *New York Times* website.
- **Medicine:** Paul Matthews and Jeffrey McQuain's *The Bard on the Brain* comments on the plague, the humors, and sources of the emotions, as seen through the framework of Elizabethan medicine.
- **Music:** The following links will take you to websites showing a rap version of *Macbeth*, how iambic pentameter is well suited for rap and hip-hop, and commentary on songs and instrumental music in Shakespeare's plays:
  https://hechingerreport.org/opinion-in-hip-hop-educators-find-a-bridge-to-shakespeare/.
  https://www.youtube.com/user/FolgerLibrary.
  https://www.kqed.org/mindshift/46215/how-hip-hop-can-bring-shakespeare-to-life
  https://dramatics.org/original-swagger/.
- **Business:** Peter Shadbolt, "The Bard's Business: Shakespeare's Economic Legacy," https://www.bbc.com/news/business-36084631.
- **Astronomy and other sciences:** Dan Falk's book *The Science of Shakespeare* argues that Shakespeare was attuned to developments in the emerging science of his time.
- **Art:** Constance McPhee's "Shakespeare and Art, 1709–1922" is a good place to begin learning about the Bard's influence on drawing and painting: https://www.metmuseum.org/toah/hd/shaa/hd_shaa.htm.
- **Linguistics:** This informative link tells about Early Modern English and the use of original pronunciation (OP) in Shakespeare's plays to understand many of his puns that don't make sense to modern ears: http://originalpronunciation.com/.
- **Shakespeare's philosophy and religion**: Contentious religious issues were always part of life's subtext in early modern England, as described in this conversation with David Scott Kastan, author of *Will to Believe: Shakespeare and Religion.* Listen online to "Shakespeare and Religion" from the Folger Shakespeare Library at https://www.folger.edu/shakespeare-unlimited/religion.
- **Sociology:** See History on the Net, "The Tudors—Society" by Salem Media at https://www.historyonthenet.com/the-tudors-society, describing the rigid social structure of England in Tudor times.
- **Law and Criminal Justice:** A good place to start is "Crime and Punishment" at Shakespeare Mag.com: http://www.shakespearemag.com/fall98/punished.asp. See also Shakespeare's own words when Portia acts as a lawyer in *The Merchant of Venice* and Angelo acts as judge and jury in *Measure for Measure*.
- **Humor:** Why does Shakespeare insert humor in his tragedies? You'll find an answer by searching online for Arthur Nason, "Comedy in Tragedy."
- **Political Science:** Have a look online at "The Progressive Case for Teaching Shakespeare," by Elizabeth Bruenig.

- **Military:** Tactics and leadership (*Henry V*), failed leadership (*Henry VI*), and the sometimes-tragic consequences of military success: *Macbeth*, *Coriolanus*, *Trolius and Cressida*, *Antony and Cleopatra*.
- **Anthropology:** Students will enjoy "Shakespeare in the Bush," a classic piece by a novice archaeologist who tried to tell the tale of *Hamlet* to a group of male elders in a remote African tribe.

4. Merriam-Webster, "What Does 'Quark' Have to Do with Finnegans Wake?," https://www.merriam-webster.com/words-at-play/quark.
5. Isaacson, *Leonardo da Vinci*.

BIBLIOGRAPHY

Ahlness, Ellen A. "The Legacy of the Ting: Viking Justice, Egalitarianism, and Modern Scandinavian Regional Governance." *World History Connected* 17, no. 1 (2020). https://worldhistoryconnected.press.uillinois.edu/17.1/forum_ahlness.html.

Ali, S., S. Jabeen, R. J. Pate, M. Shahid, S. Chinala, M. Nathani, and R. Shah. "Conversion Disorder—Mind versus Body: A Review." *Innovations in Clinical Neuroscience* 12, nos. 5–6 (May–June 2015): 27–33. https://www.ncbi.nlm.nih.gov/pmc/articles/PMC4479361/.

Alou, Yacobou. "The Relationship between Kyd's *The Spanish Tragedy* and Shakespeare's *Hamlet* and *Titus Andronicus*." *International Journal of Humanities and Social Science Invention* 6, no. 5 (May 2017): 52–57. https://www.ijhssi.org/papers/v6(5)/version-I/H0605015257.pdf"https://www.ijhssi.org/papers/v6(5)/version-I/H0605015257.pdf.

al'Thor, Rand. "Does Portia Subconsciously Influence Bassanio's Choice of Casket?" Literature Stack Exchange, April 11, 2017. https://literature.stackexchange.com/questions/2319/does-portia-subconsciously-influence-bassanios-choice-of-casket.

American Psychiatric Association. *Diagnostic and Statistical Manual of Mental Disorders, 5th ed.* Washington, DC: American Psychiatric Association, 2013.

American Psychological Association. "Multitasking: Switching Costs: Subtle 'Switching' Costs Cut Efficiency, Raise Risk." March 20, 2006. https://www.apa.org/research/action/multitask.

———. "Bias Blind Spot." *APA Dictionary of Psychology*, 2020. https://dictionary.apa.org/bias-blind-spot.

———. "Science Briefs: Evolutionary Theory and Psychology." May 2009. https://www.apa.org/science/about/psa/2009/05/sci-brief.

Andreasen, Nancy J. (1976, April 26). "The Artist as Scientist: Psychiatric Diagnosis in Shakespeare's Tragedies." *JAMA* 235, no. 17, 1868–72. doi: 10.1001. https://jamanetwork.com/journals/jama/article-abstract/345200.

Appelbaum, Robert. "War and Peace in 'The Lepanto' of James VI and I." *Modern Philology* 97, no. 3 (2000): 33–63. https://www.jstor.org/stable/439268?seq=1#metadata_info_tab_contents.

Aquinas, Thomas. Question 39, Article 2, Reply to Objection 1. In *Summa Theologica, Part 2*. Trans. Fathers of the English Dominican Province. Circa 1272; New York: Benziger Brothers, 1952. https://en.wikisource.org/wiki/Treatise_on_Human_Acts_(part_2)#QUESTION_39:_OF_THE_GOODNESS_AND_MALICE_OF_SORROW_OR_PAIN.

Arain, M., M. Haque, L. Johal, P. Mathur, W. Nel, A. Rais, R. Sandhu, and S. Sharma. "Maturation of the Adolescent Brain." *Neuropsychiatric Disease and Treatment* 9 (2013): 449–61. https://doi.org/10.2147/NDT.S39776.

Aserinsky, Eugene, and Nathaniel Kleitman. "Regularly Occurring Periods of Eye Motility, and Concomitant Phenomena, during Sleep." *Science* 118 (1953): 273–74. https://www.science.org/doi/10.1126/science.118.3062.273.

Asimov, Isaac. *Asimov's Guide to Shakespeare*. New York: Avenel Books (Random House), 1970.

Auerbach, David. "One Line from Hamlet." *Waggish* (blog), December 26, 2010. https://www.waggish.org/2010/one-line-from-hamlet/.

Barash, David P. *Homo Mysterious: Evolutionary Puzzles of Human Nature*. New York: Oxford University Press, 2012.

Barash, D. P., and N. Barash. "Biology as a Lens: Evolution and Literary Criticism." *Chronicle of Higher Education*, October 18, 2002, B7–B9.

———. *Madame Bovary's Ovaries: A Darwinian Look at Literature*. New York: Delacorte Press, 2005.

Barkan, Lou-Ellen. "What Alzheimer's Caregivers Can Learn from King Lear." *HuffPost*, September 9, 2014. http://www.huffingtonpost.com/louellen-barkan-/what-alzheimers-caregiver_b_5791078.html.

Barton, John. "Poem by Henry VI." *The Hollow Crown*. London: Samuel French, Ltd., 1962.

Bate, Jonathan. *Soul of the Age: A Biography of the Mind of William Shakespeare*. New York: Random House, 2010.

Berg, J. M. "Shakespeare as a Geneticist." *Clinical Genetics* 59, no. 3 (March 2001): 165–70. https://doi.org/10.1034/j.1399-0004.2001.590304.x.

Best, Michael. "The Age of Marriage." In *Life and Times*. Internet Shakespeare Editions, University of Victoria. https://internetshakespeare.uvic.ca/Library/SLT/society/family/marriage.html.

———. "Amleth/Hamlet." Shakespeare's Life and Times. Internet Shakespeare Editions. Posted February 29, 2020. https://internetshakespeare.uvic.ca/Library/SLT/history/prehistory/amleth.html.

Betts, Jennifer. "40 Common Words and Phrases Shakespeare Invented." Your Dictionary. http://grammar.yourdictionary.com/word-lists/list-of-words-and-phrases-shakespeare-invented.html.

Biello, David. "Does Rice Farming Lead to Collectivist Thinking?" *Scientific American*, May 12, 2014. http://www.scientificamerican.com/article/does-rice-farming-lead-to-collectivist-thinking/.

Biørnstad, Lasse. "'Twas Dangerous to Insult a Viking." *Science Norway*, August 19, 2016. https://sciencenorway.no/forskningno-norway-society--culture/twas-dangerous-to-insult-a-viking/.

Blake, Ann. "Children and Suffering in Shakespeare's Plays." *The Yearbook of English Studies* 23 (1993): 293–304. https://doi.org/10.2307/3507985.

Bloom, Harold. "The Knight in the Mirror." *Guardian*, US ed., December 13, 2003. https://www.theguardian.com/books/2003/dec/13/classics.miguelcervantes.

———. *Shakespeare: The Invention of the Human*. New York: Riverhead Books, 2003.

Bogousslavsky, J., and S. Dieguez. "Sigmund Freud and Hysteria: The Etiology of Psychoanalysis?" *Frontiers of Neurology and Neuroscience* 35 (2014): 109–225. https://doi.org/10.1159/000360244.

Bohannan, Laura. "Shakespeare in the Bush." *Natural History*, August–September 1966. http://www.naturalhistorymag.com/picks-from-the-past/12476/shakespeare-in-the-bush.

Bowling, Lawrence E. "The Theme of Natural Order in *The Tempest*." *College English* 12, no. 4 (1951): 203–9. http://links.jstor.org/sici?sici=00100994%28195101%2912%3A4%3C203%3ATTONOI%3E2.0.CO%3B2-I.

Boyce, Charles. *Shakespeare A to Z*. New York: Dell, 1991.

British Library. "Broadside Ballad on Cuckolds." https://www.bl.uk/collection-items/broadside-ballad-on-cuckolds.

———. *The Discovery of Witchcraft by Reginald Scot, 1584*. https://www.bl.uk/collection-items/the-discovery-of-witchcraft-by-reginald-scot-1584.

———. "*Gl'Ingannati*, an Italian Play about Twins and Mistaken Identity." https://www.bl.uk/collection-items/gl-ingannati-an-italian-play-about-twins-and-mistaken-identity.

———. "Saxo's Legend of Amleth in the *Gesta Danorum*." https://www.bl.uk/collection-items/saxos-legend-of-amleth-in-the-gesta-danorum.

Brown, Chip. "The Stubborn Scientist Who Unraveled a Mystery of the Night." *Smithsonian Magazine*, October 2003. https://www.smithsonianmag.com/science-nature/the-stubborn-scientist-who-unraveled-a-mystery-of-the-night-91514538/.

Bruenig, Elizabeth. "The Progressive Case for Teaching Shakespeare." *New Republic*, June 16, 2015. https://newrepublic.com/article/122040/progressive-case-teaching-shakespeare.

Bryson, Bill. *Shakespeare: The World as Stage*. New York: Harper Perennial, 2007.

Burton, Robert. *The Anatomy of Melancholy*. Originally published in 1621. Early English Books Online, STC-4159-951_05, Project Gutenberg. https://www.gutenberg.org/files/10800/10800-h/10800-h.htm.

Buss, D. "Evolutionary Psychology: A New Paradigm for Psychological Science." *Psychological Inquiry* 6, no. 1 (1995): 1–30. http://www.jstor.org/stable/1449568.

Camden, Carroll Jr. "Shakespeare on Sleep and Dreams." *Rice Institute Pamphlet* 23, no. 2 (1936): 106–33. https://scholarship.rice.edu/bitstream/handle/1911/9109/article_RI232106.pdf;sequence=5.

Carroll, Joseph. "An Evolutionary Approach to Shakespeare's King Lear." In *Critical Insights: Family*, edited by John Knapp, 83–103. Ipswich, MA: EBSCO, 2012. https://www.academia.edu/2702641/An_Evolutionary_Approach_to_Shakespeares_King_Lear?email_work_card=view-paper.

Cartwright, Mark. "Food and Drink in the Elizabethan Era." *World History Encyclopedia*. July 8, 2020. https://www.worldhistory.org/article/1578/food--drink-in-the-elizabethan-era/.

———. "Sports, Games and Entertainment in the Elizabethan Era." *World History Encyclopedia*. https://www.worldhistory.org/article/1579/sports-games--entertainment-in-the-elizabethan-era/.

Carver, C. S., and M. F. Scheier. *Perspectives on Personality*. Fourth edition. Boston: Allyn and Bacon, 2000.

Chambers, E. K. *William Shakespeare: A Study of Facts and Problems*. Oxford: Clarendon Press, 1930.

Chandler, Simon B. "Shakespeare and Sleep." *Bulletin of the History of Medicine* 29, no. 3 (1955): 255–60. http://www.jstor.org/stable/44443931.

Choi, Young Ju. "Shakespeare's Use of the Melancholy Humor." Unpublished thesis, North Texas State University, Denton, TX, 1968. https://digital.library.unt.edu/ark:/67531/metadc130981/m1/2/.

Cleaveland, Mark. "Emotions as Possession." *Mark Cleaveland.org* (blog), August 2, 2020. https://markcleaveland.org/tag/hamlet/.

Cleland, Katharine. "'Are You Fast Married?' Elopement and Turning Turk in Shakespeare's Othello." In *Irregular Unions*, 110–32. Ithaca, NY: Cornell University Press, 2021. https://doi.org/10.1515/9781501753497-007.

Cline, John. "*A Midsummer Night's Dream*: Shakespeare's Play Reminds Us of the Sweet Dreams of Midsummer." *Psychology Today*, June 28, 2019. https://www.psychologytoday.com/us/blog/sleepless-in-america/201906/midsummer-night-s-dream.

Crane, Mary Thomas. *Shakespeare's Brain: Reading with Cognitive Theory*. Princeton, NJ: Princeton University Press, 2001.

Cruttwell, Patrick. "Physiology and Psychology in Shakespeare's Age." *Journal of the History of Ideas* 12, no. 1 (1951): 75–89. doi:10.2307/2707538. https://www.jstor.org/stable/2707538.

Damasio, A. R. *Descartes' Error: Emotion, Reason, and the Human Brain*. New York: Avon Books, 1994.

———. *Looking for Spinoza: Joy, Sorrow, and the Feeling Brain*. Orlando, FL: Harcourt, 2003.

Darley, John M., and Bibb Latané. "Bystander Intervention in Emergencies: Diffusion of Responsibility." *Journal of Personality and Social Psychology* 8 (1968): 377–83. http://www.communicationcache.com/uploads/1/0/8/8/10887248/bystander_intervention_in_emergencies_diffusion_of_responsibility.pdf.

Delahoyde, Michael. "*The Tempest*." Washington State University, 2018. https://public.wsu.edu/~delahoyd/shakespeare/tempest1.html.

Delbanco, Andrew. *The Death of Satan: How Americans Have Lost the Sense of Evil*. New York: Farrar, Straus and Giroux, 1995.

Dement, William C. *Some Must Watch While Some Must Sleep*. San Francisco: San Francisco Book Company, 1980.

DeWall, C. N., J. K. Maner, T. Deckman, and D. A. Rouby. "Forbidden Fruit: Inattention to Attractive Alternatives Provokes Implicit Relationship Reactance." *Journal of Personality and Social Psychology* 100, no. 4 (2011): 621–29. https://static1.squarespace.com/static/56cf3dd4b6aa60904403973f/t/5717deb5356fb08915fab422/1461182133440/forbidden-fruit.pdf.

Dickson, Andrew. "Shakespeare's Life." British Library, March 15, 2016. https://www.bl.uk/shakespeare/articles/shakespeares-life.

Dimitrijevic, Aleksandar. "Being Mad in Early Modern England." *Frontiers in Psychology* 6 (November 19, 2015): 1740. https://www.ncbi.nlm.nih.gov/pmc/articles/PMC4652010/.

Dobzhansky, Theodosius, Arthur Robinson, and Anthony J. F. Griffiths. "Heredity." In *Encyclopedia Britannica*. https://www.britannica.com/science/heredity-genetics.

Dolan, Frances E. "Shakespeare and Marriage: An Open Question." *Literature Compass* 8–9 (2011): 620–34. https://english.ucdavis.edu/sites/english.ucdavis.edu/files/users/fdolan/Dolan,%20Shakespeare&Marriage.pdf.

Dollimore, Liz. "Shakespeare's Sources—*Othello*." *Blogging Shakespeare* (blog). https://www.bloggingshakespeare.com/shakespeares-sources-othello.

Draper, John W. *The Humors and Shakespeare's Characters*. New York: AMS Press, 1970.

Dungey, Nicholas. "Shakespeare and Hobbes: Macbeth and the Fragility of Political Order." *SAGE Open* (April–June 2012): 1–18. http://journals.sagepub.com/doi/full/10.1177/2158244012439557.

Eckert, Danny J., and Atul Malhotra. "Pathophysiology of Adult Obstructive Sleep Apnea." *Proceedings of the American Thoracic Society* 5, no. 2 (2008): 144–53. https://www.ncbi.nlm.nih.gov/pmc/articles/PMC2628457/pdf/PROCATS52144.pdf.

Edwards, Steven A. "Paracelsus, the Man Who Brought Chemistry to Medicine." AAAS. Posted March 3, 2012. https://www.aaas.org/paracelsus-man-who-brought-chemistry-medicine.

Ekström, Nelly. "Hamlet, the Melancholic Prince of Denmark." Wellcome Collection, December 16, 2016. https://wellcomecollection.org/articles/WsT4Ex8AAHruGfWf.

———. "Shakespeare and the Four Humors." Wellcome Collection, April 23, 2016. https://wellcomecollection.org/articles/the-humours-in-shakespeare/.

Elizabethan Era. "Elizabethan Tournaments." https://www.elizabethan-era.org.uk/elizabethan-tournaments.htm.

Elliot, Natalie. "Shakespeare's Worlds of Science." *New Atlantis*, no. 54 (2018): 30–50. https://www.jstor.org/stable/90021006.

English Monarchs. "The Remains of Richard III: The Archaeological Dig." http://www.englishmonarchs.co.uk/plantagenet_24.html.

Epstein, Norrie. *The Friendly Shakespeare*. London: Penguin Books, 1993.

Ergoflex. "Genius Sleep: The Sleeping Habits of History's Greatest Minds." Posted February 24, 2002. https://www.ergoflex.co.uk/blog/category/sleep-research/genius-sleep-the-sleeping-habits-of-historys-greatest-minds.

Evans, Jonathan St. B. T., and Keith E. Stanovich. "Dual-Process Theories of Higher Cognition: Advancing the Debate." *Perspectives on Psychological Science* 8, no. 3 (2013): 223–41. https://doi.org/10.1177/1745691612460685.

Falk, Dan. *The Science of Shakespeare: A New Look at the Playwright's Universe*. New York: Thomas Dunne Books (St. Martin's Press), 2014.

———. "Shakespeare, Galileo, and Cymbeline." Shakespeare in the Ruff, 2014. https://shakespeareintheruff.com/education/shakespeare-galileo-cymbeline-by-dan-falk/.

———. "What Shakespeare Knew about Science." *Scientific American*, April 2014. https://www.scientificamerican.com/article/what-shakespeare-knew-about-science-excerpt/.

Family Search. Marriage Allegations, Bonds and Licences in England and Wales. https://www.familysearch.org/wiki/en/Marriage_Allegations,_Bonds_and_Licences_in_England_and_Wales.

Fang, Edith Kuang-Lo. "The Renaissance Sense of Sleep in Shakespeare's Plays, Together with Some Modern Critical Interpretation." Doctoral thesis, University of Ottawa, 1972. Abstract: http://dx.doi.org/10.20381/ruor-17051. Full document: https://www.google.com/url?sa=t&rct=j&q=&esrc=s&source=web&cd=&ved=2ahUKEwinremfxPLwAhVLqJ4KHUA8BEEQFjABegQIBhAD&url=https%3A%2F%2Fruor.uottawa.ca%2Fbitstream%2F10393%2F10875%2F1%2FDC52346.PDF&usg=AOvVaw2txgal1vp7vUZzRbobaa7O.

Farley, Alexandra A. "A Qualitative Analysis of Superstitious Behavior and Performance: How It Starts, Why It Works, and How It Works." Thesis, Western Washington University, 2015. https://cedar.wwu.edu/cgi/viewcontent.cgi?article=1409&context=wwuet.

Fassler, Joe. "Shakespeare: One of the First and Greatest Psychologists." *Atlantic*, January 7, 2015. https://www.theatlantic.com/entertainment/archive/2015/01/by-heart-measure-for-measure/384252/.

Feminae: Medieval Women and Gender Index. "Isabella of France Meets Her Husband, Richard II, King of England." University of Iowa Libraries. Posted 2014. http://inpress.lib.uiowa.edu/feminae/DetailsPage.aspx?Feminae_ID=41702.

Ficca, Gianluca, and Piero Salzarulo. "What in Sleep Is for Memory." *Sleep Medicine* 5 (2004): 225–30. http://physiology.elte.hu/gyakorlat/cikkek/What%20in%20sleep%20is%20for%20memory.pdf.

Findlen, Paula. "The Emergence of Medicine: Middle Ages and the Renaissance: A History of the Liver, Spleen, and Gallbladder." https://web.stanford.edu/class/history13/earlysciencelab/body/liverpages/livergallbladderspleen.html.

Fleischman, John. *Phineas Gage: A Gruesome but True Story about Brain Science*. Boston: Houghton Mifflin, 2002.

Flesch, William. "Would DNA Evidence Exonerate Claudius?" Stanford Humanities Center, *Arcade*, 2022. https://shc.stanford.edu/arcade/interventions/would-dna-evidence-exonerate-claudius.

Folger Shakespeare Library. "Books and Reading in Shakespeare's England." *Shakespeare Unlimited*, episode 137, February 4, 2020. https://www.folger.edu/podcasts/shakespeare-unlimited/books-reading/.

———. "Fooles and Fricassees: Food in Shakespeare's England." Folgerpedia, https://folgerpedia.folger.edu/Fooles_and_Fricassees:_Food_in_Shakespeare%27s_England.

———. "How Shakespeare Describes Post-Traumatic Stress Disorder." *Shakespeare & Beyond* (blog), November 3, 2017. https://shakespeareandbeyond.folger.edu/2017/11/03/shakespeare-post-traumatic-stress-disorder/.

———. "The Madness of Hamlet and King Lear: When Psychiatrists Used Shakespeare to Argue Legal Definitions of Insanity in the Courtroom." *Shakespeare and Beyond* (blog), February 11, 2020. https://shakespeareandbeyond.folger.edu/2020/02/11/madness-king-lear-hamlet-insanity-courtroom-psychiatrists/#more-11302.

———. "*Measure for Measure*: Historical Background." https://shakespeare.folger.edu/shakespeares-works/measure-for-measure/historical-background/.

———. "Shakespeare and Religion." *Shakespeare Unlimited: Episode 49*, 2021. https://www.folger.edu/shakespeare-unlimited/religion.

———. "Shakespeare Documented: Family, Legal, and Property Records." https://shakespearedocumented.folger.edu/exhibition/family-legal-property-records.

———. "Shakespeare, Science, and Art." *Shakespeare Unlimited* (podcast), December 23, 2020, https://www.folger.edu/podcasts/shakespeare-unlimited/science-art-natalie-elliot/.

———. "Stephen Greenblatt: Shakespeare's Life Stories." Posted November 15, 2016. https://www.folger.edu/shakespeare-unlimited/stephen-greenblatt.

Franco, Zeno, and Philip Zimbardo. "The Banality of Heroism." *Greater Good Magazine*, September 1, 2006. https://greatergood.berkeley.edu/article/item/the_banality_of_heroism.

Franz, Julia, and Richard Paul. "Why the Moons of Uranus Are Named after Characters in Shakespeare." *The World*, January 1, 2017. https://www.pri.org/stories/2017-01-01/why-moons-uranus-are-named-after-characters-shakespeare.

Freud, Sigmund. *The Interpretation of Dreams*. New York: Modern Library, 1899/1994. Full text available online at https://en.wikisource.org/wiki/The_Interpretation_of_Dreams and https://www.google.com/url?sa=t&rct=j&q=&esrc=s&source=web&cd=&ved=2ahUKEwis7PTJi9v_AhU8IzQIHVOvAHgQFnoECEwQAQ&url=https%3A%2F%2Fpsychclassics.yorku.ca%2FFreud%2FDreams%2Fdreams.pdf&usg=AOvVaw2FNoH3Fcl5ikpU7Zp_jdJH&opi=89978449.

Friedberg, Harris. "Policing Sex: The 'Bawdy Courts.'" http://hfriedberg.web.wesleyan.edu/engl205/wshakespeare/policingsex.htm.

Friedland, G. "Discovery of the Function of the Heart and Circulation of Blood." *Cardiovascular Journal of Africa* 20, no. 3 (2009): 160.

Friedlander, Ari. "Five Myths about William Shakespeare." *Washington Post*, September 4, 2015. https://www.washingtonpost.com/opinions/five-myths-about-william-shakespeare/2015/09/04/bc986ef6-524d-11e5-9812-92d5948a40f8_story.html.

Furness, Hannah. "Could King Lear Have Suffered Lewy Body Dementia?" *Telegraph*, February 7, 2014. http://www.telegraph.co.uk/culture/theatre/william-shakespeare/10622547/Could-King-Lear-have-suffered-Lewy-body-dementia.html.

Galton, Sir Francis. *English Men of Science: Their Nature and Nurture*. London: Macmillan & Company, 1874. https://www.google.com/books/edition/English_Men_of_Science/diIJAAAAIAAJ?hl=en&gbpv=0.

Gambino, Megan. "Was Shakespeare Aware of the Scientific Discoveries of His Time?" *Smithsonian Magazine*, April 23, 2014. https://www.smithsonianmag.com/arts-culture/was-shakespeare-aware-scientific-discoveries-his-time-180951198/.

Gamblin, Hillary. "Reading between the Bloodied Lines and Bodies: Dissecting Shakespeare's *Titus Andronicus* and Vesalius's *De Humani Corporis Fabrica*." Master's thesis, Brigham Young University, 2014. https://scholarsarchive.byu.edu/etd/5249/.

Garber, Marjorie. *Shakespeare After All.* New York: Pantheon Books, 2004.

Gardner, Howard. *Creating Minds: An Anatomy of Creativity Seen through the Lives of Freud, Einstein, Picasso, Stravinsky, Eliot, Graham, and Gandhi.* New York: Basic Books, 1993.

———. *Frames of Mind: The Theory of Multiple Intelligences.* New York: Basic Books, 1985.

Ghosh, Sanjib Kumar. "Human Cadaveric Dissection: A Historical Account from Ancient Greece to the Modern Era." *Anatomy & Cell Biology* 48, no. 3 (2015): 153–69. https://doi.org/10.5115/acb.2015.48.3.153.

Goldberg, Lewis R. "The Development of Markers for the Big-Five Factor Structure." *Psychological Assessment* 4, no. 1 (1992): 26–42. https://projects.ori.org/lrg/PDFs_papers/Goldberg.Big-Five-Markers-Psych.Assess.1992.pdf.

Graham, Jesse, Jonathan Haidt, Sena Koleva, Matt Motyl, Ravi Iyer, Sean P. Wojcik, and Peter H. Ditto. "Moral Foundations Theory: The Pragmatic Validity of Moral Pluralism." In *Advances in Experimental Social Psychology 47*, edited by Patricia Devine, and Ashby Plant, 55–130. Burlington, VT: Academic Press. https://cpb-us-e2.wpmucdn.com/sites.uci.edu/dist/1/863/files/2020/06/Graham-et-al-2013.AESP_.pdf.

Grammaticus, Saxo. *Gesta Danorum (The History of the Danes).* Circa 1200 C.E. Project Gutenberg. https://www.gutenberg.org/files/1150/1150-h/1150-h.htm.

Greenblatt, Stephen. "Shakespeare Doubters." *Washington Post*, September 4, 2005. https://www.washingtonpost.com/opinions/five-myths-about-williamshakespeare/2015/09/04/bc986ef6-524d-11e5-9812-92d5948a40f8_story.html.

———. "Shakespeare's Leap." *New York Times.* Posted September 12, 2004. https://www.nytimes.com/2004/09/12/magazine/shakespeares-leap.html.

———. *Tyrant: Shakespeare's Politics.* New York: W. W. Norton, 2018.

———. *Will in the World: How Shakespeare Became Shakespeare.* New York: W. W. Norton, 2004.

Grimes, William. "Jack Yufe, a Jew Whose Twin Was a Nazi, Dies at 82." *New York Times*, November 13, 2015. https://www.nytimes.com/2015/11/14/us/jack-yufe-a-jew-whose-twin-was-a-nazi-dies-at-82.html.

Gross, Charles G. *Brain, Vision, Memory: Tales in the History of Neuroscience.* Cambridge, MA: MIT Press, 1998.

Haidt, Jonathan. *The Righteous Mind: Why Good People Are Divided by Politics and Religion.* New York: Pantheon/Random House, 2012.

Hamilton, Jon. "Think You're Multitasking? Think Again." *Morning Edition*, NPR, October 2, 2008. https://www.npr.org/templates/story/story.php?storyId=95256794.

———. "Why Brain Scientists Are Still Obsessed with the Curious Case of Phineas Gage." *Weekend Edition*, NPR, May 21, 2017. https://www.npr.org/sections/health-shots/2017/05/21/528966102/why-brain-scientists-are-still-obsessed-with-the-curious-case-of-phineas-gage.

Hammond, Claudia. "Would Shakespeare's poisons and drugs work in reality?" BBC Future, April 15, 2014. https://www.bbc.com/future/article/20140416-do-shakespeares-poisons-work.

Hanson, Marilee. "Tudor England Mental Illness Types and Facts." *English History*, March 4, 2015. https://englishhistory.net/?s=mental+illness.

Harbage, A. *Conceptions of Shakespeare*. New York: Schocken Books, 1966.

Harcum, Cornelia G. "The Ages of Man: A Study Suggested by Horace, *Ars Poetica*, Lines 153–78." *The Classical Weekly* 7, no. 15 (1914): 114–18. https://doi.org/10.2307/4386866.

Harkup, Kathryn. "Shakespeare's Remarkable Scientific Accuracy." BBC Science Focus, May 9, 2020. https://www.sciencefocus.com/the-human-body/shakespeares-remarkable-scientific-accuracy/.

Harrison, Luke, and Psyche Loui. "Thrills, Chills, Frissons, and Skin Orgasms: Toward an Integrative Model of Transcendent Psychophysiological Experiences in Music." *Frontiers in Psychology* 5 (2014): 790. https://doi.org/10.3389/fpsyg.2014.00790.

Herber, Wray. "Damned Spot: Guilt, Scrubbing, and More Guilt." APS: Association for Psychological Science, March 26, 2013. https://www.psychologicalscience.org/news/were-only-human/damned-spot-guilt-scrubbing-and-more-guilt.html.

Hirai, Hiro. "Spirit in Renaissance Medicine." In *Encyclopedia of Renaissance Philosophy*, edited by Marco Sgarbi. Cham, Switzerland: Springer Nature, 2018. https://doi.org/10.1007/978-3-319-02848-4_1107-1.

History of Royal Women. "Gruoch—The Real Lady Macbeth." November 23, 2017. https://www.historyofroyalwomen.com/gruoch/gruoch-real-lady-macbeth/.

Hobbes, Thomas. *Leviathan*. Edited by Edwin Curley. Cambridge, MA: Hackett Publishing, 1994. Originally published 1651.

Hofmann, S. A., A. Asnaani, I. J. Vonk, A. T. Sawyer, and A. Fang. "The Efficacy of Cognitive Behavioral Therapy: A Review of Meta-analyses." *Cognitive Therapy and Research* 36, no. 5 (2012): 427–40. https://www.ncbi.nlm.nih.gov/pmc/articles/PMC3584580/.

Holinshed, Raphael. *Chronicles of England, Scotland, and Ireland* (circa 1580). Project Gutenberg. https://www.gutenberg.org/ebooks/44700.

Holland, Norman N., Sidney Homan, and Bernard J. Paris, eds. *Shakespeare's Personality*. Berkeley: University of California Press, 1989.

Hook, E. B. "Shakespeare, Genetics, Malformations, and the Wars of the Roses: Hereditary Themes in *Henry VI* and *Richard III*." *Teratology* 35, no. 1 (February 1987): 147–55.

Hughes, John R. "Dictator Perpetuus: Julius Caesar—Did He Have Seizures? If So, What Was the Etiology?" *Epilepsy & Behavior* 5, no. 5 (October 2004): 756–64. https://www.sciencedirect.com/science/article/abs/pii/S152550500400160X#!.

Hurren, Elizabeth T. *Dissecting the Criminal Corpse: Staging Post-Execution Punishment in Early Modern England*. New York: Springer Nature, 2016. https://library.oapen.org/handle/20.500.12657/32102.

Imbracsio, N. M. "Corpses Revealed: The Staging of the Theatrical Corpse in Early Modern Drama." PhD dissertation, University of New Hampshire, 2010. https://scholars.unh.edu/dissertation/520/.

Ingram, Mike. "The Death of Richard III." Medievalist.net, August 2019. https://www.medievalists.net/2019/08/the-death-of-richard-iii/.

International Astronomical Union. "The Skies Are Painted with Unnumbered Sparks, They Are All Fire, and Every One Doth Shine." April 26, 2016. https://www.iau.org/news/announcements/detail/ann16019/?lang.

Isaacson, Walter. *Leonardo da Vinci*. New York: Simon & Schuster, 2017.

Jacobs, Andrew. "Tripping in the Bronze Age." *New York Times*, April 6, 2023. https://www.nytimes.com/2023/04/06/science/psychedelics-bronze-age.html?campaign_id=116&emc=edit_pk_20230411&instance_id=89944&nl=paul-krugman®i_id=53944304&segment_id=130144&te=1&user_id=5fe27f6cafec6910ade0a7b86bb59a3d.

James I, King. *Daemonologie*. Translated into modern English by Beverly Bristol. Academia. https://www.academia.edu/38053025/THE_DEMONOLOGY_OF_KING_JAMES_I. Project Gutenberg. https://www.gutenberg.org/ebooks/25929.

Jaramillo, Carina. "Macbeth the Real King of Scotland." July 18, 2015. https://www.octaneseating.com/macbeth-the-real-king-of-scotland/.

Jaworska, Natalia, and Glenda MacQueen. "Adolescence as a Unique Developmental Period." *Journal of Psychiatry & Neuroscience* 40, no. 5 (2015): 291–93. https://doi.org/10.1503/jpn.150268 and https://www.jpn.ca/content/40/5/291.

Jayatunge, Ruwan M. "Shakespearean Work and Common Mental Disorders." Lankaweb. Posted August 2, 2015. http://www.lankaweb.com/news/items/2015/08/02/shakespearean-work-and-common-mental-disorders-3/.

Johnson, S. B., R. W. Blum, and J. N. Giedd. "Adolescent Maturity and the Brain: The Promise and Pitfalls of Neuroscience Research in Adolescent Health Policy." *Journal of Adolescent Health: Official Publication of the Society for Adolescent Medicine* 45, no. 3 (2009): 216–21. https://doi.org/10.1016/j.jadohealth.2009.05.016.

Jones, Ernest. *Hamlet and Oedipus*. 1949. Reprint New York: W. W. Norton, 1976. Original work published in *The American Journal of Psychology*, January 1910. https://archive.org/details/hamletoedipus00jone.

Jones, Nolan. "In Hip-Hop, Educators Find a Bridge to Shakespeare: Rap Meets the Renaissance, and What Tupac Had to Say about the Bard." *Hechinger Report*, September 3, 2019. https://hechingerreport.org/opinion-in-hip-hop-educators-find-a-bridge-to-shakespeare/.

Jonson, Ben. "Shakespeare," in *Timber; or, Discoveries Made upon Men and Matter*. 1641. Reprint London: Leopold Classic Library, 2015.

Jorgensen, Paul A. "Hamlet's Therapy." *Huntington Library Quarterly* 27, no. 3 (1964): 239–58. https://doi.org/10.2307/3816795.

Joynes, Victoria. "Shakespeare's Friends: Burbage, Combe and Sadler." Shakespeare's Birthplace Trust. Posted August 1, 2016. https://www.shakespeare.org.uk/explore-shakespeare/blogs/shakespeares-friends-burbage-combe-and-sadler/.

Kahneman, Daniel. *Thinking Fast and Slow*. New York: Farrar, Straus and Giroux, 2013.

Karim-Cooper, Farah. "Strangers in the City: The Cosmopolitan Nature of 16th-Century Venice." Shakespeare's Globe and British Library, March 15, 2016. https://www.bl.uk/shakespeare/articles/strangers-in-the-city-the-cosmopolitan-nature-of-16th-century-venice.

Keating, Kristin M. "The Performative Corpse: Anatomy Theatres from the Medieval Era to the Virtual Age." PhD dissertation, University of California, Irvine, 2014. https://escholarship.org/uc/item/9th868k6.

Keidel, James L., Philip M. Davis, Victorina Gonzalez-Diaz, Clara D. Martin, and Guillaume Thierry. "How Shakespeare Tempests the Brain: Neuroimaging Insights." *Cortex* 49, no. 4 (April 2012): 913–19. https://www.researchgate.net/publication/224912467_How_Shakespeare_tempests_the_brain_Neuroimaging_insights.

Kelly, Philippa. "Division, Harmony, and Medical Mistakes: Twins in Shakespeare." CalShakes, February 18, 2014. https://calshakes.org/division-harmony-and-medical-mistakes-twins-in-shakespeare/.

Kettt, Joseph F. "Discovery and Invention in the History of Adolescence." *Journal of Adolescent Health* 14 (1993): 605–12. https://www.sciencedirect.com/sdfe/pdf/download/eid/1-s2.0-1054139X9390193S/first-page-pdf.

Killgrove, Kristina. "Julius Caesar's Health Debate Reignited: Stroke or Epilepsy?" *Forbes*, May 15, 2015. https://www.forbes.com/sites/kristinakillgrove/2015/05/15/julius-caesars-health-debate-reignited/#41caa8fc10a3.

King, Turi E., Gloria Gonzalez Fortes, Patricia Balaresque, et al. "Identification of the Remains of King Richard III." *Nature*, December 2, 2014. https://www.nature.com/articles/ncomms6631.pdf.

Kipen, David. "The Secret Connection between Cervantes and Shakespeare." *Los Angeles Times*, April 14, 2016. https://www.latimes.com/books/la-ca-jc-cervantes-shakespeare-connection-20160417-story.html.

Korbey, Holly. "How Hip-Hop Can Bring Shakespeare to Life." *Mind/Shift*, KQED, September 1, 2016. https://www.kqed.org/mindshift/46215/how-hip-hop-can-bring-shakespeare-to-life.

Kreider, P. V. "The Mechanics of Disguise in Shakespeare's Plays." *Shakespeare Association Bulletin* 9, no. 4 (1934): 167–80. http://www.jstor.org/stable/23675558.

Kreps, Barbara. "The Paradox of Women: The Legal Position of Early Modern Wives and Thomas Dekker's 'The Honest Whore.'" *ELH* 69, no. 1 (2002): 83–102. http://www.jstor.org/stable/30032012.

Lagay, Faith. "The Legacy of Humoral Medicine." *Virtual Mentor* 4, no. 7 (2002). https://doi.org/10.1001/virtualmentor.2002.4.7.mhst1-0207.

Lantern Theater Company. "The Venetian Mercantile Empire." Searchlight, February 24, 2020. https://medium.com/lantern-theater-company-searchlight/the-venetian-mercantile-empire-93a4c6f8ceed.

Lever, J. W. "The Date of Measure for Measure." *Shakespeare Quarterly* 10, no. 3 (1959): 381–88. https://doi.org/10.2307/2866859.

Levy, Eric. "The Problematic Relation between Reason and Emotion in Hamlet." *Renascence* 53, no. 2 (Winter 2001): 83–95. http://www.houseofideas.com/mscorneli us/resources/hamlet/hamlet_vol_71__eric_levy_essay_date_winter_2001_276895-.pdf.

Lewis, Liz. "The Mixture of Styles in Shakespeare's Last Plays: *The Winter's Tale* and *The Tempest*." Literature-Study-Online. http://www.literature-study-online.com/essays/shakespeare_last_plays.html.

Liu, Yongsheng. "A New Perspective on Darwin's Pangenesis." *Biological Reviews of the Cambridge Philosophical Society* 83, no. 2 (April 21, 2008): 142–49. https://onlinelibrary.wiley.com/doi/full/10.1111/j.1469-185X.2008.00036.x.

Lyon, Karen. "Wooing and Wedding: Courtship and Marriage in Early Modern England." *Folger Magazine*, June 8, 2018. Folger Shakespeare Library. https://www.folger.edu/folger-story/wooing-and-wedding-courtship-and-marriage-in-early-modern-england.

Mabillard, Amanda. "*Julius Caesar* Character Introduction." Shakespeare Online. Last modified August 20, 2000. http://www.shakespeare-online.com/plays/juliuscaesar/juliuscaesarcharacters.html.

———. "Shakespeare and the Gunpowder Plot." Shakespeare Online. Last modified January 21, 2022. http://www.shakespeare-online.com/biography/gunpowderplot.html.

———. "Words Shakespeare Invented." Shakespeare Online. Posted August 20, 2000. http://www.shakespeare-online.com/biography/wordsinvented.html.

Machovec, Frank J. "Shakespeare on Hypnosis: *The Tempest*." *American Journal of Clinical Hypnosis* 24, no. 2 (1981): 73–88. https://doi.org/10.1080/00029157.1981.10403293.

Mandel, Jerome. "Dream and Imagination in Shakespeare." *Shakespeare Quarterly* 24, no. 1 (1973): 61–68. https://doi.org/10.2307/2868739.

Marcaggi G., and F. Guénolé. "Freudarwin: Evolutionary Thinking as a Root of Psychoanalysis." *Frontiers in Psychology* 9 (June 19, 2018): 892. doi: 10.3389/fpsyg.2018.00892.

Markley, Robert. "Shakespeare in the Little Ice Age." In *Early Modern Ecostudies: From the Florentine Codex to Shakespeare*, edited by Thomas Hallock, Ivo Kamps, and Karen Raber. New York: Palgrave Macmillan, 2008. https://www.academia.edu/20566800/Summers_Lease_Shakespeare_in_the_Little_Ice_Age.

Matthews, Paul M., and Jeffrey McQuain. *The Bard on the Brain*. New York: The Dana Press, 2003.

Matusiak, Carol. "Was Shakespeare 'Not a Company Keeper'? William Beeston and MS Aubrey 8, fol. 45v." *Shakespeare Quarterly* 68, no. 4, 2017: 351–73. https://doi.org/10.1353/shq.2017.0040.

McAdams, Dan P. "The American Identity: The Redemptive Self." *General Psychologist* 43, no. 1 (Spring 2008): 20–27. https://www.sesp.northwestern.edu/docs/publications/2094657112490a0f25ec2b9.pdf.

———. "Exploring Psychological Themes through Life Narrative Accounts." In *Varieties of Narrative Analysis*, edited by John A. Holstein and Jaber F. Gubrium, 15–32. London: Sage, 2012.

———. *The Redemptive Self: Stories Americans Live By*. New York: Oxford University Press, 2005.

McAdams, Dan P., and Jennifer L. Pals. "A New Big Five: Fundamental Principles for an Integrative Science of Personality." *American Psychologist* 61, no. 3 (2006): 204–17. https://doi.org/10.1037/0003-066X.61.3.204. Full text available online at: https://www.sesp.northwestern.edu/docs/publications/89090097490a0624369b9.pdf.

McPhee, Constance C. "Shakespeare and Art, 1709–1922." *The Met*, 2021. https://www.metmuseum.org/toah/hd/shaa/hd_shaa.htm.

McWhorter, John. "How the N-Word Became Unsayable." *New York Times*, April 30, 2021. https://www.nytimes.com/2021/04/30/opinion/john-mcwhorter-n-word-unsayable.html.

Merriam Webster Dictionary. "What Does 'Quark' Have to Do with Finnegans Wake?" https://www.merriam-webster.com/words-at-play/quark.

Messbarger, Rebecca. "Anna Morandi's Wax Self-Portrait." In *The Fine Arts, Neurology, and Neuroscience: Neuro-Historical Dimensions*, edited by S. Finger, D. Zaidel, F. Boller, and J. Bogousslavsky, 75–94. Amsterdam: Elsevier, 2013. https://www.researchgate.net/publication/256664249_Anna_Morandi%27s_Wax_Self-Portrait_with_Brain.

Milgram, Stanley. *Obedience to Authority: An Experimental View*. New York: HarperCollins, 1974.

Miller, Marcy J. "The Four Humors and the Integrated Universe: A Medieval World View." *Owlcation*, July 20, 2022. https://owlcation.com/humanities/The-Four-Humors.

Minton, Eric. "Every Man in His Humor: Watching the Other Side of Shakespeare." Shakespeareances.com. Posted April 28, 2015. http://www.shakespeareances.com/willpower/nonshakespeare/Every_Man_In_Humour-03-ASC15.html.

Misra, Ria. "No, William Shakespeare Did Not Really Invent 1,700 English Words." *Gizmodo*, April 24, 2015. https://io9.gizmodo.com/no-william-shakespeare-did-not-really-invent-1-700-eng-1700049586.

Montuori, Alfonso, and Ronald E. Purser. "Deconstructing the Lone Genius Myth: Toward a Contextual View of Creativity." *Journal of Humanistic Psychology* 35, no. 3 (Summer 1995): 69–112. https://www.academia.edu/168677/Deconstructing_the_Lone_Genius_Myth_Towards_a_Contextual_View_of_Creativity or http://jhp.sagepub.com/cgi/content/abstract/35/3/6.

Morris, Sylvia. "Shakespeare's Minds Diseased: Mental Illness and Its Treatment." *The Shakespeare Blog*, March 12, 2012. http://theshakespeareblog.com/2012/03/shakespeares-minds-diseased-mental-illness-and-its-treatment/.

Mowat, Barbara A. "*The Tempest*: A Modern Perspective." In William Shakespeare, *The Tempest*, edited by B. A. Mowat and P. Werstine, 185–99. Folger Shakespeare Library. New York: Washington Square Press, 1994.

Mullan, John. "*Measure for Measure* and Punishment." British Library: *Discovering Literature: Shakespeare and Renaissance*, March 15, 2016. https://www.bl.uk/shakespeare/articles/measure-for-measure-and-punishment.

Myers, Seth. "Difference between the Psychopath and So-Called Sociopath: Persistent Misinformation about the Psychopath Reinforces Public Confusion." *Psychology Today*, December 12, 2018. https://www.psychologytoday.com/us/blog/insight-is-2020/201812/difference-between-the-psychopath-and-so-called-sociopath.

Nason, Arthur Huntington. "Shakespeare's Use of Comedy in Tragedy." *Sewanee Review* 14, no. 1 (1906): 28–37. http://www.jstor.org/stable/27530731.

National Institute of Mental Health. *Genetics and Mental Disorders: Report of the National Institute of Mental Health's Genetics Workgroup*. Rockville, MD: National

Institute of Mental Health, September 19, 1997. https://www.nimh.nih.gov/about/advisory-boards-and-groups/namhc/reports/genetics-and-mental-disorders-report-of-the-national-institute-of-mental-healths-genetics-workgroup.shtml.

National Library of Medicine. "And There's the Humor of It: Shakespeare and the Four Humors." NLM exhibition, January 30–August 17, 2012. https://www.nlm.nih.gov/exhibition/shakespeare/fourhumors.html.

———. "Melancholy in Age: The Case of Shylock." In *And There's the Humor of It: Shakespeare and the Four Humors.* NLM exhibition, January 30–August 17, 2012. https://www.nlm.nih.gov/exhibition/shakespeare-and-the-four-humors/index.html#section5.

———. "Melancholy Virgins: The Case of Ophelia." In *And There's the Humor of It: Shakespeare and the Four Humors.* NLM exhibition, January 30–August 17, 2012." https://www.nlm.nih.gov/exhibition/shakespeare/hamlet.html.

Neely, Carol Thomas. "'Documents in Madness': Reading Madness and Gender in Shakespeare's Tragedies and Early Modern Culture." *Shakespeare Quarterly* 42, no. 3 (1991): 315–38. https://www.jstor.org/stable/2870846?seq=1#metadata_info_tab_contents.

Neill, Michael. "A Modern Perspective: *Hamlet.*" The Folger Shakespeare Library. https://shakespeare.folger.edu/shakespeares-works/hamlet/hamlet-a-modern-perspective/.

Nelson, Walter. "Elizabethan Incomes." *Mass Historia* (blog). http://walternelson.com/dr/elizabethan-incomes.

———. "Elizabethan Money." *Mass Historia* (blog). http://walternelson.com/dr/elizabethan-money.

Norwood, Arlisha. "Rosa Parks." Women's National History Museum, 2017. https://www.womenshistory.org/education-resources/biographies/rosa-parks.

O'Driscoll, Kieran, and John Paul Leach. "'No Longer Gage': An Iron Bar through the Head." *British Medical Journal* 317, no. 7174 (December 19, 1998): 1673–74. doi: 10.1136/bmj.317.7174.1673a. https://www.ncbi.nlm.nih.gov/pmc/articles/PMC1114479/.

Oleynick, Victoria C., Colin G. DeYoung, Elizabeth Hyde, et al. "Openness/Intellect: The Core of the Creative Personality." In *The Cambridge Handbook of Creativity and Personality Research*, edited by Gregory J. Feist, Roni Reiter-Palmon, and James C. Kaufman, 9–27. Cambridge: Cambridge University Press, 2017. https://scottbarrykaufman.com/wp-content/uploads/2017/03/Oleynick-et-al.-2017.pdf.

Orbell, John. "Hamlet and the Psychology of Rational Choice under Uncertainty." *Rationality and Society* 5, no. 1 (January 1993): 127–40. https://www.academia.edu/926307/Hamlet_and_the_psychology_of_rational_choice_under_uncertainty.

Original Pronunciation.com. "Original Pronunciation." http://originalpronunciation.com/GBR/Home.

Ottilingam, Somasundaram. "The Psychiatry of King Lear." *Indian Journal of Psychiatry* 49, no. 1 (2007): 52–55. https://www.ncbi.nlm.nih.gov/pmc/articles/PMC2900000/.

Owens, Rebecca. "Shakespeare and Medicine: Mental Health in Tudor Times." Shakespeare Birthplace Trust, March 8, 2018. https://www.shakespeare.org.uk/explore-shakespeare/blogs/shakespeare-and-medicine-mental-health-tudor-times/.

PAEI—Structures of Concern. "Hippocrates and Galen—The Four Humors." Last edited December 23, 2008. http://paei.wikidot.com/hippocrates-galen-the-four-humors.

Parvini, Neema. "How Noble in Reason: Shakespeare Reveals the Primacy of Emotions in Human Nature." *This View of Life*, April 11, 2017. https://thisviewoflife.com/how-noble-in-reason-shakespeare-reveals-the-primacy-of-emotions-in-human-nature/.

———. *Shakespeare and Cognition: Thinking Fast and Slow through Character.* New York: Palgrave Macmillan, 2015.

Paster, G. K. "*Much Ado About Nothing*: A Modern Perspective." In *Much Ado About Nothing*. The Folger Shakespeare Library, edited by B. A. Mowat and P. Werstine. New York: Washington Square Press, 213–30.

Paulhus, Delroy L., Paul D. Trapnell, and D. Chen. "Birth Order Effects on Personality and Achievement within Families." *Psychological Science* 10, no. 6 (1999): 482–88.

Penfield, Wilder. "The Interpretive Cortex." *Science* 129, no. 3365 (June 26, 1959): 1719–25. doi:10.1126/science.129.3365.1719.

Pinker, Steven. *The Better Angels of Human Nature: Why Violence Has Declined.* New York: Penguin Books, 2012.

———. *The Blank Slate: The Modern Denial of Human Nature.* New York: Penguin, 2002.

Plato. *The Republic.* Book VII. Translated by B. Jowett (updated 2021). Project Gutenberg eBook. https://www.gutenberg.org/files/1497/1497-h/1497-h.htm.

Pollard, Tanya. "'A Thing Like Death': Sleeping Potions and Poisons in 'Romeo and Juliet' and 'Antony and Cleopatra.'" *Renaissance Drama* 32, new series (2003): 95–121. JSTOR. http://www.jstor.org/stable/41917377.

Pray, L. A. "Discovery of DNA Structure and Function: Watson and Crick." *Nature Education* 1, no. 1 (2008): 100. http://www.nature.com/scitable/topicpage/discovery-of-dna-structure-and-function-watson-397.

Pressley, J. M. "Mrs. Shakespeare: Anne Hathaway." *The Shakespeare Resource Center.* http://www.bardweb.net/content/ac/hathaway.html.

PsychCentral. "The Forbidden Fruit in Relationships." (Blog post). 2011. https://psychcentral.com/blog/the-forbidden-fruit-in-relationships#1.

Psychology Today. "The Invention of Adolescence." January 1995. https://www.psychologytoday.com/us/articles/199501/the-invention-adolescence.

———. "Six Degrees of Separation: Two New Studies Test 'Six Degrees of Separation' Hypothesis." November 2003. https://www.psychologytoday.com/us/articles/200311/six-degrees-separation.

Pumfrey, Stephen. "'Your Astronomers and Ours Differ Exceedingly': The Controversy Over the 'New Star' of 1572 in the Light of a Newly Discovered Text by Thomas Digges." *British Journal for the History of Science* 44, no. 1 (2011): 29–60. http://www.jstor.org/stable/41241533.

Quealy, Gerit, and Sumie Hasegawa Collins. *Botanical Shakespeare: An Illustrated Compendium of All the Flowers, Fruits, Herbs, Trees, Seeds, and Grasses Cited by the World's Greatest Playwright.* New York: HarperCollins, 2017.

Rafiq, Muhammad. "What Is a Shakespearean Tragedy?" *Owlcation*. https://owlcation.com/humanities/Shakespearean-Tragedy-Definition-and-Characteristics-of-Shakespearean-Tragedy.

Ramachandran, Vilayanur. "Adventures in Behavioral Neurology—or—What Neurology Can Tell Us about Human Nature: A Talk with Vilayanur Ramachandran." *Edge*, February 21, 2012. https://www.edge.org/conversation/vilayanur_ramachandran-adventures-in-behavioral-neurology-%E2%80%94-or-%E2%80%94-what-neurology-can.

Rasmussen, Eric. "Marriage and Courtship." Discovering Literature: Shakespeare and Renaissance. British Library, March 15, 2016. https://www.bl.uk/shakespeare/articles/marriage-and-courtship.

Reiss, Benjamin. *Theaters of Madness: Insane Asylums and Nineteenth-Century American Culture*. Chicago: University of Chicago Press, 2008.

Reuters. "Radar Scan of Shakespeare's Grave Confirms Skull Apparently Missing." March 24, 2016. http://www.reuters.com/article/us-britain-shakespeare-idUSKCN0WQ192.

Reuven, Orna, Nira Liberman, and Reuven Dar. "The Effect of Physical Cleaning on Threatened Morality in Individuals with Obsessive-Compulsive Disorder." *Clinical Psychological Science* 2, no. 2 (2013): 224–29. https://journals.sagepub.com/doi/pdf/10.1177/2167702613485565.

Rex, Richard. "The English Campaign against Luther in the 1520s: The Alexander Prize Essay." *Transactions of the Royal Historical Society* 39 (1989): 85–106. https://doi.org/10.2307/3678979.

Risen, Clay. "The Lady Macbeth Effect." *New York Times*, December 10, 2006. https://www.nytimes.com/2006/12/10/magazine/10Section2a.t-9.html.

Roe, Richard Paul. *The Shakespeare Guide to Italy: Retracing the Bard's Unknown Travels*. New York: Harper Perennial, 2011.

Romm, Cari. "Rethinking One of Psychology's Most Infamous Experiments." *Atlantic*, January 2015. https://www.theatlantic.com/health/archive/2015/01/rethinking-one-of-psychologys-most-infamous-experiments/384913/.

Rosen, Marty. "The History Behind Shakespeare's Henriad Series." *Leo Weekly*. July 14, 2021. https://www.leoweekly.com/2021/07/history-behind-shakespeares-henriad-series/.

Rowan-Robinson, Michael. "Shakespeare's Astronomy." *Brewminate: A Bold Blend of News and Ideas*, February 15, 2017. https://brewminate.com/shakespeares-astronomy/.

Rowse, A. L. "The Personality of Shakespeare." *Huntington Library Quarterly* 27, no. 3 (1964): 193–209. https://doi.org/10.2307/3816792.

Sabio, Anna Fluvià. "The Mask of Madness: Identity and Role-Playing in Shakespeare's *Hamlet*." Thesis, University of Barcelona, 2016. https://ddd.uab.cat/pub/tfg/2016/169519/TFG_annafluvia.pdf.

Salem Media. "The Tudors—Society." History on the Net. https://www.historyonthenet.com/the-tudors-society.

Schieffelin. (2017, January 29). "'Lady Macbeth Effect' and Obsessive-Compulsive Disorder." *Challenging Conventions: EN4HD* (blog), January 29, 2017. http://schieffelinen4hd.blogspot.com/2017/01/lady-macbeth-effect-and-obsessive.html.

Schmidt, Alexander. *Shakespeare Lexicon and Quotation Dictionary: A Complete Dictionary of All the English Words, Phrases and Constructions in the Works of the Poet.* Vol. 1. Third edition. New York: Dover, 1971.

Schuessler, Jennifer. "Shakespeare: Actor. Playwright. Social Climber." *New York Times*, June 29, 2016. https://www.nytimes.com/2016/06/30/theater/shakespeare-coat-of-arms.html.

Shadbolt, Peter. "The Bard's Business: Shakespeare's Economic Legacy." *BBC News*, April 21, 2016. https://www.bbc.com/news/business-36084631.

Shakespeare Birthplace Trust. "Method in the Madness: Treating Mental Illness." http://collections.shakespeare.org.uk/exhibition/exhibition/method-in-the-madness/page/2#method-in-the-madness-treating-mental-illness.

Shakespeare Lives. "Shakespeare Lives in Science; Poisons, Potions and Drugs: Do Shakespearean Concoctions Really Work?" 2015. https://www.shakespearelives.org/poisons-potions/.

Shakespeare Mag.com. "Crime and Punishment." Fall 1998. http://www.shakespearemag.com/fall98/punished.asp.

Shakespeare Online. "Preface to The First Folio (1623)." http://www.shakespeare-online.com/biography/firstfolio.html.

———. "Words That Shakespeare Invented." http://www.shakespeare-online.com/biography/wordsinvented.html.

Shapiro, James. *1599: A Year in the Life of William Shakespeare*. New York: HarperCollins, 2005.

———. *Contested Will: Who Wrote Shakespeare?* New York: Simon and Schuster, 2010.

Sheir, Rebecca. "Shakespeare and Insane Asylums." *Shakespeare Unlimited*, Episode 9. Interview with Benjamin Reiss. August 27, 2014. https://www.folger.edu/shakespeare-unlimited/shakespeare-insane-asylums.

Sherrington, C. *Man on His Nature*. Cambridge: Cambridge University Press, 1942.

Siev, Jedidiah, Shelby E. Zuckerman, and Joseph J. Siev. "The Relationship between Immorality and Cleansing: A Meta-Analysis of the Macbeth Effect." *Social Psychology* 49 (2018): 303–9. https://doi.org/10.1027/1864-9335/a000349.

Simon, Edward. "What's So 'American' about John Milton's Lucifer?" *Atlantic*, March 16, 2017.

Skulsky, Harold. "Pain, Law, and Conscience in *Measure for Measure*." *Journal of the History of Ideas* 25, no. 2 (1964): 147–68. https://doi.org/10.2307/2708009.

Smith, Roff. "Richard III Killed by Sustained Attack, Suffering 9 Wounds to Head." National Geographic Society, September 18, 2014. https://www.nationalgeographic.com/news/2014/9/140917-richard-cause-death-helmet-forensic-science/.

Snipes, Shakiya. "Treatment of the Elderly in Shakespeare." *Articulāte* 7 (2002): 31–33. http://digitalcommons.denison.edu/articulate/vol7/iss1/6.

SparkNotes. *Titus Andronicus*. https://www.sparknotes.com/shakespeare/titus/summary/.

Stables, Andrew. "The Unnatural Nature of Nature and Nurture: Questioning the Romantic Heritage." *Studies in Philosophy and Education* 28, no. 1 (2009, January): 3–14. https://www.researchgate.net/publication/225759970_The_Unnatural_Nature_of_Nature_and_Nurture_Questioning_the_Romantic_Heritage.

Stigler, Brett. "Gender Swaps in Shakespeare Plays." Thirteen PBS. Posted April 26, 2016. https://www.thirteen.org/blog-post/gender-swaps-in-shakespeare-plays/.

Stofan, Ellen. "Where Shakespeare Meets Science." *NASA Leadership, 2014–2016* (blog), April 21, 2016. https://blogs.nasa.gov/leadership/2016/04/21/where-shakespeare-meets-science/.

Strickland, Agnes. "The Seven Ages of Woman and Other Poems." Google Books. https://www.google.com/books/edition/The_Seven_Ages_of_Woman/rORJAAAAIAAJ?hl=en.

Sulloway, Frank J. *Born to Rebel: Birth Order: Family Dynamics, and Creative Lives*. New York: Pantheon, 1996.

Swift, Deborah. "A History of the Cuckold's Horns." https://englishhistoryauthors.blogspot.com/2017/02/a-history-of-cuckolds-horns.html.

Takaki, Ronald. "*The Tempest* in the Wilderness: The Racialization of Savagery." *Journal of American History* 79, no. 3 (1992): 892–912. https://doi.org/10.2307/2080792.

Tasca, Cecilia, Mariangela Rapetti, Mauro Giovanni Carta, and Bianca Fadda. "Women and Hysteria in the History of Mental Health." *Clinical Practice & Epidemiology in Mental Health* 8 (2012): 110–19. https://clinical-practice-and-epidemiology-in-mental-health.com/VOLUME/8/PAGE/110/.

Tate, Adam. "The Ancient Planets." Medicine Traditions, 2019. https://www.medicinetraditions.com/7-ancient-planets.html.

Tavris, Carol. *The Mismeasure of Woman*. New York: Simon & Schuster, 1992.

Texas State University. "Researchers Say Star in Hamlet May Be Supernova of 1572." September 6, 1998. https://news.txst.edu/about/news-archive/press-releases/1998/09/supernova092898.html.

Than, K. "A Brief History of Twin Studies." *Smithsonian Magazine*, March 4, 2016. http://www.smithsonianmag.com/science-nature/brief-history-twin-studies-180958281/.

Thorpe, Vanessa. "Alas, Poor Hamnet: Spotlight Falls on Shakespeare's Tragic Only Son." *Guardian*, February 22, 2020. https://www.theguardian.com/culture/2020/feb/22/alas-poor-hamnet-shakespeare-tragic-son-finally-steps-into-the-spotlight.

Todorov, Lubomir. "Shakespeare, Cervantes, and the Evolution of Civilizational Thinking." April 23, 2016. https://universalfuture.org/shakespeare-cervantes-and-the-evolution-of-civilizational-thinking-e3dc66714ac.

Tosh, Will. "Shakespeare and Madness." British Library. Posted March 15, 2016. https://www.bl.uk/shakespeare/articles/shakespeare-and-madness.

Turnbull, Rio. "The Underrated Heroines of Shakespeare." Shakespeare Birthplace Trust. Last modified March 8, 2018. https://www.shakespeare.org.uk/explore-shakespeare/blogs/underrated-heroines-shakespeare/.

Turner, Sam. Aspects of the Development of Public Assembly in the Danelaw. Archaeology Data Service, 2000. https://archaeologydataservice.ac.uk/archives/view/assemblage/html/5/turner.html.

Usher, Peter. "Hamlet and Infinite Universe." *Penn State News*, September 1, 1997. https://news.psu.edu/story/140839/1997/09/01/research/hamlet-and-infinite-universe.

———. "Hamlet's Love Letter and the New Philosophy." *The Oxfordian* 8 (2005): 93–109. https://shakespeareoxfordfellowship.org/wp-content/uploads/Oxfordian2005_Usher-Hamlets_Poem.compressed.pdf.

———. "Jupiter and Cymbeline." *The Shakespeare Newsletter* 53, no. 1 (Spring 2003): 7–12. https://scholarsphere.psu.edu/resources/6aa2541c-ff75-41e4-98d8-c4827de82dc4.

Waardt, Hans de. "Witchcraft, Spiritualism, and Medicine: The Religious Convictions of Johan Wier." *Sixteenth Century Journal* 42, no. 2 (2011): 369–91. http://www.jstor.org/stable/23076788.

Wagenknecht, E. *The Personality of Shakespeare*. Norman: University of Oklahoma Press, 1972.

Walthall, Bill. "The Lunatic, the Lover, and the Poet." The Bill / Shakespeare Project, April 21, 2010. https://thebillshakespeareproject.com/2010/04/the-lunatic-the-lover-and-the-poet/.

Warren, William S. "The Biology in Shakespeare." *Bios* 15, no. 1 (1944): 21–36. http://www.jstor.org/stable/4604801.

Watson, Robert N. "False Immortality in *Measure for Measure*: Comic Means, Tragic Ends." *Shakespeare Quarterly* 41, no. 4 (1990): 411–32. https://doi.org/10.2307/2870774.

Wheeler, Richard P. "Deaths in the Family: The Loss of a Son and the Rise of Shakespearean Comedy." *Shakespeare Quarterly* 51, no. 2 (2000): 127–53. https://www.jstor.org/stable/2902129?seq=1.

Whitbourne, Susan K. "Shakespeare's Seven Ages of Man, Revisited." *Psychology Today*, November 10, 2009. https://www.psychologytoday.com/us/blog/fulfillment-any-age/200911/shakespeares-seven-ages-man-revisited.

Wikipedia. "Moons of Uranus." Last modified June 29, 2023. https://en.wikipedia.org/wiki/Moons_of_Uranus.

———. "Sophia Brahe." https://en.wikipedia.org/wiki/Sophia_Brahe.

———. "Thing (assembly)." Last modified March 28, 2023. https://en.wikipedia.org/wiki/Thing_(assembly).

Will I Am Shakespeare. "Dr. William Shakespeare, M.D." http://will-i-am-shakespeare.weebly.com/mental-illness-in-shakespearean-tragedy.html.

Williams, George Walton. "Fastolf or Falstaff." *English Literary Renaissance* 5, no. 3, 1975: 308–12. https://www.jstor.org/stable/43446825.

Wilson, Jeffrey R. "'When Evil Deeds Have Their Permissive Pass': Broken Windows in William Shakespeare's *Measure for Measure*." *Law and Humanities* 11, no. 2, 160–83. https://doi.org/10.1080/17521483.2017.1371953.

Winkler, Elizabeth. *Shakespeare Was a Woman and Other Heresies*. New York: Simon and Schuster, 2023.

Woike, Barbara A. "The State of the Story in Personality Psychology." *Social and Personality Psychology Compass* 2, no. 1 (2008): 434–43. https://doi.org/10.1111/j.1751-9004.2007.00070.x.

Woo, Elaine. "Jack Yufe Dies at 82; He Was Raised Jewish, His Identical Twin as a Nazi." *Los Angeles Times*, November 11, 2015. http://www.latimes.com/local/obituaries/la-me-jack-yufe-20151111-story.html.

Wood, Janice. "Sleepwalking Linked to Depression, Anxiety." *PsychCentral*, May 16, 2012. https://psychcentral.com/news/2012/05/16/sleepwalking-linked-to-depression-anxiety/38671.html.

Wood, Michael. *Shakespeare.* New York: Basic Books, 2003.

Zajonc, Robert B., and Gerald B. Markus. "Birth Order and Intellectual Development." *Psychological Review* 82 (1975): 74–88.

Zimbardo, Philip G. *The Lucifer Effect: Understanding How Good People Turn Evil.* New York: Random House, 2007.

———. "Revising the Stanford Prison Experiment: A Lesson in the Power of Situation." *The Chronicle Review*, 2007, B6. https://eric.ed.gov/?id=EJ766029.

Zimbardo, P. G., Robert Johnson, and Vivian McCann. *Psychology: Core Concepts*. New York: Pearson, 2017.

Zlatar, Alexandria. "When the Body Is Ill, The Mind Suffers: Shakespeare's Unravelling of Women's Hysteria and Madness in the Elizabethan Era." The Folger Shakespeare Library. August 9, 2022. https://www.folger.edu/blogs/collation/when-the-body-is-ill-the-mind-suffers/.

INDEX